Writing Organic Reaction Mechanisms:
A Practical Guide

*To my family
past, present and future*

Writing Organic Reaction Mechanisms: A Practical Guide

by

Michael Edenborough

Taylor & Francis
Publishers since 1798

UK	Taylor & Francis Ltd, 4 John St, London WC1N 2ET
USA	Taylor & Francis Inc., 1900 Frost Road, Suite 101, Bristol PA 19007

British Library Cataloguing in Publication Data

A catalogue record for this book is available from the British Library

ISBN 0 7484 0171 7 (paper)

Library of Congress Cataloging in Publication Data are available

Cover design by Linda Wade

Typeset by Santype International Limited, Netherhampton Road, Salisbury, Wilts SP2 8PS

Printed in Great Britain by Burgess Science Press, Basingstoke on paper which has a specified pH value on final paper manufacture of not less than 7.5 and is therefore 'acid free'.

Contents

Part II: Mechanisms

Preface

This book is designed to give the student a solid foundation in the basic principles involved in writing organic mechanisms so that further study may be confidently undertaken.

In Part I the basic principles that are common to all mechanisms are explained in full. Some of the exercises may seem very easy, but they are all important, because they instil a precision and discipline that is essential in order to deal with more complicated mechanisms. Fluency in writing mechanisms comes from practice, and in particular from reecognising when a proposed intermediary is impossible or highly unlikely.

In Part II examples of the main mechanistic pathways are given. No attempt has been made to cover exhaustively all the synthetic routes or reactions that could be encountered. However, by reaching an understanding of why a particular mechanism is followed in any given situation, an appreciation is necessarily developed of the major synthetic methods that may be deployed in producing organic chemicals. There is an emphasis upon understanding the stereochemistry of the interactions between reagents and substrate and the consequences on the geometry of the product, because such an appreciation is of fundamental importance in comprehending an organic mechanism fully.

Part III consists of several appendices which cover all the essential information and simple odds and ends that are often left untaught because it has been erroneously assumed that the student already knows the point in question. In general the appendices will answer all those questions that are too embarrassing to ask to ask either your peers or your teachers.

There may only be one name on the front cover, but every book is the result of the hard work of many people. Firstly, I gratefully acknowledge the large amount of time and effort that resulted in the invaluable comments made by my friends Drs. Ruth M. Dixon and Deborah J. G. Mackay, and thank my parents for the valuable criticisms which they made. Without their efforts this book would have been the poorer. I would also like to thank my publishers who have been of great assistance at every stage in the production of this book. In particular I would like to mention Ms Ann Berne, Dr. Jeremy Lucas and Miss Wendy Mould.

Even after every effort has been made to eliminate all possible errors, ambiguities and omissions, it is still likely that some such faults remain within the text. I would welcome any comments from readers, in particular those that point out mistakes so that they may be rectified, but also those comments that suggest additions or other alterations that would make the book of greater value to the reader.

Michael Edenborough
Saint Valentine's Day, 1994
Covent Garden, London

Part I:
Basic Principles

1

Introduction

1.1 The Aim of this Book

By the time you have worked your way through this book, you will be able to write a plausible mechanism for any organic reaction that you may meet on an undergraduate chemistry course.

For the more complicated mechanisms there may be several possible choices, but you will be able to give reasons as to why your mechanism is reasonable. In order to choose between the various theoretical possibilities, experimental data is required; but once this data has been supplied you should be able, if required, to modify your proposal accordingly so as to fit the experimental observations. However, regardless of which mechanistic route you have proposed, the various species will be reacting in accordance with their known characteristics, and for each step you will be able to justify the hypothesized electron movements. You will be able to do all this with confidence, because of an understanding of the electronic characteristics of different systems and from experience of other reactions with which you have become familiar.

To be able to achieve this, only a few fundamental ideas need to be learnt and then practised thoroughly so that you may apply these basic tools fluently in a variety of situations. Many of the errors that students make when writing organic mechanisms are very simple and stem from a lack of understanding of the basic skills of the art. Once these skills have been learnt then they may be applied to a wide range of reactions.

The basic skills include being able to count correctly the number of electrons around any given atom in a molecule, and being aware of the position of one group relative to another during a reaction. Some topics, such as thermodynamics and kinetics, will be introduced without any detailed explanation, because a full understanding is not needed in order to use the idea effectively. An analogy may be drawn with the photographer, who uses a special lens to achieve a certain effect. He may know how to use the lens and how to arrange matters to obtain the effect which he desires, but he has no detailed knowledge of the physics of refraction upon which the lens depends

1

in order to work. This latter knowledge, although interesting, is not essential to produce a successful picture. Similarly, a detailed understanding of thermodynamic or kinetics is not needed in order to utilise some of the basic points profitably.

The point of this book is to give you confidence in writing organic mechanisms, so that you can write down something sensible when needed. This will only come from the practice which results from writing down what you think is the answer to each problem in the text and then comparing your answer to the solution which is given later. Once you have worked your way through this book you should know the majority of what has been covered, because you will have been using the information and ideas at every stage. Thus there should be less need for revision! The glossaries and other appendices will form a useful reference point for the basic definitions and ideas that have been covered in this book, as well a answering those questions which may to too embarrassing to ask.

1.2 What is Organic Chemistry?

Organic chemistry originally referred to the study of chemicals that were derived from nature, *i.e.* an organic source. It was believed that these chemicals were fundamentally different from those that were mineral in origin, *i.e.* the inorganic chemicals: so different in fact, that it would be impossible ever to make an organic chemical from component parts that were derived only from mineral sources, because the latter would not contained the essential "essence of life" which was required to form organic compounds.

The fallacy of this argument was demonstrated in 1828 when Wöhler heated ammonium cyanate, an "inorganic salt", and produced urea, hitherto considered purely an "organic compound". This showed that there was no essential difference between organic and inorganic compounds. There was no "vital force" or "essence of life" which was needed to make organic compounds.

The two major branches of chemistry, however, continue to be studied separately. This is because there is still a useful distinction to be drawn between the two areas. No longer is this distinction based upon the old erroneous idea of the presence of a "vital force", but instead it is now based upon an appreciation that an understanding of the chemistry of carbon-containing compounds is necessary in order to comprehend the fundamental biochemical processes of life. In contrast, inorganic chemistry is often concerned with the structure and utility of inanimate compounds. So the distinction is the same, but the justification is different.

A present day definition of organic chemistry would be one that includes the study of carbon compounds, and in particular those compounds which possess covalent carbon/hydrogen bonds. This definition is very general and as such there are many exceptions to it. For example, carbonates and carbides are normally considered under the chemistry of the cation with which they are associated. This is because even though they contain covalent bonds they are often ionic solids with high melting points. Furthermore, some examples of these compounds are commonly found as inanimate species, such as calcium carbonate which is the main component of limestone.

To highlight the pragmatic approach of this definition the compounds which contain the cyanide unit are studied under whichever heading is most appropriate at the time. So ionic cyanides are studied under inorganic chemistry while compounds in which the cyanide group is covalently bonded to the rest of the molecule are usually studied under organic chemistry.

1.3 Organic Synthesis

Industrial organic chemistry is primarily concerned with the manufacture of useful compounds such as pharmaceuticals or commercial chemicals like glues or plastics. All these products were made from simpler compounds, *i.e.* they were synthesized. Synthesis may be defined as the rearrangement of the constituent atoms within the starting materials to give the target molecule. This involves the making and breaking of covalent bonds. Not only is the elemental composition (*i.e.* the molecular formula) of the resultant product important, but so is the spatial arrangement of the atoms (*i.e.* the structural formula).

In addition, many organic compounds are chiral, that is, they can exist in one of two forms which are called enantiomers. Enantiomers have identical chemical and physical properties and only differ in that one is the mirror image of the other. In biological systems such chiral molecules abound. Many of them are only active when they exist as the correct enantiomer: so that even though a compound may have exactly the correct elemental composition, its biological activity may be reduced from that which was expected because it is not enantiomerically pure. Sometimes the wrong enantiomer may just have no activity, but on other occasions the enantiomer with the incorrect stereochemistry may have harmful effects. For example, the drug thalidomide was produced with a small contamination of the incorrect enantiomer and it was this minor side product which led to the adverse effects upon the developing foetus within a pregnant woman who took the drug. The consequence of this is that those reactions which control the precise spatial arrangement of the constituent atoms within the target molecule are of great commercial value, not only because they avoid waste, *i.e.* more of the product of the correct enantiomer is produced, but also because they ensure that there are no contaminants which may have adverse side effects.

1.4 The Need for Mechanisms

The covalent bonds that are made and broken in the synthesis of any organic product are formed by pairs of electrons which occupy molecular orbitals. Organic mechanisms chart the rearrangement of the constituent atoms of the starting materials into the target molecule by indicating the movement of the electrons from one molecular orbital to another. This allows the mechanistic chemist not only to predict the molecular formula of the new product, but also to predict the final spatial arrangement of the atoms relative to one another, *i.e.* the stereochemistry of the product.

By studying these mechanisms an understanding of the principle synthetic routes may be gained, which in turn may lead to an optimization of the existing synthesis and the development of new synthetic methods.

1.5 Examples

This section gives a foretaste of the types of reaction that you should be able to tackle once you have worked through this book. The two examples will probably appear somewhat daunting at this stage, but they are given here to introduce some of the concepts that will be explained in more detail later, and to show how an apparently complicated reaction may be broken down into a sequence of simpler steps.

At an elementary level organic chemistry is often explained by allowing the adjacent groups of the two dimensional representations of the reagents to interact with each other. A common example of this is the formation of an ester from the reaction of an alcohol with a carboxylic acid. The hydrogen of the alcohol group and the hydroxyl group of the carboxylic acid are encircled and so bind together to form water, while leaving the truncated alcohol and acid to fuse together to form an ester. This is often referred to as "lasso" chemistry.

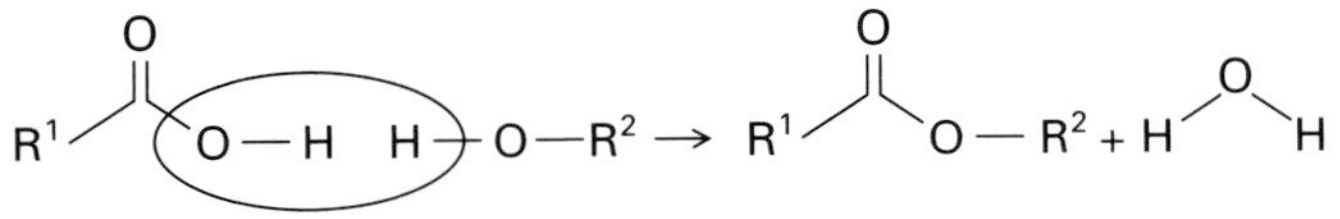

The simplicity of the idea that the most proximate groups react with each other is appealing, however, it does not led to a very sound understanding of how the bonds within the molecules rearrange themselves in order to form new compounds. Also it is of little help in anticipating what will happen when there is a choice of groups which are all near enough to react with each other. For example, in the following reaction sequence it is not immediately obvious how the epoxide ring has been formed.

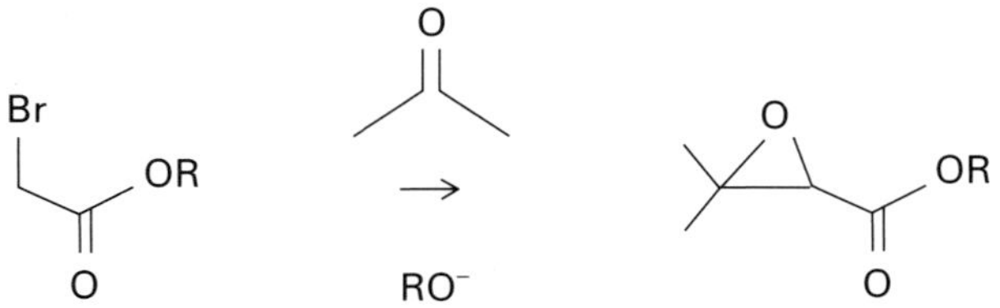

Here an α-bromocarboxylate, which has been treated with a base, reacts with a carbonyl containing compound to form an epoxide species.

It is traditional in organic chemistry to name a reaction after the person who first published it or the reagents which first characterized the reaction, and then to use that name as a convenient way of referring to it on all future occasions. This is a bit daunting to the uninitiated, but it does serve the useful purpose that a reaction type may be quickly identified. In this book we will use the appropriate name to identify reactions, but also you will find a full list in the first appendix for reference. The reaction we have just looked at is called the Darzen's condensation.

If we now look at each step of the reaction sequentially, showing the individual movement of electron pairs within the reacting molecules some understanding of what is actually happening may be gleaned.

Here the base has removed the hydrogen atom that was attached to the carbon bearing the bromine and the ester group. The resultant anion, which is stabilized by the ester group, then attacks the slightly positive, δ+, carbon of the carbonyl group.The movement of the electron pairs is indicated by the use of a curved arrow, which is why this type of representation is often referred to as "curly arrow" chemistry.

The last part of this reaction is the attack by the oxygen anion on the δ+ carbon which has bromine, which in turn is leaves so that the carbon never has more than four covalent bonds around its nucleus at any one time. The attack causes the three membered epoxide ring to be formed.

In summary, there was an initial acid/base reaction which resulted in the formation of an anion, with the negative charge on a carbon atom, which was stabilized by a neighbouring group. This anion then attacked another molecule to form a further anion, which in turn reacted, but this time internally to form a cyclic compound.

When this epoxide product is treated with aqueous acid a rearrangement of the carbon skeleton occurs. This is the glycidic acid rearrangement.

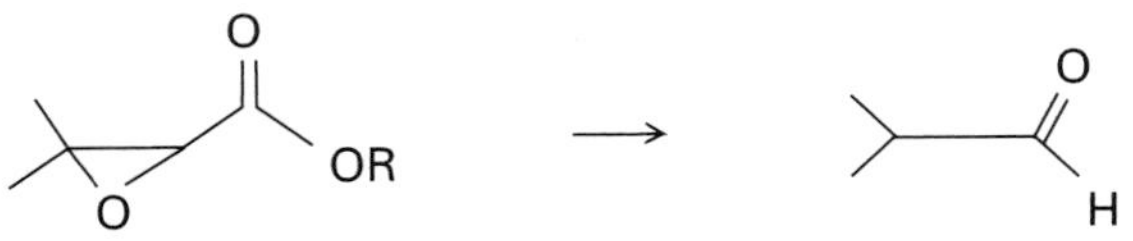

The second part of the synthetic sequence is slightly more complicated. Firstly, the ester is hydrolysed to give the carboxylate anion which then undergoes a rearrangement of the carbon skeleton that involves the migration of part of the molecule. However, each step follows the principles introduced above, *i.e.* an initial acid/base reaction followed by the stabilization of the charged species formed, then an attack by that charged species with a final reorganization.

The second example highlights the attention that must be paid to the exact spatial arrangement of the atoms within the starting materials and in particular to any intermediate products which may be formed.

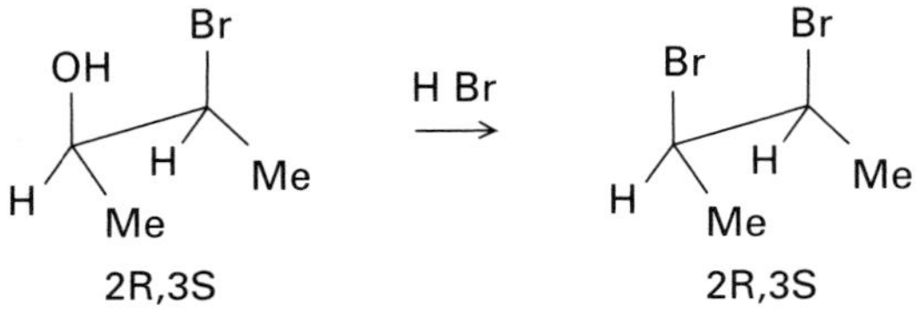

Lasso chemistry seems to be able to solve this problem well:

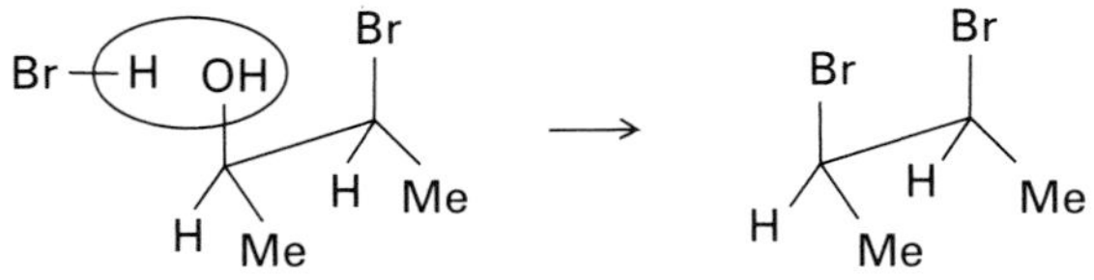

However, this analysis would predict that when (2*S*, 3*S*)-3-bromobutan-2-ol reacts with HBr the steroechemistry of the resultant product would be (2*S*, 3*S*)-2,3-dibromobutane, *i.e.* an optically active product. In fact the product is a racemic mixture of (2*S*, 3*S*)-2,3-dibromobutane and (2*R*, 3*R*)-2,3-dibromobutane. The suggested product and the product which is actually formed have exactly the same elemental composition, but the detailed stereochemistry is different.

The exact stereochemistry of the product indicates that so long as the two reactions follow the same mechanism, then the mechanism cannot proceed along the pathway indicated by the lasso approach.

The following representation of the mechanism accounts for the stereo-chemistry in both cases.

In this case there is an initial acid/base reaction in which the hydroxyl oxygen is protonated and this in turn activates the carbon to which the oxygen is attached. Then there is an internal attack by the bromine group which forms a bridged bromonium ion as an intermediate. This may then be attacked by the incoming bromide ion at either end to give the desired products. It is the formation of the bridged bromonium ion and the stereo-chemical demands which that places on any further attack which explains the stereochemistry of the products.

1.6 How to use this Book

This book is divided into three parts: Basic Principles, Mechanisms and Appendices. Within the first part all the basic ideas that are needed in order to write organic mechanisms will be discussed and explained. The areas covered are electron counting; covalent bonding and polarization; shapes of molecules; stabilization of charged species; thermodynamic and kinetic con-siderations; and lastly acid/base characteristics. In each case the underlying principles will be highlighted and many of the common errors and mis-understandings will be explained so that not only do you know what to do, but also you know what not to do and why.

The second part of the book uses the basic principles that have been intro-duced in the first part and applies them to the major mechanistic types, so that you may see how these ideas are used in practice. The mechanisms will be considered in increasing order of complexity: substitution reactions; addition reactions; elimination reactions; rearrangement reactions; and finally, redox reactions.

The majority of the text in the first two parts will be in the form of an interactive question and answer format. A point will be introduced in an instructional frame and then a question will be posed around that information. Maximum benefit will be gained if each question is attempted and your proposed answer is written down before moving onto the next frame where a suggested solution will be given. Also there will be some further elaboration that will include pointers on how to avoid the common errors or misunderstandings that are encountered when dealing with that particular sort of problem.

Sufficient information will always be given to ensure that the correct solution may be deduced without having to guess, so long as you have understood what has been taught and you have remembered what has gone before. The information will continually be used in future frames, as the later frames build upon what has gone before, and so you should know all that has been taught in the book by the time you have worked your way through it. In the introduction to some sections a little more background information will be given before the interactive question and answer sequence is commenced.

This book is not intended to cover comprehensively either every synthetic method which is available or every mechanism that is possible. Rather the object is to teach the basic principles of the tools that are required to write the fundamental organic mechanisms so that they may be written with confidence and free from errors. This will allow you to understand and appreciate the chemistry that is contained in the standard textbooks on organic chemistry.

In the third part there are various appendices, within which you will find a working definition and explanation of every common term, notation or abbreviation that you will have encountered in this book and most of those terms which you will meet in an undergraduate chemistry course. There is also a brief overview of stereochemical terminology and a short explanation as to the use of oxidation numbers in covalent molecules. These sections may be used as a quick source of information while using the main text, or later as a ready reference.

2

Electron Counting

2.1 Introduction

The ability to count correctly the number of electrons around any given atom in an organic molecule is central to being able to write a plausible mechanism. Many of the errors that are found in the scripts of first year university students, and even more in the case of second year A-level students, result from either placing too many electrons around an atom, *e.g.* pentavalent carbons, or by placing the wrong charge on an atom, *e.g.* R_3NH. Both of these errors stem from an inability to count the number of electrons on the central atom and then to realize the mistake involved, namely in the first case that a carbon atom may never have ten electrons around it; and in the second case, that a nitrogen atom which is sharing eight electrons between four covalent bonds must bear a single positive charge.

So to start with we are going to learn how to count the electrons around just a single atom or ion. This is very simple but you must be absolutely fluent in doing this flawlessly before you progress to molecules in which the electrons are shared between different atoms.

2.2 Atoms

Atoms are formed from protons, neutrons and electrons. The protons and neutrons together constitute the nucleus, while the electrons occupy orbitals which are centred on the nucleus. The number of protons within the nucleus determines the element. This number is called the atomic number. The sum of the number of neutrons and protons is called the atomic mass number. The number of neutrons within the nucleus of a particular element is not fixed, though usually there are approximately the same number of neutrons as there protons. For a given element, atoms which contain different numbers of neutrons are called isotopes and they have different physical properties, *e.g.* a different atomic mass number, and a different stability of the nucleus. However, most importantly, the chemical properties of different isotopes are the same. The smallest unit of an element is the atom.

9

Neutrons are electrically neutral, while protons carry a single positive charge and electrons carry a single negative charge. The number of electrons always equals the number of protons in an atom. Thus every element in the atomic state is electrically neutral.

The element which has only one proton is hydrogen, *i.e.* its atomic number is one. The commonest isotope of hydrogen does not possess a neutron in its nucleus, but only contains a solitary proton. More strictly this isotope should be called protium in order to distinguish it from the other isotopes of hydrogen with one or two neutrons within the nucleus and which are called deuterium and tritium respectively. Draw a diagram which indicates the arrangement of the protons, neutrons and electrons of these three isotopes of hydrogen.

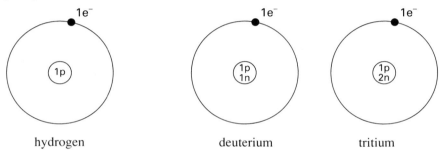

hydrogen   deuterium   tritium

What you drew should have shown for each isotope a single electron in an orbital centred on the nucleus; while within the nucleus there was always one proton with zero, one or two neutrons depending on whether the diagram was meant to represent hydrogen, deuterium or tritium.

Normally, the various isotopes of an element are not distinguished by separate names, but instead are distinguished by having the different atomic mass numbers as a raised prefix to the atomic symbol. Hydrogen is unusual in that each isotope has a separate name and also may be represented by a separate symbol, namely H, D or T, although the conventional ^1H, ^2H or ^3H may also be used. When referring to an element it is usual to imply that the most stable isotope, or the naturally occurring mixture of isotopes, is meant unless otherwise stated.

Helium is the element which has the atomic number two. There are two common isotopes which contain one and two neutrons respectively. Draw two diagrams to represent these two isotopes and also write down the conventional atomic symbols.

You should have drawn a nucleus which contained two protons and either one or two neutrons depending on the isotope, and that in each case two electrons occupied an orbital centred on the nucleus so as to ensure that the atom was electrically neutral. The conventional symbols are ^3He and ^4He respectively.

The chemical properties of an element are determined by its electrical properties and so chemists are particularly interested in the placement of

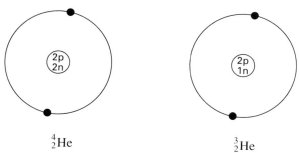

$$^4_2He \qquad ^3_2He$$

positive and negative charges. If an element loses or gains one or more electrons then that element will no longer be electrically neutral as the number of protons will not equal the number of electrons that surround the nucleus. The resultant charged species is called an ion. If the ion is negative then it is an anion, while if it has a positive charge it is a cation.

The hydrogen atom may lose or gain one electron to form either a cation or an anion respectively. Draw two diagrams to represent the resultant anion and cation of hydrogen.

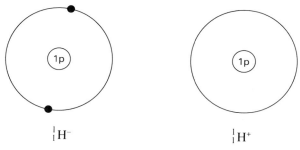

$$^1_1H^- \qquad ^1_1H^+$$

The cation of hydrogen is a bare proton, while the anion of hydrogen has one proton and two electrons, and is called a hydride ion. Ions of elements are represented by the atomic symbol of the element with the appropriate overall charge as a raised postscript. So these ions would be H^+ and H^- respectively.

The commonest isotope of helium is ^4He, and this may lose one or two electrons to form two different cations. Draw a diagram to represent these two cations.

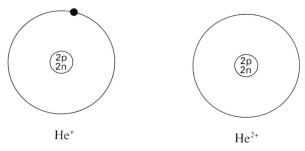

$$He^+ \qquad He^{2+}$$

The cation of helium which results from the loss of one electron carries a single positive charge as there is only one electron around a nucleus which contains two neutrons and two protons. The cation which results from the loss of two electrons carries a double positive charge and is a bare nucleus consisting of two neutrons and two protons. This species is called an alpha particle.

If there is an odd number of electrons in an atom or ion, then it is called a radical. Draw the electronic configurations of all the radicals of hydrogen and helium.

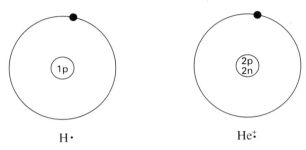

H·   He⁺

The hydrogen atom has only one electron and so is a radical; while the unipositive cation of helium is the only species which has an odd number of electrons. All the other species have either none or two electrons and so are not radicals.

The pairing of electrons is important because it confers a degree of stability on the species concerned. Electrons in an atom are contained within atomic orbitals, each of which may only hold a maximum of two electrons. The electron in a hydrogen atom is contained in the first principal quantum shell, which is indicated by "1". Within this principal quantum shell there is only one type of subshell that is represented by the letter "s". The electronic configuration of hydrogen may be written as $1s^1$, while the raised postscript indicates that there is only one electron.

In helium the two electrons are both contained within the first principal quantum shell. Write down the electronic configuration of helium.

The electronic configuration of helium is $1s^2$.

Each subshell is an orbital and so may contain a maximum of tow electrons. Write down the electronic configurations of the ions of hydrogen and helium that have been discussed above, *i.e.* $^1H^+$, $^1H^-$, $^4He^+$ and $^4He^{2+}$.

The cation of hydrogen is $1s^0$, and the anion of hydrogen is $1s^2$, and the cations of helium are $1s^1$ and $1s^0$ for the unipositive and bipositive cations respectively.

As each orbital may contain a maximum of two electrons the helium atom has a full complement at $1s^2$ as does the hydrogen anion. There is only one subshell in the first principal quantum level and hence there can only be two elements which have electrons only in this level. That is why there are only two first row elements of the periodic table.

In the second quantum level there are four subshells, one "s" and three "p" subshells, each of which may contain two electrons, which gives a total of eight electrons in all at the second level. This is why there are eight second row elements. It is possible to add or remove electrons to or from a principal quantum level which is different from that of the neutral atom. However, this requires a lot of energy and so does not usually occur. Write down the electronic configuration of the uninegative anion of helium and hence suggest why this species is not readily formed.

He$^-$ has the electronic configuration of $1s^2, 2s^1$.

The electron that was added to the helium atom to form the anion could not fit into the "s" subshell of the first principal quantum shell because it already had two electrons and so was full. Instead this extra electron occupied the next available orbital, which was the "s" subshell of the second principal quantum level. This requires a lot of energy and so it is difficult. Hence the He$^-$ anion is not normally found.

The second row elements, Li to Ne, sequentially fill the available four sub-shells, starting with the "s" subshell and then progressing to the three "p" subshells, the latter each have the same energy, *i.e.* they are degenerate. The three "p" subshells are distinguished by the subscripts x, y and z. If electrons are placed in degenerate orbitals they fill them so as to minimize the doubling up of electrons, so as to reduce the electron/electron repulsion. This is called Hund's rule. Write down the electronic configurations of the second row elements.

Li $1s^2, 2s^1$; Be $1s^2, 2s^2$; B $1s^2, 2s^2, 2p^1$; C $1s^2, 2s^2, 2p_x^1, 2p_y^1$; N $1s^2, 2s^2, 2p_x^1, 2p_y^1, 2p_z^1$; O $1s^2, 2s^2, 2p_x^2, 2p_y^1, 2p_z^1$; F $1s^2, 2s^2, 2p_x^2, 2p_y^2, 2p_z^1$; Ne $1s^2, 2s^2, 2p_x^2, 2p_y^2, 2p_z^2$

As the loss or gain of electrons normally only occurs from the highest principal quantum level which is occupied in the atom, this is the level which is concerned with the chemistry of the element, and is called the valence shell. Usually only the electrons within the valence shell are indicated and the electrons within the core shells are assumed to be in order.

Suggest the electronic configurations for the anions of the second row elements formed by the gain of one electron.

Li$^-$ $1s^2, 2s^2$; Be$^-$ $1s^2, 2s^2, 2p^1$; B$^-$ $1s^2, 2s^2, 2p_x^1, 2p_y^1$; C$^-$ $1s^2, 2s^2, 2p_x^1, 2p_y^1, 2p_y^1$; N$^-$ $1s^2, 2s^2, 2p_x^2, 2p_y^1, 2p_z^1$; O$^-$ $1s^2, 2s^2, 2p_x^2, 2p_y^2, 2p_z^1$; F$^-$ $1s^2, 2s^2, 2p_x^2, 2p_y^2, 2p_z^2$

In each case the ion is uninegative because one electron has been added to the neutral atom and so the number of electrons now exceeds the number of protons by one. The electronic configurations of the uninegative ions corresponds to that of the element one to the right in the Periodic Table. Notice that as neon already has eight electrons so no more may be added to it without a great deal of energy being expended thus the Ne$^-$ ion is very difficult to form. Now write down the electronic configuration of the ion which results from the addition of two electrons to the oxygen atom, and the electronic configuration of the two ions which result from the removal of one and two the beryllium atom.

O^{2-} $1s^2, 2s^2, 2p_x^2, 2p_y^2, 2p_z^2$
Be^+ $1s^2, 2s^1$
Be^{2+} $1s^2, 2s^0$

The electron configuration of the binegative anion corresponds to that of the neutral element two positions to the right of the original element *i.e.* in this case the fluorine atom. In contrast the electronic configurations of the cations correspond to those of the elements to the left of the original element, and carry a single or double positive charge depending on how many electrons have been lost.

It is difficult to form certain ions of the second row elements, *e.g.* Ne^-, F^{2-} and Li^{2+}, because in each case it would be necessary either to remove (in the case of lithium), or to add (in the case of neon and fluorine), an electron from a different principal quantum level than that in which the electrons of the neutral atom were accommodated.

As each element in the second row has four subshells in the second principal quantum level and as each subshell may only accommodate a maximum of two electrons, the result is that no element in the second row may have more than eight electrons in the second principal quantum shell. This is called the "Octet Rule", and is one of the most fundamental rules concerning the electron distribution around second row elements. Under normal conditions is it never violated.

The third row elements have more subshells than the second row elements, *namely* one "s" subshell, three "p" subshells and 5 "d" subshells. This means that they may accommodate a maximum of 18 electrons. As far as the organic chemist is concerned the important consequence of this is that the elements phosphorus and sulphur may expand their octet and have five or six bonds, instead of being limited to four bonds, as is carbon.

Now that we can count the unshared electrons around an atom or ion, we will look at molecules in which the electrons are shared between more than one atom.

2.3 Uncharged Molecules

In an atom the charge that is resident may easily be determined by the difference between the number of electrons that are accommodated in orbitals around it and the number of protons within the nucleus. For molecules the matter is more complicated because there are two or more nuclei to consider. Yet in principal the same counting process is involved.

The simplest molecule is bihydrogen, H_2. This consists of two hydrogen nuclei joined by one covalent bond which contains two electrons, one donated from each hydrogen atom. Each hydrogen atom now has two electrons accommodated within a molecular orbital. Each molecular orbital, like each atomic orbital, may only hold two electrons. What is the charge on each hydrogen atom?

Each hydrogen atom is neutral, because each hydrogen nucleus sees the two electrons on average for half of the time. That is, in a single covalent bond between identical atoms, each bonded atoms shares the electrons equally with the other nucleus. Thus 2 electrons shared by two nuclei gives rise to an effective single negative unit. So subtracting the one positive charge of the proton leaves an overall zero charge in electronic terms. Hydrogen is only ever involved in one covalent bond.

The overall charge on an atom is, of course, the difference between the number of protons in its nucleus and the total number of electrons in the core and valence shell. When discussing atoms in molecules, however, it is easier to consider only the electrons in the valence shell, it being assumed that the electrons in the core shell are fully occupied. The the charge on any given atom can then be calculated according to the following formula.

Firstly, find the number of electrons that would surround the uncharged, isolated atom in its valence shell. Then subtract the total number of electrons that are actually in the valence shell of that atom within the molecule. Lastly, add half of the number of electrons which that atom shares with any other atom in the molecule. The result is the charge on that atom within the molecule.

For the elements Li to Ne, write down the number of electrons that surround the neutral atom in the valence shell.

For Li to Ne the number increases form one to eight sequentially.

In the diatomic molecule of fluorine, there is one covalent bond joining the two fluorine atoms. Draw the dot and cross structure of this molecule, using a dot to represent the electrons in the valence shell of one fluorine atom, and a dot to represent the electrons from the other fluorine atom. Then calculate the charge on each fluorine atom within this molecule.

The charge on each fluorine atom in the diatomic fluorine molecule is zero. Normally every fluorine atom has 7 electrons in the valence shell, which are combined into three lone pairs and one unpaired electron. It is this unpaired electron which combines with the unpaired electron of the other fluorine atom to form the single covalent bond. Each fluorine atom now has eight electrons around its nucleus. So now each fluorine nucleus is involved in one single covalent bond which contains two electrons within a molecular orbital which is accommodated in a molecular orbital between the two nuclei. These two electrons are shared equally between the two fluorine nuclei. Using the above formula one has seven electrons normally in the valence shell, minus the eight that are actually present, plus half of the two that are shared in the single covalent bond, to result in an overall zero charge.

Neon has eight electrons in its valence shell and chemically it is a very stable species. This stability is due to the atom having a full valence shell. When atoms combine to form molecules, they obtain a full complement of electrons in the valence shell and hence obtain this stability. Carbon has four electrons in its valence shell and needs four more electrons to reach the stable octet of electrons. This it may do by combining with four hydrogen atoms to form a methane molecule. Write down a dot and cross representation of this molecule and calculate the charge that is present on each of the five atoms.

$$\begin{array}{c} \text{H} \\ {}_{\text{x}}\ {}^{\bullet} \\ \text{H}\ {}^{\bullet}_{\text{x}}\ \text{C}\ {}^{\text{x}}_{\bullet}\ \text{H} \\ {}_{\bullet}\ {}^{\text{x}} \\ \text{H} \end{array}$$

All the atoms have zero charge. The four hydrogens are in identical environments. Hydrogen has one electron in its valence shell in its uncharged, atomic state. Now there is a total of two electrons in one covalent bond accommodated in a molecular orbital around the nucleus, both electrons being shared with the central carbon atom. Thus each hydrogen atom bears a zero charge. The carbon nucleus has a total of eight electrons accommodated in four molecular orbitals. All of the electrons are shared with the hydrogen atoms. It has four electrons in the atomic state in its valence shell. Hence it bears a zero overall charge.

Draw the dot and cross representation of the dihydride of oxygen, *i.e.* water, and so calculate the charge that is resident on each of the three atoms.

$$\begin{array}{c} {}^{\text{x x}} \\ {}^{\text{x}}\ \text{O}\ {}^{\bullet}_{\text{x}}\ \text{H} \\ {}_{\text{x}}\ {}^{\text{x}}_{\bullet} \\ \text{H} \end{array}$$

Each atom is neutral. The same counting procedure applies to the hydrogens as it did in methane. Oxygen has six electrons in its atomic state in its valence shell. In the water molecule the oxygen atom has a total of eight electrons accommodated around its nucleus in the valence shell, and so it has a stable octet. Four of these electrons are involved in the two single covalent bonds with the hydrogen atoms. Hence the charge on the oxygen atom is zero. Notice that of the eight electrons that surround the oxygen nucleus, four are in two molecular orbitals forming the two covalent bonds, while the other four are in two lone pairs. The latter four electrons are not shared with any other atom.

Draw the dot and cross representation of the trihydride of nitrogen, *i.e.* ammonia, and so calculate the charge that is resident on each of the four atoms.

$$ H \overset{\bullet}{\underset{x}{x}} N \overset{x}{\underset{\bullet\ x}{x}} H $$
$$ H $$

Each atom is neutral. For the nitrogen atom there are three molecular orbitals which contain six electrons and also one lone pair of electrons, making a total of eight electrons. In the atomic state of nitrogen there are five electrons in the valence shell. The nitrogen atom is involved in three covalent bonds which require six electrons, hence giving a resultant charge of zero, again!

Now that you are fully familiar with the number of electrons that accommodate the valence shell of the common elements when they are combined in neutral species, we may proceed to the next step which is the counting of electrons in molecular species that carry a whole unit of charge and try to find which atom has gained or lost an electron.

2.4 Molecules with Whole Charges

This is the area in which most mistakes are made. Most mechanisms that will be examined in the second part of this book deal with intermediates which carry one or more whole charges. This arises because most mechanisms start by splitting a covalent bond unequally so that one part has an extra electron while the other part has a corresponding deficiency, which means that each part carries a charge.

When drawing atoms or molecules which carry a whole electric charge the charge itself is indicated by a plus or minus sign which in some texts is enclosed within a circle. Here, however it will be just the plain +/- sign, which is placed as raised postscript to the atomic symbol. You are already familiar with the notation that an electron is represented by a dot, "•". Sometimes a short solid bar, "-", is used to represent a pair of electrons, though this notation may be easily confused and thus is best avoided. These bars are placed around the atomic symbol to form a box which is not joined up at the corners, *e.g.* for neon which as four electron pairs:

$$ \overset{x\,x}{\underset{x\,x}{{}_{x}^{x}Ne_{x}^{x}}} \quad \equiv \quad \overline{|\,Ne\,|} $$

When a molecular species is indicated with a net charge shown, then often the charge is shown at the end of the group symbol and so does not necessarily indicated the atom or atoms on which the charge resides.

So long as the electrons are carefully counted on every atom then mistakes may be easily avoided. With practice it will become second nature as to what should be the correct charge on each species.

A molecule of water may be ionized to give a˙ hydroxyl ion and a hydrogen ion. Write down the equation for this reaction and so calculated the charge on each species.

$$H_2O = OH^- + H^+$$

The hydroxyl ion bears a single negative charge, while the hydrogen ion bears a single positive charge. We have already looked at how to calculate the charge on the hydrogen ion. Within the hydroxyl group, the hydrogen is involved in one single covalent bond which is occupied by two electrons, hence it is neutral. The uncharged atomic oxygen would have six electrons in its valence shell. However in the hydroxyl group, eight electrons are accommodated around it, which are involved in only one single covalent bond with the hydrogen atom. Hence there is a single unit of negative charge resident upon the oxygen atom. Draw a dot and cross representation of the hydroxyl ion.

$$\left[\begin{array}{c} \overset{\times\ \times}{\underset{\times\ \blacksquare}{\times}} \ \overset{}{O}\ \overset{\bullet}{\underset{\times}{}}\ H \end{array} \right]^{-} \qquad \blacksquare \ \text{extra electron}$$

There are three lone pairs on the oxygen and one single covalent bond, and the oxygen atom bears a negative charge. The oxygen when it is involved in two single covalent bonds has two lone pairs of electrons. When it only has one single bond and three lone pairs there is in effect an extra electron, which accounts for the single negative charge.

Write down the equation for the addition of a hydrogen cation to a water molecule to give a hydroxonium ion.

$$H_2O + H^+ = H_3O^+$$

Now draw a dot and cross representation of the hydroxonium ion.

$$\left[\begin{array}{c} H\ \overset{\times\ \times}{\underset{\times\ \bullet}{\times}}\ \overset{}{O}\ \overset{\bullet}{\underset{\times}{}}\ H \\ H \end{array} \right]^{+}$$

Calculate the charge which is resident on each atom in the hydroxonium ion.

For each of the hydrogen atoms the charge is zero, while on the oxygen it is plus one. There are eight electron around the oxygen, compared with the six which there are in the valence shell of the uncharged atom. The oxygen atom is involved in three single covalent bonds, which each accommodate two electrons. This means that there is a deficit of one electron on the oxygen atom and so it carries a plus one charge. Note that there is now only one lone pair of electrons. The second lone pair that normally exists in involved in the dative bond to the hydrogen cation.

Write the equation which represents the further addition of another hydrogen cation to the hydroxonium ion. Draw a dot and cross structure for the resultant ion.

$$\left[\begin{array}{c} H \\ {}^{x\,x} \\ H \ {}^{x}_{x}O{}^{\bullet}_{x}H \\ {}^{x\ \bullet} \\ H \end{array} \right]^{2+}$$

$$H_3O^+ + H^+ = H_4O^{2+}$$

Calculate the charge which is resident on each atom in this ion.

For each of the hydrogens the charge is zero, while on the oxygen it is plus two. There are eight electrons accommodated in the valence shell of the oxygen atom. These eight electrons are involved in four single covalent bonds with the four hydrogen atoms. Thus there is a deficit of two electrons and so the oxygen carries a plus two charge. Note that there are no lone pairs of electrons left as they are both now involved in dative bonds to the two extra hydrogen cations.

For each of the three equations that have just been given add up the charges on the left hand side of the equation and those on the right.

$H_2O = OH^- + H^+$	0	0
$H_2O + H^+ = H_3O^+$	$+1$	$+1$
$H_3O^+ + H^+ = H_4O^{2+}$	$+2$	$+2$

On every occasion the total charge is the same before the reaction as it is after the reaction. This observation is embodied in the principle called the Conservation of Charge, which is never broken, even in exotic situations like subatomic physics and astrochemistry. It is a very simple principle, but it is a common error to violate this principle by losing or gaining charges.

Another facet of this principle is placing the wrong charge on a molecular species. We will now look at a few common mistakes. Consider what is wrong with the charged species: H_4O^+.

If you are having problems, try to write a balanced equation to form this species from known species, and then try to draw the dot and cross structure.

Using the principle of the Conservation of Charge:

$$H_4O^{2+} + e^- = H_4O^+ \qquad\qquad +1 \qquad\qquad +1$$

$$\begin{bmatrix} H \\ H \overset{x\,x}{\underset{x\ \bullet}{\overset{x}{\underset{x}{O}}}} H \\ H \end{bmatrix}^{2+} \xrightarrow{\ \beta^-\ } \begin{bmatrix} H \\ H \overset{x\,x}{\underset{\blacksquare\,x\,\bullet}{\overset{x}{\underset{x}{O}}}} H \\ H \end{bmatrix}^{+} \qquad \blacksquare \ \text{extra electron}$$

If an electron, e⁻, is added to the H_4O^{2+} species in order to reduce the overall charge to plus one, on drawing out the required electronic structure it becomes apparent that there are now nine electrons attempting to be accommodated around the oxygen. This would violate the Octet Rule. Thus this species cannot exist under normal circumstances. When calculating the charge on a molecular species both the Octet Rule and the principle of the Conservation of Charge must be followed.

If instead of using water, one or both of the hydrogens is replaced by a general alkyl group, R, then a general alcohol or ether respectively is formed. Write down these two general formulae.

| alcohol | ROH |
| ether | R_2O or ROR |

How many lone pairs are there on the oxygen in the alcohol and the ether?

There are two lone pairs in each case on the oxygen atom.

When a hydrogen ion is added to a species, the process is called protonation. Write a balanced equation for the protonation of a general alcohol and a general ether.

$$ROH + H^+ = ROH_2^+$$
$$R_2O + H^+ = R_2OH^+$$

Notice that in each case the oxygen now has three groups joined to it by single covalent bonds and there is only one lone pair of electrons remaining. Suggest a generalized formula for the ionic and neutral compounds of oxygen which contain eight electrons around the oxygen atom.

$$O^{2-},\ XO^-,\ X_2O,\ X_3O^+ \text{ and } X_4O^{2+}.$$

These general formulae describe the ionic and neutral states of all the common specie of oxygen compound. The middle three species are the ones which are common in organic chemistry. The dianion is common in inorganic compounds as the ionic oxide. The dication is rare in all circumstances since the high charge density upon the oxygen reduces its stability.

In summary if oxygen has one covalent bond, then it carries a minus one charge; with two covalent bonds it is neutral; and with three covalent bonds it has a plus one charge.

The next element to the left of oxygen in the Periodic Table is nitrogen. This has five electrons in its valence shell and so in order to make up the octet it may become involved in three covalent bonds. Draw a dot and cross structure for the trihydride of nitrogen, and calculate the charge which is resident on each atom.

$$H \overset{x}{\underset{\bullet}{}} \overset{\times\times}{N} \overset{\bullet}{\underset{\times}{}} H$$
$$\overset{\times\,\bullet}{\underset{H}{}}$$

The charge is zero on each atom. This compound is called ammonia. Note that there is only one lone pair of electrons.

Write an equation of the protonation of ammonia, and then calculate the charge on each atom. Draw the resultant ion.

$$\left[\begin{array}{c} H \\ H \overset{x}{\underset{\bullet}{}} \overset{\times\times}{N} \overset{\bullet}{\underset{\times}{}} H \\ \overset{\times\,\bullet}{H} \end{array} \right]^{+}$$

$$NH_3 + H^+ = NH_4^+$$

The charge on the nitrogen is now plus one, while it is still zero on each of the hydrogen atoms. The lone pair of the ammonia is now utilized in the dative bond to the hydrogen ion. This cation is called the ammonium ion. Note that the principle of Conservation of Charge has been obeyed in the equation, and also that the nitrogen has only eight electrons around it and so still conforms with the Octet Rule.

Write an equation for the deprotonation of the ammonia molecule. Calculate the charge on each atom and draw a dot and cross representation for the resultant species.

$$\left[\begin{array}{c} \overset{\times\times}{N} \overset{\bullet}{\underset{\times}{}} H \\ \overset{x}{\underset{\blacksquare}{}} \\ \overset{\times\,\bullet}{H} \end{array} \right]^{-}$$ ■ extra electron

$$NH_3 = NH_2^- + H^+$$

The charge on each of the hydrogen atoms is still zero, while on the nitrogen it is now minus one. This ion is called the amide ion and has two lone pairs of electrons. Again notice that the overall charge has been conserved and that the nitrogen still has eight electrons accommodated in orbitals around it. However, as there are now only two single covalent bonds, there is an overall excess of one electron and so the molecule bears a single negative charge.

Gives reasons why the NH_5^{2+} ion does not exist under normal circumstances.

By using the principle of the Conservation of Charge, the following equation may be suggested for the formation of the NH_5^{2+} ion:

$$NH_4^+ + H^+ = NH_5^{2+}$$

However, adding a hydrogen cation does not add any electrons, and so there are still eight electrons left around the nitrogen. Yet these eight electrons are required to bond five hydrogen atoms to the central nitrogen atom. This would make available less than two electrons per bond. Even though this is possible, it does not occur under normal circumstances and so for the present purposes we may assume that this molecule does not exist because there are too few electrons to hold all the hydrogens in place.

Suggest why the species NH_5^+ does not exist under normal circumstances.

By using the principle of the Conservation of Charge, the following equation may be suggested for the formation of the NH_5^+ ion:

$$NH_4^+ + H = NH_5^+$$

If a hydrogen atom is added to the ammonium ion then an extra electron would be added to the eight that already are accommodated around the nitrogen and this would bring the total to nine which would violate the Octet Rule. Thus we have shown why nitrogen has a maximum of four covalent bonds radiating from it.

Now suggest why the neutral species NH_4 does not exist.

By using the principle of the Conservation of Charge again, the following equation may be suggested for the formation of the NH_4 species:

$$NH_4^+ + e^- = NH_4$$

However, adding one electron to the eight that already are accommodated around the nitrogen would violate the Octet Rule; so for this reason this species does not exist.

One or more of the hydrogen atoms may be replaced by alkyl groups. Write down the general formulae of the neutral nitrogen compounds with one, two and three alkyl groups.

$$RNH_2, R_2NH \text{ and } R_3N$$

Notice that there are only three groups in total, and each species has one, and only one, lone pair. The univalent $\leftarrow NH_2$ group is called the amino group. Unlike the alkyl derivatives of oxygen where the mono- and di-substituted oxygen compounds were given different names, *i.e.* alcohols and ethers respectively, all three types of substituted nitrogen species are classed as amines. However, they are sub-classed into primary, secondary and tertiary amines depending on whether the nitrogen bears one, two or three alkyl groups.

Write equations for the syntheses of the products which result from the mono-protonation of each type of amine.

$$RNH_2 + H^+ = RNH_3^+$$
$$R_2NH + H^+ = R_2NH_2^+$$
$$R_3N + H^+ = R_3NH^+$$

Write equations for the syntheses of the products which result from the mono-deprotonation of each type of amine.

$$RNH_2 = RNH^- + H^+$$
$$R_2NH + H^+ = R_2N^- + H^+$$

The tertiary amine cannot lose a proton from the nitrogen, it can only be protonated.

Write down the generalized formulae for the ionic and neutral compounds of nitrogen which contain eight electrons around the nitrogen atom.

$$N^{3-}, XN^{2-}, X_2N^-, X_3N \text{ and } X_4N^+$$

The trianion is found in inorganic compounds, while the last three species are common in organic chemistry.

In summary, if nitrogen is involved in two covalent bonds then it carries a minus one charge; if it is involved in three covalent bonds then it is neutral, and if it is involved in four covalent bonds then it bears a plus one charge.

The element to the left of nitrogen is carbon. This has four electrons in its valence shell in the uncharged atomic state. Thus in order to achieve the octet it must combine with four other atoms to form four single covalent bonds. Draw the dot and cross structure of the tetrahydride of carbon, *i.e.* methane.

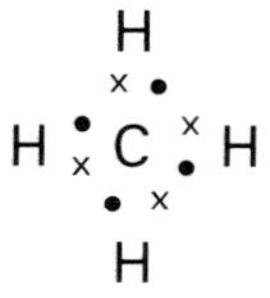

Now write an equation for the deprotonation of methane, and draw the structure of the resultant species.

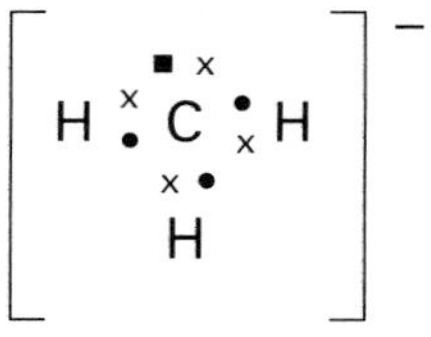

■ extra electron

$$CH_4 = CH_3^- + H^+$$

Notice that the carbon atom has one lone pair of electrons and is involved in three single covalent bonds. This carbon species is called a methyl carbanion.

If instead of a proton being removed from the methane molecule, a hydride anion is removed, then a positive carbon species will be formed. Write an equation for its formation and draw its structure.

$$\left[\begin{array}{c} H \overset{\times}{\underset{\bullet}{}} C \overset{\bullet}{\underset{\times}{}} H \\ \overset{\times\;\bullet}{H} \end{array} \right]^{+}$$

$$CH_4 = CH_3^{+} + H^{-}$$

This carbon species is called a methyl carbonium ion or a carbocation. It is different from the charged species that we have looked at before, because it only has six electrons accommodated around the central atom. These electrons occupy three single covalent bonds. The central atom is thus electron deficient.

Suggest reasons why the species CH_5^{+} and CH_5 do not exist under normal circumstances.

By using the principle of the Conservation of Charge, the following equations may be suggested for the formation of the CH_5^{+} ion and CH_5 species:

$$CH_4 + H^{+} = CH_5^{+}$$
$$CH_4 + H = CH_5$$

In the first case the carbon would now have nine electrons accommodated around it. In the second case it would have ten electrons. In both cases this would violate the Octet Rule. So under no normal circumstances do we find pentavalent carbon atoms, regardless of whether they are charged or neutral.

So far we have looked at the CH_3^{-} and CH_3^{+} ions, the first has eight electrons and the second has six electrons accommodated around the carbon. Suggest a structure for the trihydride of carbon which has seven electrons around the carbon, and calculate the charge which is resident on the carbon. Suggest an equation for its formation from methane.

$$H \overset{\times}{\underset{\bullet}{}} \overset{\bullet}{C} \overset{\bullet}{\underset{\times}{}} H$$
$$\overset{\times\;\bullet}{H}$$

$$CH_4 = CH_3\bullet + H\bullet$$

The $CH_3\bullet$ species is neutral. It has seven electrons accommodated around the carbon, six of which are in three single covalent bonds to the three hydrogens. This leaves a single unpaired electron on the carbon. Molecules which have unpaired electrons are called radicals. This particular radical is called the methyl radical.

Write down the general formulae of the charged and neutral trisubstituted carbon species that have been discussed so far and indicate the number of electrons around each carbon.

$$CX_3^-, \ CX_3 \bullet \ \text{and} \ CX_3^+$$

The carbon atoms bear eight, seven and six electrons respectively. These are the three principal carbon species that are found in organic chemistry.

There is one more carbon species that is encountered less frequently. Write down the product which results from the loss of the X^- anion from the CX_3^- species. Draw its dot and cross structure.

$$H \ {}^{x}_{\bullet} \ C \ {}^{x}_{x} \\ {}^{x}_{} \ {}_{\bullet} \\ H$$

$$CX_3^- = {:}CX_2 + X^-$$

The carbon atom has only six electrons accommodated around it, four of which are involved in two single covalent bonds, while the other two are in a lone pair. The carbon atom is electron deficient but uncharged. This type of carbon species is called a carbene and is highly reactive.

We have looked at species which have gained or lost electrons and so the charge that they bear must be either zero or an integral positive or negative number. There have been no fractional charges. Also we have used the principle of the Conservation of Charge to ensure that all the electrons have been accounted for when suggesting equations for the formation of the different species. We have also found that elements in the second row of the periodic table cannot accommodate more than eight electrons in the valence shell, although in some circumstances they can have fewer than eight valence electrons.

In the next chapter we will look at partial charges.

3

Covalent Bonding and Polarization

3.1 Introduction

In this chapter we will look at covalent bonds and the manner in which they may be broken. Prior to discussing the complete breaking of a bond it is natural to investigate bonds which exhibit a degree of polarization.

3.2 Partially Charged Species

In a homoatomic bond, *i.e.* one between two atoms which are the same, the electrons in the covalent bond are equally shared between the two nuclei. This means that there is no permanent charge separation, *i.e.* one end is not permanently positive nor the other permanently negative. Suggest two homoatomic molecules from the first two rows of the Periodic Table.

e.g. H-H and F-F

There may be no permanent separation of charge in such homoatomic molecules, but there may be a temporary separation of electronic charge in a covalent bond. The electron density within the molecular orbital is subject to random fluctuations which result in a random, temporary separation of charge called the London effect. These fluctuations average to zero over time.

There is another way in which temporary charge separation may occur. Suggest what would happen if a homoatomic molecule was brought close to a fixed positive charge.

The positive charge would attract the electrons within the bond and hence the end closer to the fixed positive charge would become negative. The other end of the bond would become positive as its is now electron deficient. Here a charge separation has been induced by the approach of a fixed charge.

A charge separation is called a dipole, because there are two poles, one positive and the other negative. If the molecule is neutral then the negative

27

and positive charges that have been separated must sum to zero. Also note that only small amounts of charge are separated, not full units of one electron. The sign δ is used to indicate this small amount of charge. It is used as a raised postscript with the appropriate sign. Draw the dihydrogen molecule which is next to a fixed positive charge, and indicate the charge separation that has been induced.

$$H \overset{\delta+}{\rule{3cm}{0.4pt}} H^{\delta-} \quad [+]$$

The hydrogen closest to the fixed positive charge will bear a small negative charge, while the other hydrogen will bear an equally small positive charge. The sum of these charges is obviously zero; while the sum of the modulus, *i.e.* the magnitude of the charge without consideration of whether it is positive or negative, of these charges is less than one.

When there are two different atoms in a diatomic molecule is a heteroatomic bond exists. Each element has a different affinity for electrons, and this is called its electronegativity. The greater an element's attraction for electrons, the greater its electronegativity. So far we have ignored the difference in the electronegativities of the different atoms when we have looked at the hydrides of carbon, nitrogen and oxygen, because the effect is small. In the diatomic molecule of hydrogen fluoride, suggest which atom carries the small negative charge, bearing in mind that fluorine is the most electronegative element in the periodic table.

$$H \overset{\delta+}{\longrightarrow} F^{\delta-}$$

The fluorine atom is more electronegative than the hydrogen atom, and thus attracts the electrons within the single covalent bond more strongly than the hydrogen atom. This dipole is permanent and is indicated on a diagram of the molecule by an arrowhead on the line which represents the single covalent bond. The arrowhead points towards the negative end of the bond. This attraction of electrons towards one end of a covalent bond is called induction. A group that attracts electrons is said to exert a negative inductive effect and a *vice versa*. The symbol for this property is I.

Show the inductive effect on the following bonds: N-H, O-H and C-H.

$$H \overset{\delta+}{\longrightarrow} N^{\delta-}$$

$$H \overset{\delta+}{\longrightarrow} O^{\delta-}$$

$$H \overset{\delta+}{\longrightarrow} C^{\delta-}$$

In all cases the hydrogen is the positive end of the covalent bond. The numbers which represent the relative magnitudes of the electronegativities of oxygen, nitrogen, carbon and hydrogen are 3.5, 3.0, 2.5 and 2.1 respectively.

Hence suggest which of the above bonds is the most polarized and which is the least.

The O-H bond is the most, while the C-H bond is the least polarized. In each case the charge separation in still only a small fraction of an electrical unit and these small charges sum to zero. The difference between the electronegativities of carbon and hydrogen is fairly small and for most purposes the polarization of the C-H bonds may be ignored.

The electronegativities of sulphur and chlorine are 2.5 and 3.5 respectively. Suggest what the polarization will be in the S-H bond and the H-Cl molecule.

$$\overset{\delta+}{H} \xrightarrow{\hspace{2cm}} \overset{\delta-}{S}$$

$$\overset{\delta+}{H} \xrightarrow{\hspace{2cm}} \overset{\delta-}{Cl}$$

The degree of charge separation in the S-H bond is similar to that which is found in the C-H bond, while the H-Cl bond is quite strongly polarized.

The strong polarization of the single covalent bond which is found in the H-F, H-Cl and H-O bonds means that the hydrogen is quite positive. In each of these cases there is a lone pair of electrons which exists on the other atom, *i.e.* the heteroatom. (A heteroatom is any other atom apart from hydrogen or carbon.) A lone pair of electrons is an area which is rich in electrons. Suggest how in the case of HCl there may be an interaction between two different molecules of HCl.

$$\overset{\delta+}{H} \xrightarrow{\hspace{2cm}} \overset{\delta-}{Cl} \cdots\cdots\cdots \overset{\delta+}{H} \xrightarrow{\hspace{2cm}} \overset{\delta-}{Cl}$$

The dotted line indicates a weak intermolecular coulombic interaction, *i.e.* an electrostatic attraction which is called a hydrogen bond, between the $\delta+$ hydrogen and the electron rich lone pairs of the chlorine atom of another HCl molecule.

For the single covalent bonds of carbon with nitrogen, oxygen and chlorine indicate the polarization of each bond.

$$\overset{\delta+}{C} \xrightarrow{\hspace{2cm}} \overset{\delta-}{O}$$

$$\overset{\delta+}{C} \xrightarrow{\hspace{2cm}} \overset{\delta-}{N}$$

$$\overset{\delta+}{C} \xrightarrow{\hspace{2cm}} \overset{\delta-}{Cl}$$

In all these cases the carbon is at the positive end. This means that it is liable to be attacked by a negative species, or that it is on the way to becoming a carbonium ion, which is a carbon species that bears a positive charge.

Carbon sometimes forms covalent bonds with metals such as lithium and magnesium, the electronegativities of which are 1.0 and 1.2 respectively. Indicate the polarization that is present in these bonds.

$$C \xrightarrow{\delta-\qquad\qquad\delta+} Li$$

$$C \xrightarrow{\delta-\qquad\qquad\delta+} Mg$$

In these cases the carbon atom is at the negative end of the bond. This means that it is liable to attack by positive species, or that the carbon is on the way to becoming a carbanion, which is a carbon species that bears a negative charge. This is the case for organometallic bonds in general.

So far we have only considered each bond in isolation, but in all except the simplest of molecules there will be a chain of atoms which will interact. Suggest what will happen in the $CH_3CH_2CH_2Cl$ molecule.

$$CH_3 \relbar\joinrel\relbar CH_2 \xrightarrow{\delta\delta+} CH_2 \xrightarrow{\delta+} Cl^{\delta-}$$

The C-Cl bond is polarized as expected, but now the first carbon in the chain bears a permanent partial positive charge. This attracts electrons from the next C-C bond in the chain. However, this attraction is much smaller than the initial one and to indicate this the symbol $\delta\delta$ is used, *i.e.* very small. The total of all the small and very small charges must still sum to zero. The transmission of this effect is very inefficient and so after the second carbon in the chain the effect is negligible. Generally you need only consider the effect on the carbon which is bonded directly to the polarizing element.

The polarization along the bond is characterized by both its magnitude and its direction. This may be illustrated by looking at a molecule of trichloroborane, which has a symmetrical planar trigonal shape. Firstly consider an individual B-Cl bond and determine the direction of polarization, given the information that the electronegativity of boron is 2.0. Then suggest what the overall polarization is for this molecule.

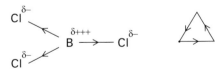

Each B-Cl bond is polarized towards the chlorine atom. However, by vector addition of the three dipoles, one discovers that they cancel each other out so that there is no overall polarization of the molecule.

If we now look at the water molecule, which by experiment has been shown to exist with a bond angle, HOH, of about 109°, suggest what will be the overall polarization of this molecule.

In this case the dipoles created by the polarization of the two oxygen/hydrogen bonds does not sum to zero, but instead there is a resultant vector which bisects the HOH bond angle and is directed towards the oxygen.

The single bond between the elements that has been considered so far is called a σ bond. So far we have only looked at the polarization of this σ bond, which in all cases remained intact. This was the case even if there was a permanent charge separation. Now we will look at what happens when the charge separation is taken a stage further and the bond breaks.

3.3 Bond Fission

In a homoatomic bond such as between two fluorine atoms there is no permanent charge separation. If this bond broke what do you think would be the likely product? Write an equation for your suggested answer.

$$F_2 = 2F\bullet$$

It is not unreasonable to suppose that the symmetrical fluorine molecule will break into two equal parts, each of which is a fluorine atom.

What is the charge on each fluorine atom after the bond has broken, and how many electrons are there around each atom?

The charge is zero, and there are seven electrons around each atom. Using the principle of the Conservation of Charge the products must have a net zero charge as that is the charge on the fluorine molecule. Another way of deriving this result is that as two identical products are formed, which means that each part must be neutral. The fluorine atom has seven electrons in its valence shell, six of which exist in three lone pairs while the last is an unpaired electron and so the fluorine atom is a radical.

This type of bond breakage is called homolytic cleavage, because the bond breaks equally between the atoms which formed it. This type of cleavage is represented by two arrows, each with only half of the arrowhead present. The tail of the arrow indicates the source of the electrons which are moving and is usually electron rich, *i.e.* a bond or a lone pair; while the head of the arrow is the target to which the electrons are moving and is usually deficient in electrons, *i.e.* an empty orbital or a positive charge, or an electronegative atom. Half an arrowhead is used to represent one electron, and two electrons are represented by a complete arrowhead.

Draw the appropriate arrows with a fluorine molecule, which represent the homolytic cleavage of the fluorine molecule.

$$F \overset{\frown}{\underset{\smile}{\rule{2cm}{0.4pt}}} F \quad \rightarrow \quad F\bullet \qquad F\bullet$$

Here there are two arrows of which each has its tail in the middle of the bond which is going to break. One arrow with half an arrowhead is going to one fluorine atom, while the other is going to the second. The half arrowhead indicates that only one electron is being considered. This shows that the two electrons which originally formed the single bond between the atoms will be shared equally between them, with one electron going to each atom.

For the hydrogen chloride molecule draw the bond polarization that would be expected.

$$\overset{\delta+}{H} \xrightarrow{\hspace{2cm}} \overset{\delta-}{Cl}$$

If the polarization continued to the limit, which element would gain the electrons of the bond and which would lose them?

The chlorine atom would gain them, while the hydrogen atom would lose them.

What would be the nature of the ions that are formed in each case?

The chlorine would form the chloride anion, while the hydrogen would form the hydrogen cation.

Draw a representation of this bond cleavage using the curly arrow notation.

$$H \overbrace{\hspace{2cm}}^{} Cl \quad \longrightarrow \quad H^+ \quad + \quad Cl^-$$

Note that only one arrow is used and that it has a complete arrowhead, which represents the two electrons in the bond. It is not necessary to use two arrows, each with only half a head. The tail of the arrow is on the bond which is about to break, and the head is on the chlorine atom which is where the electrons are about to reside. Note the Conservation of Charge in the formation of a monocation and a monoanion. This type of bond cleavage is called heterolytic fission.

Draw the bond fission that would occur in chloromethane.

$$CH_3 \overset{\delta+}{\xrightarrow{\hspace{1.5cm}}} \overset{\delta-}{Cl} \quad \longrightarrow \quad CH_3^+ \quad + \quad Cl^-$$

The most polarized bond is the C-Cl bond, with the chlorine being the more negative end. If this bond broke it is reasonable to suppose that it would break by continuing to polarize in the original direction, and so the electrons in the bond would go to the chlorine, and the carbon would be left as the cation.

Suggest what will be the nature of the bond fission in a water molecule.

$$H_2O = OH^- + H^+$$

$$\overset{\delta=}{O} \overset{\delta+}{\underset{H}{\xleftarrow{\hspace{1cm}}}} H \quad \longrightarrow \quad HO^- \quad + \quad H^+$$

The O-H bonds are polarized towards the oxygen and if this polarization is continued until the bond is broken, then after the heterolytic cleavage, a hydroxyl anion and a hydrogen cation will result.

Suggest why it would be unlikely that a hydride anion would be produced from the heterolytic cleavage of water.

Using the principle of the Conservation of Charge, then if a heterolytic fission occurred which resulted in a hydride ion then the other ion must be an OH^+ species. This is unlikely because it would mean that the O-H bond had cleaved in the opposite direction to the original polarization.

In summary, homolytic cleavage usually occurs in homoatomic bonds, or when the degree of polarization is small. The result of such a cleavage is two radicals. Heterolytic cleavage occurs when the degree of polarization is large, or when the resultant ionic species are particularly stable. The result of the two electrons in the original bond both going to one atom means that ions are formed, one of which has an extra lone pair and the other an empty orbital. In using curly arrows to indicate the movement of the electrons you have written out the mechanism of these bond cleavages.

So far we have only looked at single covalent bonds. In organic chemistry there are many compounds which contain double or even triple bonds between atoms.

3.4 Isolated Multiple Bonds

The nitrogen atom has five electrons in its valence shell, so it needs to be involved in three covalent single bonds in order to obtain the stability which is conferred by having an octet of electrons. Nitrogen combines with itself to form a diatomic molecule where each atom provides the three extra electrons which are needed by the other to satisfy the octet.

Draw a dot and cross structure for the diatomic molecule of nitrogen.

$$\overset{x}{\underset{x}{x}} N \overset{x \ x \ x}{\underset{\bullet \ \bullet \ \bullet}{}} N \overset{\bullet}{\bullet}$$

Each nitrogen has one lone pair and is involved in three bonds, each of which has two electrons. The second and third bonds in the dinitrogen molecule are different from the first bond which, it will be recalled, is called a σ bond. These multiple bonds are called π bonds.

Calculate the charge which is resident on each nitrogen atom.

The charge on each is zero. There are a total of eight electrons around each nitrogen in the valence shell, which normally has five accommodated there, and each nitrogen atom has a half share in the six electrons which are involved in the three bonds between the atoms. This results in a zero net charge on each atom.

The oxygen atom, which has six electrons in its valence shell, requires two covalent single bonds in order to obtain the stability of the octet of electrons. Oxygen combines with itself to form a diatomic molecule where each atom provides the two extra electrons which are needed by the other to satisfy the octet requirement.

Draw a dot and cross structure for the diatomic molecule of oxygen.

Each oxygen has two lone pairs and is involved in two bonds, each of which has two electrons.

Calculate the charge which is resident on each oxygen atom.

The charge on each is zero. There are a total of eight electrons around each oxygen in the valence shell, which normally has six, and each oxygen atom has a half share in the four electrons which are involved in the bonds between the atoms. This results in a zero net charge on each atom.

Diatomic oxygen is quite reactive. Suggest a reason for this.

Generally paired electrons are less reactive than unpaired electrons. In oxygen under normal circumstances one of the bonds between the atoms breaks and so forms a diradical species. Draw the mechanism for this fission.

One of the bonds undergoes homolytic fission and so forms the diradical dioxygen. Radicals possess unpaired electrons which would be more stable if they combined and became paired, thus they tend to be quite reactive. Hence the diradical of dioxygen is quite reactive as it has two radical parts.

In the molecule HN=NH there is a double nitrogen/nitrogen bond. Draw the dot and cross structure for this molecule.

Suggest the polarization that is present between the central atoms of the three molecules that have just been discussed.

In each case of nitrogen and oxygen the multiple bonds were homoatomic and thus there is no permanent polarization across the central bond. In the case of HN=NH, exactly the same atoms appear on each side of the central bond, and so again there is no permanent charge separation. However, temporary and induced polarizations may occur as in homoatomic single bonds.

Carbon may form double or triple bonds with itself. Draw a dot and cross representation for these species.

$$H \overset{x}{\underset{\bullet}{\cdot}} C \overset{x \, x}{\underset{\bullet}{\cdot}} \overset{x}{\cdot} C \overset{x}{\underset{\bullet}{\cdot}} H$$

Again there is no permanent polarization of the multiple bond and each carbon is electrically neutral.

Carbon may also form multiple bonds with many other elements. It is this ability which is one of the reasons for the richness of the chemistry of carbon. One of the commonest heteroatomic multiple bonds in which carbon is involved is with oxygen. Draw the dot and cross structure of this carbon/oxygen double bond system.

Calculate what is the charge on the carbon and the oxygen and also suggest whether there is any bond polarization.

here are no whole units of charges resident on either the carbon or oxygen atom. However, the bond between the carbon and oxygen has a permanent polarization due to the difference in electronegativities, with the oxygen atom being negative and the carbon being positive. The sum of this charge separation is zero. The carbon/oxygen double bond system is called a carbonyl group.

Draw the mechanism for the heterolytic fission of one of the bonds within the carbonyl group.

Note that a single arrow with a complete arrowhead is used, with the tail on the bond which is breaking and the head on the oxygen atom. This results in one ionic species which has both positive and negative poles. This heterolytic fission is very important in the chemistry of the carbonyl group. The attraction of electrons along the multiple bond network is called the mesomeric effect: those atoms or groups which attracts electrons exert a negative mesomeric effect and *vice versa*. The symbol is M, with the appropriate plus or minus sign.

Draw a dot and cross representation of the cyanide group. Then suggest the direction of polarization, if any, of the bond.

$$H \overset{x}{\underset{\blacksquare}{:}} C \overset{x\ x\ x}{\underset{:\ :}{:}} N \overset{\bullet}{\underset{\bullet}{:}} \qquad\qquad H \!-\!\!-\!\!-\!\! C \overset{\delta+}{=\!=\!=\!\Longrightarrow} N^{\delta-}$$

The bond is slightly polarized towards the nitrogen, with each atom bearing only a small partial unit of electric charge.

Draw the dot and cross structure of the $RC=NX$ group, and again suggest the direction of polarization, if any. Note the R is used to indicate a general alkyl group, while X is used to indicate a group which exerts a negative inductive effect, *i.e.* attracts electrons.

Again the bond is slightly polarized towards the nitrogen, but this time the degree of polarization depends on the nature of X. If X is a strongly electronegative element or group then the degree of polarization will be higher. The $C=N$ bond is called an imine bond.

In a general compound which contains a carbonyl group, $R_2C=O$, the oxygen has two lone pairs which may be protonated. Write an equation for the protonation of the carbonyl group, and draw a dot and cross structure for the product.

$$R_2C=O + H^+ = R_2C=OH^+$$

Calculate the charge of the carbon and oxygen of the product.

The carbon is still neutral, while the oxygen now carries a plus one charge. This is to be expected as the carbon is involved in four covalent bonds, while the oxygen is involved in three.

The oxygen now carries a positive charge and so it attracts electrons even more than it did when it was neutral. Has the inductive effect of the oxygen along the single bond system increased or decreased?

Increased. What about the mesomeric effect along the double bond system?

This has increased as well. Draw the result of the heterolytic fission of the double bond.

The positive charge now resides on the carbon atom.

If the nitrogen in an imime bond is protonated, draw the structure of the protonated product and the product after the carbon/nitrogen double bond has been broken heterolytically.

Initially the nitrogen carried the positive charge and then, after the heterolytic fission of the double bond, the carbon bears the charge.

In summary, the inductive effect operates along the single bond or σ system and results in partial charge separation. In contrast, the mesomeric effect operates along the multiple bond system or π system and results in whole units of charge separation.

3.5 Conjugated Multiple Bonds

When multiple bonds alternate with single bonds the whole arrangement is called a conjugated system. It has certain properties which are of great importance in organic chemistry.

In the molecule $R_2C = CH-CH = CR_2$, there are two carbon/carbon double bonds which are separated by a carbon/carbon single bond. Draw the result of homolytically cleaving one of the carbon/carbon bonds.

$$R_2C = CH - CH = CR_2 \longrightarrow R_2C = CH - \overset{\bullet}{C}H - \overset{\bullet}{C}R_2$$

Draw the result of breaking the other double bond in the same manner.

$$R_2C = CH - \overset{\bullet}{C}H - \overset{\bullet}{C}R_2 \longrightarrow R_2\overset{\bullet}{C} - \overset{\bullet}{C}H - \overset{\bullet}{C}H - \overset{\bullet}{C}R_2$$

Now, reform the double bond, but not between the terminal atoms, but instead between the central atoms.

$$R_2\overset{\bullet}{C} - \overset{\bullet}{C}H - \overset{\bullet}{C}H - \overset{\bullet}{C}R_2 \longrightarrow R_2\overset{\bullet}{C} - CH = CH - \overset{\bullet}{C}R_2$$

The result is a di-radical, but the unpaired electrons are in the 1,4 positions and not the 1,2 or adjacent positions. The numbers refer to the carbon atoms in the chain. Try and draw a single mechanistic step which will result in this 1,4 di-radical starting from the molecule with the two double bonds.

$$R_2C = CH - CH = CR_2 \longrightarrow R_2\overset{\bullet}{C} - CH = CH - \overset{\bullet}{C}R_2$$

If we now move to the heteroatomic conjugated system of $R_2C = CH-CH = O$ draw the product which results from the heterolytic cleavage of the carbonyl bond.

$$R_2C = CH - CH = O \longrightarrow R_2C = CH - \overset{+}{C}H - \overset{-}{O}$$

The carbon atom which carries the positive charge will exert a strong inductive effect. More importantly though, it will also exert a strong mesomeric effect on the carbon/carbon double bond. We saw earlier that it is through the π system that whole units of charge may be separated. Draw the result of the heterolytic cleavage of the carbon/carbon bond and then the subsequent formation of a new carbon/carbon double bond between the central carbon atoms.

$$R_2C = CH - \overset{+}{C}H - \overset{-}{O} \longrightarrow R_2\overset{+}{C} - \overset{-}{C}H - \overset{+}{C}H - \overset{-}{O}$$

$$\longrightarrow R_2\overset{+}{C} - CH = CH - \overset{-}{O}$$

Notice that the terminal carbon/carbon bond cleaves in the direction which places the newly formed carbanion next to the existing carbonium ion. If it cleaved the other way then two negative charges would be adjacent to each other which would be unstable. Hence by using the mesomeric effect the charge has been moved a further two atoms down the chain in a very efficient manner, *i.e.* retaining the full unit of the charge. This is in contrast with the inductive effect, which it will be remembered, only effectively transmitted a partial charge to the adjacent atom and no further.

In the triene system, *i.e.* one with three double bonds: $R_2C = CH-HC = CH-CH = CR_2$, the same type of rearrangement of the double bonds may occur. Draw a mechanism for the rearrangement of all the double bonds.

$$R_2C = CH - CH = CH - CH = CR_2$$

$$\longrightarrow R_2\overset{\bullet}{C} - CH = CH - CH = CH - \overset{\bullet}{C}R_2$$

Now let us consider the cyclic version of this linear triene, which would be 1,3,5-cyclohexatriene. Draw the rearrangement of the electrons for this compound.

This time three new double bonds may be formed because the two unpaired electrons which were at the terminals of the chain are now adjacent and so may interact. The new molecule is the same as the old one except that the double bonds have moved one position around the ring. This cyclic triene molecule is benzene. Draw a representation for the time averaged rearrangement of these three double bonds within this cyclic molecule.

In the time averaged molecule every carbon/carbon bond has both single and double bond character, and each bond is the same length. This cyclic system is called an aromatic ring. The representation of benzene which shows each bond as single or double is called a canonical structure and represents a hypothetical structure which does not exist. The representation that indicates every bond as half way between a single and double is called the resonance structure and is closer to the true picture of the molecule. This smearing out of the electrons is called delocalization. The aromatic ring is often represented by a regular hexagon with a circle inside it. This is convenient when the aromatic ring itself is not involved in the reaction, but when drawing mechanisms that involve the interaction of electrons within the aromatic ring, the canonical structure is used in order to keep track of the electrons.

After the rearrangement of the π electrons in the $R_2C{=}CH\text{-}CH{=}O$ molecule, the carbonyl group ended up with more electrons than it initially had, *i.e.* it was an electron sink. There are many other groups which have a similar property. Suggest a rearrangement of the double bonds in the molecule $R_2C{=}CH\text{-}NO_2$ which would enrich the nitro group with electrons.

$$R_2C = CH - \overset{+}{N}\!\!\begin{array}{c}\diagup O \\ \diagdown O^-\end{array} \quad \longrightarrow \quad R_2\overset{+}{C} - CH = \overset{+}{N}\!\!\begin{array}{c}\diagup O^- \\ \diagdown O^-\end{array}$$

Here the second oxygen in the nitro group has taken the extra electron and so acted as an electron sink in this system.

Suggest a mechanism whereby the cyano group may act as an electron sink.

$$R_2C = CH - C \equiv N \longrightarrow R_2\overset{+}{C} - CH = C = \overset{-}{N}$$

In general groups which have multiple bonded heteroatoms may act as electron sinks, because the multiple bond may cleave heterolytically and leave the extra electron pair on the heteroatom.

If a group with a lone pair of electrons is placed at the other end of the chain, then this electron pair may be donated into the conjugated system. Such a group may, for example, be an amino group. suggest a possible rearrangement of the π electrons in the molecule $H_2N\text{-}RC\text{=}CH\text{-}NO_2$ of the double bonds.

$$H_2\ddot{N} - CR = CH - \overset{+}{N}\underset{O^-}{\overset{O}{\diagup}} \longrightarrow H_2\overset{+}{N} = CR - CH = \overset{+}{N}\underset{O^-}{\overset{O^-}{\diagup}}$$

Here the amino group has provided the extra electrons, *i.e.* it has acted as an electron source. The hydroxyl group or halide atom may act in a similar manner to the amino group. If an electron source and an electron sink are conjugated together then a permanent dipole may be formed due to the resultant polarization along the multiple bond system as a result of the mesomeric effect.

4

Shapes of Molecules

4.1 Introduction

So far we have only considered the electronic properties of the atoms within a molecule. We will now turn our attention to the spatial arrangement of those atoms within the molecule.

There are two main types of bonding, ionic and covalent. Ionic bonding is characterized by the non-directional nature of the Coulombic attractions between ions, *i.e.* the force is felt equally in all the directions which radiate from the central ion. The main factor which influences the structure of the crystal lattice is the relative sizes of the cations and anions, because this affects how the ions will pack together in the lattice.

Covalent bonds are different from ionic bonds, in that they are directional in nature. Furthermore each covalent bond has a particular length. The consequence of this is that when one atom is covalently bonded to another atom then the relative position in space of these two atoms is fixed.

This means that in a molecule which is held together by covalent bonds, each and every atom has a defined geometric relationship to the other atoms within the molecule. In simple ionic compounds, each ion only occupies one type of environment, with all the ions of the same type having exactly the same geometric relationship to all the other ions in the crystal lattice. In more complicated ionic compounds it is possible for ions of one species to exist in one of limited number of environments, but this is the exception rather than the rule. In the case of covalent compounds, the situation is reversed, and it is now the norm for every atom to be considered individually in respect of its geometric relationship to the other constituent atoms.

We will now look more closely at certain types of covalent bonding, before examining some of the more common geometric possibilities.

4.2 Types of Bonding

Each covalent bond is formed from a molecular orbital which may accommodate two electrons. Molecular orbitals are formed from atomic orbitals

41

each of which may accommodate two electrons. The same number of molecular orbitals are produced as there were atomic orbitals originally. In each hydrogen atom there is one atomic orbital, which holds the single 1s electron. When two such atoms combine to form the hydrogen molecule, then two molecular orbitals are formed from the two atomic orbitals of the original hydrogen atoms. One molecular orbital is shaped like a cigar tube around the two hydrogen nuclei. This is the bonding orbital, because it allows the electrons which occupy it to reside between the two positive nuclei. Thus there is some electron density between the nuclei which supplies the attractive force to keep the two nuclei together and so form a stable molecule. The other molecular orbital is the anti-bonding orbital. Its shape is such that there is no electron density between the two nuclei and so there is no electron density between the two nuclei to hold them together. The shapes of each of these orbitals is shown below.

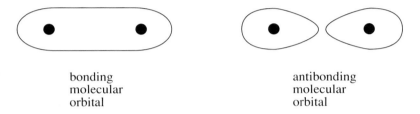

bonding	antibonding
molecular	molecular
orbital	orbital

For every bonding molecular orbital that is formed there is an anti-bonding molecular orbital as well. In certain circumstances non-bonding orbitals may be formed. If an anti-bonding orbital is occupied it destabilizes the molecule, hence its name. The two important points about anti-bonding orbitals is that, firstly, they project in the opposite direction in space to the bonding orbitals to which they are related, and secondly, they are usually empty in an otherwise stable molecule.

The reason why we must consider them at all is because electrons must move from one orbital to another. If the electrons are to remain attached to an atom or molecule they cannot just float about in space. As we saw above, each orbital may only accommodate two electrons, so if an electron rich species is going to attack another species which has a full complement of electrons there needs to be an empty orbital which can accommodate the incoming electrons before the electrons which were originally there have departed. The empty orbital is often an anti-bonding orbital. The geometry of the anti-bonding orbital will determine the line of approach for the incoming reagent, and so will impose certain restrictions on the way a mechanism must proceed. This in turn will affect the stereochemistry of the product.

The shape of organic molecules is determined by the electron repulsion theory which, in summary, states that any given electron pair in an orbital will repel all other electron pairs. Thus the electron pair will adopt a position in space which is furthest from all the other electron pairs. To a first approx-imation the geometries adopted are regular in shape.

Suggest the shapes that will be adopted by a triatomic molecule of general formula AB_2, which only has the electrons in the two molecular orbitals required to bond the B atoms to the central A atom, *i.e.* there are no other lone pairs of electrons to consider.

$$B \longrightarrow A \longrightarrow B$$

The geometry will be linear, *i.e.* the bond angle is 180°. Even though this is rather rare in organic molecules, there are important classes of compounds which have this geometry, *e.g.* alkynes and cyanides. Now suggest the geometry of the molecular of general formula AB_3.

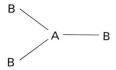

This geometry is called trigonal planar, and the bond angle is 120°. This geometry is important in all compounds which contain a double bonded carbon, or a carbonium ion. Now suggest the geometry of the molecular type AB_4.

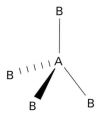

This adopts a regular tetrahedron, and is the first structure which requires the third dimension, *i.e.* it is not a flat structure. The bond angle is about 109°. This is the fundamental building block of organic molecules. Now suggest the geometry of the molecular type AB_5.

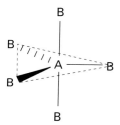

This structure is called a trigonal bipyramid. There are two different bond angles in this structure. Between the atoms in the trigonal plane the angle is

120°, while between the axial and equatorial atoms the angle is only 90°, *i.e.*
a right angle. This is an important shape in the transition of one saturated
carbon centre to another.

We will now look at some of these geometries in a little more detail,
starting with that of the tetrahedron, *i.e.* AB_4 compounds, because this the
most important shape in organic chemistry.

4.3 Tetrahedral Geometry

This is the single commonest geometry which is encountered in organic
molecules, because this is the shape which is adopted by a saturated carbon,
i.e. one which is bonded to four other groups by four single bonds.

Write down the electronic configuration of the carbon atom in the ground
state.

$$1s^2, 2s^2, 2p_x^{1}, 2p_y^{1}$$

There are four electrons in the valence shell of carbon, so in order to
achieve an octet it requires four more. The shape of the s atomic orbital is
spherical, while the shape of the p orbital is like two spheres which just
touch at the nucleus. Each of the p orbitals is orthogonal, that is at right
angles, to each of the others. More strictly, there is no shape to an orbital, as
it is just a wave function; however, that is of little use to most organic
chemists. Thus a shape is allocated to each orbital which approximates to
volume in space around each nucleus where the electron in question is most
likely to be found. If a carbon atom was bonded only by means of the p
orbitals, what would be the bond angle?

As each p orbital is orthogonal to the others, the bond angle must be 90°.
In practice this is not observed, instead the bond angle is about 109°. In
order to explain this observation, it is proposed that one electron from the
valence shell, *i.e.* one of the $2s^2$ electrons, is promoted to the empty $2p_z$
orbital to give $2s^1, 2p_x^{1}, 2p_y^{1}, 2p_z^{1}$. Then all four atomic orbitals are mixed
to give four equal hybrid orbitals called sp^3 type, *i.e.* formed from one s
orbital and 3 p orbitals. This is the hybridization theory.

If these four hybrid orbitals are the same, suggest what their relative
energy level to each other is, and also the directions in space which they
occupy with respect to each other.

As they are the same, then the energy levels must be equal, *i.e.* they are
degenerate. As they are equal in energy and also as they have the same

shape, each one points towards a different corner of a regular tetrahedron. These four hybrid orbitals may then combine with four other atomic orbitals to form four bonding and four antibonding molecular orbitals.

Suggest the shape of the tetrahydride of carbon. Draw it using a stereo projection.

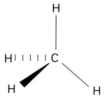

The tetrahydride of carbon is called methane and is a regular tetrahedron. If one of the hydrogen atoms is replaced by another methyl group, CH_3, then what will be the shape of the resultant molecule? Again draw this using a stereo projection.

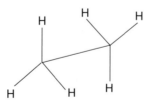

Around each carbon the bonds still are arranged in a tetrahedral fashion. Now redraw this using the sawhorse projection.

Now that the three dimensional structure of the molecule is important, the dot and cross notation that was so useful when counting electrons around each atom is of less utility. Firstly, it is rather cumbersome, and secondly, it does not give the spatial information which we now require. It is still useful, however, in ensuring that all the electrons have been included, and thus in ensuring that any necessary lone pairs of electrons are present.

In the nitrogen atom the same hybridization of the atomic orbitals occurs and so four sp^3 hybrid orbitals are formed. These are distributed around the central atom in the same tetrahedral manner as in carbon. Suggest the shape of the ammonia molecule.

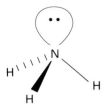

This shape is often called a trigonal pyramid because that is the shape that is mapped out by the hydrogen and nitrogen atoms. However, the true geometry is related to the tetrahedron, the fourth corner of which is occupied by the lone pair. The nitrogen is bonded to three hydrogens, but this only accounts for six of the eight electrons which surround the nitrogen. The other two are in the lone pair. The lone pair has similar spatial requirements as the pair of electrons in each of the bonding orbitals, and so exerts a similar repulsive force on the three bonding orbitals as do the latter on each other. Hence even though, initially this looks like an AB_3 geometry because there are only three hydrogen atoms joined to the central nitrogen, it is really an AB_4 geometry because of the presence of the lone pair.

Suggest the shape of the dihydride of oxygen.

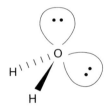

This molecule appears bent if one only looks at the atoms. The true geometry, however, is based on a tetrahedron, with two corners occupied by the hydrogen atoms and the other two occupied by the lone pairs. So again it is based on the AB_4 geometry and not the AB_2 geometry as it might at first sight appear; this is because of the presence of the two lone pairs.

The lone pairs are held slightly closer to the central atom than the electrons which are involved in the bonding orbitals. Suggest a reason for this.

Simply in a bonding orbital there are two nuclei trying to share the electrons, while in a lone pair the attraction of the nucleus comes only from one side.

If the lone pairs are held slightly closer to the nucleus than the bonding pair, then what are the consequences for the bond angle between the hydrogen atoms in the methane, ammonia and water molecules.

In methane the HCH bond angle is that which is found in a perfectly regular tetrahedron, about 109°. However, in ammonia the HNH angle is reduced slightly because the lone pair, being closer to the nucleus, repels the bonding electrons slightly more strongly and thus forces the bonding electron orbitals together. The resultant bond angle is about 107°. In water the effect is greater because there are now two lone pairs, and so the HOH angle is about 105°.

For molecules which have lone pairs there is the possibility of protonation. In ammonia the protonated species is the ammonium ion. Draw the shape of this ion.

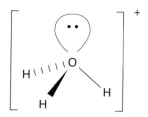

This molecule is a perfectly regular tetrahedron.

Suggest what will be the shape of the hydrazine molecule, N_2H_4, and its mono-protonated ion, $N_2H_5^+$.

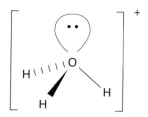

Hydrazine is similar to ethane, but with a lone pair replacing a hydrogen on each of the central atoms. There is free rotation around the nitrogen/ nitrogen single bond. In the protonated form the geometry around each nitrogen atom is still based on the tetrahedron.

What is the shape of the molecule which results from the mono-protonation of water?

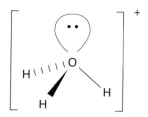

This molecule is similar in shape to the ammonia molecule, and so is still based on the tetrahedron.

What is the shape of the molecule which results from the deprotonation of water?

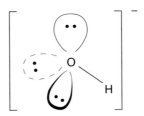

This ion, only having two atoms, must be linear, but the three lone pairs project towards the corners of tetrahedron which is centred on the oxygen anion, and so still forms the basic template for the orientation of the lone pairs. The position in space of these lone pairs is important because that will determine the direction of approach of the protonating reagent.

Methane may be deprotonated under extreme conditions to give the carbanion, CH_3^-. What is the shape of this ion?

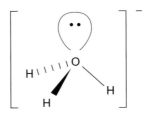

Again this ion is based upon a tetrahedron, with three of the corners occupied by the hydrogen atoms and the last one occupied by the lone pair.

When one of the hybrid orbitals, sp^3, of carbon combines with a s type atomic orbital of a hydrogen, then two molecular orbitals must be formed, one of which is the bonding while the other is the anti-bonding orbital. The bonding orbital extends from the carbon towards the hydrogen atom which is placed at the corner of a tetrahedron. In which direction do you think the related anti-bonding orbital is pointed?

This anti-bonding orbital points in the opposite direction from the bonding one, and so is directed between the three remaining carbon/hydrogen bonds. The same hold true for the other three anti-bonding orbitals. If a carbon centre which has a full complement of eight electrons is attacked by an electron rich species, from what direction do think that this species will approach?

Electrons may only move from orbital to orbital, and as each orbital may only contain two electrons then it cannot place more electrons in a bonding orbital, hence it must approach along the line of the empty anti-bonding orbital and may then transfer electrons into this. Thus the direction not only of the bonding orbitals but also of the anti-bonding orbitals is very important in understanding mechanistic organic chemistry. An anti-bonding orbital is indicated by a raised postscript asterisk, *e.g* sp^{3*}.

4.4 Trigonal Planar Geometry

The hybridization theory may be used to account for the direction of the bonds in an atom which is involved in three single bonds only, and which does not have any lone pairs. If the s orbital and only two of the three p orbitals are hybridized, leaving the p_z orbital untouched, then three hybrid orbitals are formed which are called sp^2 orbitals.

If these three orbitals are degenerate, what configuration relative to each other do you think they will adopt?

These three orbitals will form a trigonal planar shape, with a bond angle of 120°. This leaves the last p orbital to project above and below the plane formed by the three hybrid orbitals, *i.e.* it is orthogonal to the plane.

Suggest the shape of the molecule BCl_3.

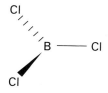

In the isolated molecule of boron trichloride there are only three electron pairs around the boron, and so they adopt the trigonal planar geometry. The last orbital, p_z, is left unused and so is an empty orbital. Draw a diagram of the BCl_3 molecule and incorporate into that diagram the unused p_z orbital.

Now draw the shape of the CH_3^+ ion.

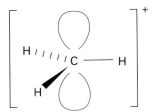

Again this is a trigonal planar species, with an empty orbital which projects above and below the plane of the C-H bonds.

The carbanion CCl_3^-, which may be formed by the deprotonation of trichloromethane, $CHCl_3$, may under favorable conditions lose a chloride anion. Write down the balanced equation for the lose of the chloride anion from CCl_3^-.

$$CCl_3^- = :CCl_2 + Cl^-$$

The CCl_2 must be neutral in order that charge is conserved in the reaction. Draw the dot and cross structure of the CCl_3^- ion and so deduce the electronic configuration of the $:CCl_2$ species, and hence suggest the shape of this molecule.

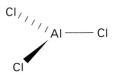

The value of the dot and cross structure here is to help to ensure that all the electrons are accounted for before attempting to deduce the structure. The electron repulsion theory gives the result that the $:CCl_2$ molecule has a trigonal planar shape. There are two C-Cl single bonds and also a lone pair. It will be recalled that this carbon species is called a carbene. It is a most interesting species because it has both a lone pair and an empty orbital, and yet is electrically neutral.

Earlier we looked at the BCl_3 molecule and observed that it was trigonal planar. Suggest what the structure of $AlCl_3$ in the gas phase will be.

The aluminium trichloride monomer is trigonal planar, with an empty orbital which extends above and below the plane in which the Al-Cl bonds are located. However, aluminium trichloride forms a dimer at room temperature and pressure. Suggest a structure for this dimer.

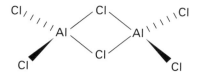

One of the lone pairs from a chlorine atom forms a dative bond with the aluminium atom. This lone pair was able to attack the aluminium atom because of the presence of the empty orbital. Now each aluminium atom has four single bonds around it and so adopts a tetrahedral configuration. Notice that the non-bridging chlorine atoms now lie in the same plane which is orthogonal to the plane which the bridging chlorines occupy.

The molecular orbitals that have been formed between the sp^3 or sp^2 hybrid orbitals and the atomic orbitals of the other atoms have all had circular symmetry about the axis connecting the atoms which are bonded together. In other words there has not been a nodal plane which passes through both of the bonded nuclei. A molecular orbital with this type of symmetry is called a σ bond. A diagram of such a bond between a sp^3 carbon and a s type atomic orbital of a hydrogen atom is shown below. This is a normal carbon/hydrogen single bond.

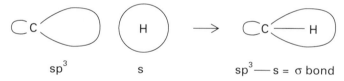

Notice that the lettering used for atomic orbitals is the lower case Roman alphabet, while for molecular orbitals it is the Greek alphabet.

If two carbons atoms are hybridized into the sp^2 configuration, then each one will have a spare p_z orbital remaining. A normal σ type carbon/carbon bond may be formed by using two sp^2 hybrid orbitals. Furthermore, the other two sp^2 orbitals on each carbon may be used to bond to hydrogen to form four more σ type carbon/hydrogen bonds. Draw the ethene molecule, representing the five σ type bonds that have been discussed as lines, and drawing out the remaining p_z orbital on each carbon.

The σ bonds framework gives the trigonal planar structure to the ethene molecule. The two p_z orbitals may interact by overlapping side on, *i.e.* by overlapping each lobe, one above and the other below the plane of the molecule. This is the second carbon/carbon bond in ethene. Notice that in this type of bond there is now a nodal plane which transects the two nuclei. A molecular orbital with this type of symmetry is called a π bond.

Draw the result of rotating one carbon atom by 90°.

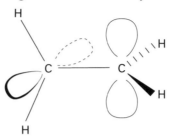

Now the p_z orbitals of each carbon atom cannot overlap successfully and hence there is no bond formed. So by rotating around the C-C axis the π bond has been broken. This cleavage would require energy. The consequence is that rotation about an axis which has a π bond is very difficult.

Suggest the position of the anti-bonding orbital which results from the formation of the σ bond between the two carbons.

This empty orbital projects out, away from the carbon/carbon bond, between the remaining two hydrogen atoms. The π bond also has an anti-bonding orbital associated with it. Suggest where this may be in space with respect to the bonding orbital.

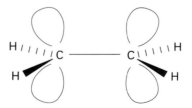

This empty orbital projects away from the double bond, but unlike the σ anti-bonding orbital, it is above and below the plane of the atoms. This makes it is more approachable to an incoming attacking species.

Carbon may form double bonds with other atoms apart from itself, such as oxygen and nitrogen. Suggest the shape of the general ketone $R_2C=O$.

Notice that the three groups are in the same plane. Draw in the position of the lone pairs on the oxygen and on the carbon draw in the anti-bonding orbital which is associated with the π bonding orbital.

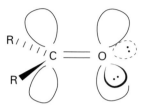

The lone pairs are in the same plane as the other groups, while the anti-bonding orbital is projected above and below this plane and away from the lone pairs.

The carbon/oxygen double bond is polarized by both the inductive and mesomeric effects of the oxygen. Draw a diagram which shows the inductive effect.

$$
\begin{array}{c}
R \\
 \\
R
\end{array}
> C \overset{\delta+}{\Longrightarrow} O^{\delta-}
$$

Draw another diagram showing the mesomeric effect.

$$R_2C = O \quad \longleftrightarrow \quad R_2\overset{+}{C} - \overset{-}{O}$$

Suggest the shape of the general imine, $R_2C = NX$.

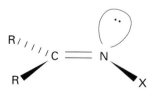

If the two R groups are different, then the X group may be on the same side as one or the other. This gives rise to the possibility of geometric isomers due to the lack of free rotation around the carbon/nitrogen double bond.

4.5 Linear Geometry

If only one p orbital and the s orbital are hybridized together, then two sp hybrid orbitals are formed. How will these two hybrid orbitals be orientated with respect to each other?

The bond angle will be 180°, so that they will be diametrically opposed to each other. There are also two remaining p orbitals, the p_z and the p_y orbital. What position will these two orbitals adopt?

They will be orthogonal to each other and the two sp hybrid orbitals. If two carbon atoms were hybridized into the sp configuration, then the remaining p orbitals could form two π bonds, each orthogonal to the other. Draw the structure of the ethyne molecule, in which the two carbons are joined by one σ bond and two π bonds, indicating the σ bonds with a line and drawing out the p orbitals which form the π bonds.

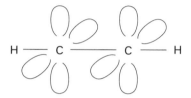

The ethyne molecule is linear. There are two π bonds which restrict rotation, but in this case it is not noticeable because of the circular symmetry around the σ bond.

If the ethyne molecule is deprotonated it gives rise to an anion, C_2H^-. Draw the shape of this anion and in particular indicate the direction of the lone pair.

$$\left[\; H - C \equiv C\!\!\bigcirc :\; \right]^-$$

The lone pair projects directly away from the triple bond, and so the whole molecule is still linear. If this anion is used as a reagent, it is possible for it to attack centres which are quite crowded because it is long and thin. For this reason it is sometimes called the "Heineken" nucleophile as it can reach the parts that other nucleophiles cannot reach!

The cyanide group is another group in which there is a triple bond, this time between carbon and nitrogen. Draw the shape of the related anion, CN^-, indicating the lone pairs.

$$\left[\; :\!\bigcirc C \equiv N\!\bigcirc :\; \right]^-$$

Two atoms must form a linear unit, however, both lone pairs project directly out from the triple bond. Thus suggest the shape of the protonated form, HCN.

$$H - C \equiv N \overset{\textstyle :}{\bigcirc}$$

The anion is protonated on the carbon end to give rise to a linear molecule, which is hydrogen cyanide or hydrocyanic acid. A reagent which has lone pairs on two or more different atoms with which is may attack another compound is called ambident. Which atom acts as the nucleophile often depends upon the nature of the atomic centre which is being attacked. In the case of protonation, the hard, less easily polarized atom, reacts with the proton, which is itself a hard species. This type of interaction will be studied in more detail in the chapter on acid/base characteristics.

There is a rather unusual functional group which is related to the cyanide group. It contains a carbon atom which is only doubly bonded to a nitrogen atom, but which is not involved in any further bonding. This is the isonitrile group, R-NC. Draw the dot and cross structure of this grouping.

$$R \quad \overset{\bullet\bullet}{\underset{\bullet\bullet}{:N}} \overset{x\ x}{\underset{\bullet\ \bullet}{:}} C \overset{x}{\underset{x}{}}$$

The nitrogen atom is involved in one single and one double bond. Furthermore there is a lone pair on that atom. What positions would you expect these three groups to adopt around the central nitrogen atom?

This is based on the trigonal planar geometry, AB_3, with one B group being the lone pair, another the R group and the last the doubly bonded carbon atom. However, this functional group is linear and not bent at an angle around the nitrogen atom. Suggest a rearrangement of the electrons which would account for this observation.

$$R - \overset{\bullet\bullet}{N} = C \quad \longleftrightarrow \quad R - \overset{+}{N} \equiv \overset{-}{C}$$

If the nitrogen donates its lone pair into the empty orbital of the carbon atom and so forms a dative bond, then there is now a triple bond between the carbon and the nitrogen atoms. What is the shape of the group in this electronic configuration and in what direction does the lone pair on the carbon atom project?

$$R \overset{+}{\longrightarrow} \overset{+}{N} \equiv \overset{-}{C} \left(\quad : \right)$$

The group is now co-linear with the R group and the lone pair on the carbon projects directly away from the carbon/nitrogen triple bond. This is the favored electronic configuration of the isonitrile functionality as shown by the fact that this the shape that it adopts.

4.6 Further Examples

In this section we will look at a few examples which will develop the basic ideas that have been introduced in the previous sections. When confronted with a molecule of unknown structure it is often useful to build towards it from segments which are well known, maybe using hypothetical inter-mediates to help on the way.

The first example is quite simple but will introduce the idea. The structure of water has already been discussed. You will recall that it is a bent triatomic molecule based on a tetrahedron, with the two lone pairs of the oxygen pointed towards the unoccupied corners of the tetrahedron. We will use the example of hydrogen peroxide, H_2O_2, to demonstrate the principle of building towards an unknown structure from known molecules. Diatomic oxygen must be linear because there are only two atoms in it. Write down its electronic configuration.

$$\overset{x}{\underset{x}{\overset{x}{O}}} \overset{x \ x}{\cdots} \overset{\bullet\bullet}{\underset{\bullet}{O}}.$$

Each oxygen has a full octet so if an electron is to be added then at least one of the oxygen/oxygen bonds must be broken. Write down a mechanism whereby one of the oxygen/oxygen bonds is homolytically broken and an electron is added to form the O_2^- anion.

$$\overset{x \ x}{\underset{x}{\overset{x}{O}}} = \overset{\bullet\bullet}{O}: \quad \rightarrow \quad \overset{x \ x}{\underset{x}{\overset{x}{O}}} - \overset{\bullet\bullet}{\underset{\bullet}{O}}: \quad \overset{\beta^-}{\rightarrow} \quad \left[\overset{x \ x}{\underset{x}{\overset{x}{O}}} - \overset{\bullet\bullet}{\underset{\bullet\bullet}{O}}: \right]^- \quad \blacksquare \text{ extra electron}$$

Adding another electron forms the O_2^{2-} anion. Write down the electron configuration of this anion.

$$\left[\overset{x \ x}{\underset{x \ \blacksquare}{\overset{x}{O}}} - \overset{\bullet\bullet}{\underset{\blacksquare\bullet}{O}}: \right]^{2-}$$

In this anion the lone pairs are arranged around each oxygen atom in a tetrahedral manner. Draw the shape of this ion.

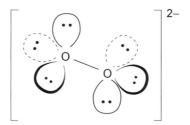

If two protons are added we form the desired molecule, H_2O_2. Now draw the shape of this molecule.

This is a bent molecule. Hydrogen peroxide is a relatively easy example, but it illustrates the method that could be used to find the structure of more complicated compounds. The next example follows the same pathway, but is slightly more difficult.

The structure of NH_3 has been discussed previously; however, the structure of HN_3 is less obvious. Here we will start with diatomic nitrogen. Write down the dot and cross structure of diatomic nitrogen.

$$\overset{x}{\underset{x}{}}N \,\overset{x\ x\ x}{\underset{\bullet\ \bullet\ \bullet}{}}\, N \overset{\bullet}{\underset{\bullet}{}}$$

There is a triple bond between the two nitrogen atoms, which is composed of one σ bond and two orthogonal π bonds. Furthermore, there is a lone pair which is pointed away from the multiple bond. Each nitrogen has a full octet. Now form the hypothetical anion N_2^-, by the addition of an electron. Write down the electronic structure of this intermediate.

$$:N \equiv N: \;\rightarrow\; :N = N: \;\overset{\beta^-}{\rightarrow}\; \left[:\overset{\bullet\bullet}{N} = \overset{\bullet\bullet}{N}\!\bullet \right]^{-}$$

Firstly one of the π bonds must be broken to give the diradical, and then secondly, an electron is added to one of the nitrogen atoms. This nitrogen now has two lone pairs. Draw the structure of this anion.

$$\left[\overset{\curvearrowleft}{N} = N \right]^{-}$$

If the unpaired electron on the nitrogen of this anion now forms a covalent bond with one of the unpaired electrons on a nitrogen atom we form the N_3^- anion, which is the deprotonated form of the molecule in which we are interested. Draw a dot and cross structure of this anion, and then draw a structure for it.

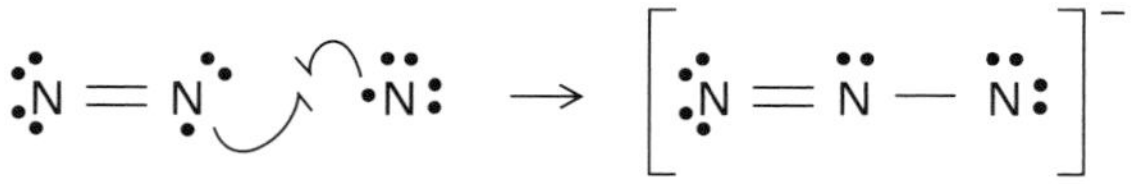

$$\ddot{:}N = N \cdot \quad \overset{\cdot\cdot}{\cdot N} : \quad \rightarrow \quad \left[\ddot{:}N = \ddot{N} - \ddot{N} : \right]^-$$

At the moment the ion is asymmetrical. One terminal nitrogen has two lone pairs and bears a negative charge and is doubly bonded to the central atom. The other terminal nitrogen has only one covalent bond and two lone pairs and is thus electron deficient and yet neutral. Suggest a rearrangement of the electrons so as to make the anion symmetrical.

$$N = \ddot{N} \quad N \quad \leftrightarrow \quad \overset{-}{N} = \overset{+}{N} = \overset{-}{N}$$

Now each terminal nitrogen is doubly bonded and carries a negative charge, while the central atom is doubly bonded to each of the two terminal nitrogen atoms and carries a single positive charge. Suggest a shape for this molecule.

$$\left[\overset{-}{:N} = \overset{+}{N} = \overset{-}{N:} \right]^-$$

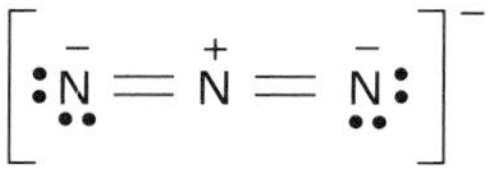

It is linear. If you have not already, now indicate for this anion the direction in which the lone pairs point.

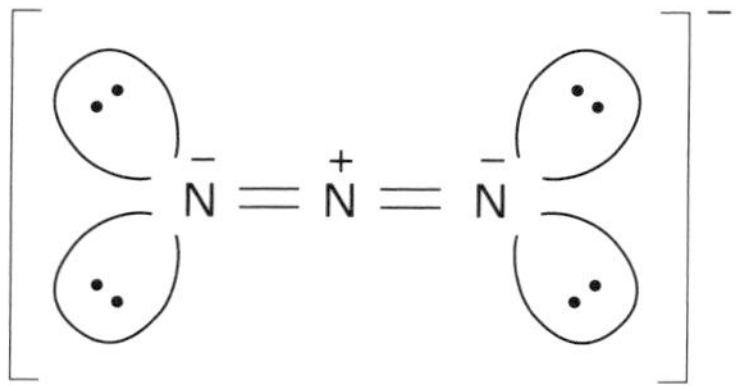

Each of the two lone pairs on the terminal nitrogens adopts a trigonal planar configuration. If this anion is now protonated then hydrogen azide is formed, HN_3. Suggest a shape for this molecule.

$$\overset{H}{\diagup} N = \overset{+}{N} = \overset{-}{N}$$

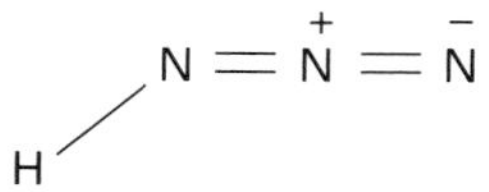

One of the lone pairs of the anion forms a dative bond with the proton and hence the final molecule is bent, with the three nitrogen atoms in a straight line, and the hydrogen off to one side. So by systematically building up the molecule from smaller units of known structure, the geometry of an unknown molecule has be deduced.

The shape of a general carbonyl containing compound, $R_2C=O$, has been discussed above. If one of the general alkyl groups is replace by an electronegative group such as a hydroxyl group, then the properties of the carbonyl group change. Write down the general formula for such compounds.

$$\underset{R \qquad X}{\overset{O}{\underset{\|}{C}}}$$

Now suggest some possibilities for X, and name the resultant compounds.

If X equals:

OH	carboxylic acid
Cl	acid chloride
OR	ester
NH_2	primary amide
NHR	secondary amide
NR_2	tertiary amide
OCOR	acid anhydride

Draw out the structure of the carboxylic acid.

$$R - C \overset{\displaystyle O}{\underset{\displaystyle O-H}{\big<}}$$

Write down a balanced equation for the deprotonation of this molecule using a general base, B, as the reagent.

$$RCOOH + B = RCOO^- + BH^+$$

Draw out the structure of the resultant anion, indicating the lone pairs on the oxygen atoms.

$$R - C \overset{\displaystyle O:}{\underset{\displaystyle O:}{\big<}}$$

The carbonyl group is an electron sink and the oxygen anion is a source of electrons and so the latter may donate electrons to the former. Suggest a possible movement of electrons that may occur.

$$R - \overset{O}{\underset{O^-}{\big<}} \quad \leftrightarrow \quad R - \overset{O^-}{\underset{O}{\big<}}$$

In this new species the bonding to the oxygen atoms is reversed, otherwise the molecule is the same. In reality the electronic configuration is a time weighted average of these two hypothetical, canonical structures. The electrons are said to be delocalized over the three atoms to give rise to the resonance form of the carboxylate anion. This resonance interaction affects the geometry which is adopted by some of the derivatives of the carboxylic acids.

Draw out the structure of a general carboxylic ester.

There is a single bond between the carbonyl carbon and the bridging oxygen bond and so the groups may rotate with respect to each other around this bond and thus give rise to many different conformers. There are two conformers of particular interest. Draw the two conformers of a general ester which have all the groups in one plane.

trans   *cis*

In essence these are the *cis* and *trans* conformer of the ester, as the ester alkyl group is either adjacent to the carbonyl alkyl group or it is opposite it. Suggest which one is the favored conformer.

The *trans* conformer is the favored one. Suggest two reasons which may account for this observation.

The first is that there is less steric interference between the two alkyl groups if they are in the *trans* conformation. The other reason is that while in the *trans* position the lone pairs on the oxygen may interact with the carbonyl group. Draw the mechanism for this interaction.

As a result of this interaction, the *trans* conformer is favored. In a secondary amide the same considerations may apply. Hence suggest the favored conformer of a general secondary amide.

In biological systems this is important, because small energy differences often determine the overall geometry of interacting species.

5

Stabilization of Charged Species

5.1 Introduction

When two molecules react with each other new bonds are made while others are broken. At the half way point between the making and breaking of any given bond there is a intermediary stage between the original compounds and the products which result from the bond rearrangement. This intermediary stage is called a transition state, and represents a species that is very short lived. It does not have an independent existence and may not be isolated like a true intermediate in a reaction pathway. In all transition states the distribution of charge is different from that which is to be found in the original molecules. This is because of the movement of the electrons that has occurred in the bonds that are breaking and forming.

The immediate result of the bond rearrangement may not be the final product that is to be produced from the reaction, but instead it may be a true intermediate. Such intermediates may be neutral, but more often they bear an electrical charge. These intermediates may exist long enough to undergo bond rotation or other bond flexing movements which may affect the eventual stereochemistry of the product. It may also be possible to isolate or trap these intermediates.

In order to facilitate the reaction, the charges that have been created by the reagents must be stabilized. It is the methods that are available for such stabilization that will be covered in this chapter.

The two main forms of charge stabilization have already been introduced in the context of polarized bonds, *i.e.* the inductive and mesomeric effects. Other mechanisms exist and may be of overriding importance when they are applicable. However, for most reactions it is the effects of inductive and mesomeric groups which have most influence upon the stability of the charged species, and so we will start with an investigation of these methods.

63

5.2 Inductive Effects

5.2.1. Charged Hydrocarbon Species

The simplest charged hydrocarbon species is the methyl cation, CH_3^+. Draw a dot and cross structure of this ion, and suggest a shape for it.

There are six electrons around the carbon in three single σ bonds. There is an empty p_z orbital. The structure adopted is a trigonal planar, with the empty orbital at right angles to this plane.

Suggest the direction of polarization along these three single σ bonds. Indicate this polarization on a diagram of the charged species.

The central carbon bears the single positive charge, and hence its effective electronegativity is increased, *i.e.* it attracts electrons more strongly. It exerts this attractive force upon the electrons that are within the adjacent carbon/ hydrogen bonds. This reduces the charge that the carbon is carrying, and in effect places a δ+ charge on each of the three hydrogens. This in effect distributes the charge and reduces the charge density on the carbon.

Suggest the shape of the ethyl cation, $CH_3CH_2^+$, and indicate the charge polarization which is present.

The positive carbon adopts the trigonal planar conformation, while the other carbon is in the normal tetrahedral configuration. The carbon bearing the positive charge may inductively draw towards itself the electrons within the two σ bonds to the hydrogens, and so place a δ+ charge on each of them. It may furthermore draw towards itself the electrons within the σ bond to the second carbon atom, which in turn may attract the electrons in the three carbon/hydrogen bonds of the terminal methyl group and so spread out the positive charge over another three atoms. This lowers still further the charge density on the original charged carbon and so stabilizes it even further. This species is called a primary carbonium ion because there is one carbon attached to the carbon which bears the charge.

In the propyl cation with the charge upon a terminal carbon, $CH_3CH_2CH_2{}^+$, suggest the polarization which is present and indicate this on a diagram of the molecule.

Here the addition of the extra methyl group is of little further assistance in the stabilization of the charge, because of the inefficiency of the transmission of the charge by the inductive effect.

In the propyl cation with the charge upon the central carbon, $CH_3CH^+CH_3$, suggest the polarization which is present and indicate this on a diagram of the molecule.

In this case there is a further diminution of the charge upon the central carbon, because the extra atoms are that much closer, *i.e.* only two bonds away and the electrons within their σ bonds may be draw towards the charge by induction. This species is called a secondary carbonium ion because there are two carbons attached to the carbon which bears the charge.

Suggest the order of stability of the four cations that have been discussed so far.

$$CH_3CH^+CH_3 > CH_3CH_2CH_2{}^+ > CH_3CH_2{}^+ > CH_3{}^+$$

Suggest where the *t*-butyl cation, $(CH_3)_3C^+$, should be placed in the above list of stability of the cations. This species is called a tertiary carbonium ion because there are now three carbons attached to the carbon which bears the charge.

It would be the most stable cation, because it has the largest number of bonds within two bond lengths.

Earlier we mentioned the shape of the methyl cation; suggest the shape of the *t*-butyl cation, $(CH_3)_3C^+$.

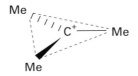

This molecule is trigonal planar around the central charged carbon. It is observed experimentally that carbonium ions are more stable if it is possible for them to adopt a planar configuration. One reason for this is that it allows for the maximum separation of the substituents, and so reduce steric crowding to a minimum. Hence, carbonium ions at a bridgehead are less stable than would be expected purely using the considerations that have been mentioned already.

Hydrocarbon species may also bear a net negative charge and thus be anions. Draw the methyl anion, CH_3^-.

This species is tetrahedral in shape as there are eight electrons around the carbon in three σ single covalent bonds to the hydrogens and one lone pair.

Suggest the direction of polarization within the bonds in this species.

The carbon atom is electron rich and this now decreases its electronegativity, which was initially greater than that of the hydrogen atoms. Now, however, electrons are being pushed towards the hydrogen atoms; yet each one already has a full complement of two electrons accommodated around it, so the negative charge is not effectively shared onto the neighbouring hydrogen atoms. Furthermore, each hydrogen prefers to donate its electrons rather than acquire more. The overall result is that the anion is destabilized.

In the ethyl anion, $CH_3CH_2^-$, suggest the shape of this charged species and also indicate the polarization of the bonds which may occur. Comment on the degree of stabilization relative to the methyl anion.

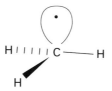

Each carbon is tetrahedral: one has three carbon/hydrogen σ single bonds, and the other has only two such bonds and a lone pair. This ion is even less stable than methyl anion because there are even more hydrogens which would prefer to donate electrons.

Suggest the relative order of stability of the following anions: $CH_3CH_2^-$, CH_3^-, $CH_3CH^-CH_3$ and $CH_3CH_2CH_2^-$.

$$CH_3^- > CH_3CH_2^- > CH_3CH_2CH_2^- > CH_3CH^-CH_3$$

The relative stability of the hydrocarbon anions is the opposite of that which was proposed for the related cations.

The last hydrocarbon species which we will consider is the free radical. Suggest a shape for the methyl radical, $CH_3\bullet$.

The overall shape is a flattened tetrahedron. The central carbon has six electrons in three covalent bonds to the three hydrogen atoms, and also there is an unpaired electron which occupies the p_z orbital. However, because there is only one electron in that orbital it does not have the same repulsive effect as a pair of electrons, and so the carbon/hydrogen bonds are not forced together as much as they are in, for example, the methane molecule which results in a regular tetrahedron.

Suggest the direction and degree of polarization which is present in the methyl radical.

The carbon is electrically neutral, but it does have a greater electro-negativity than hydrogen and so it attracts the electrons within the carbon/hydrogen bonds slightly.

Suggest the relative order of stability of the primary, secondary and tertiary carbon radicals.

$$\text{tertiary} > \text{secondary} > \text{primary}$$

The order is the same as that of the carbonium ions, because they share the same electronic property in that electrons are attracted towards the carbon which bears the unpaired electron from the adjacent hydrogen atoms.

Draw out the methyl radical using the stereochemical notation and also draw out the methyl cation indicating the empty p_z orbital.

There are two forms of the methyl radical, one with the unpaired electron above the plane of the other atoms, and the other with it below. In the methyl cation all the atoms are in the same plane with one lobe of the p_z orbital above this plane and the other below. It is observed that the carbon radical species inverts its configuration very readily and presumably it passes through a shape which is closely related to that adopted by the methyl cation as the unpaired electron goes from above to below the plane of the atoms.

5.2.2 Charged Non-Hydrocarbon Species

Electronegative atoms draw charge towards them. Indicate the direction of the polarization along a carbon/chlorine σ bond.

$$\overset{\delta+}{C} \longrightarrow \overset{\delta-}{Cl}$$

The chlorine exerts a negative inductive effect and so pulls the electrons within the σ bond towards itself. This in turn makes the carbon atom slightly positive.

If one of the hydrogen atoms within the methyl cation is replaced by a chlorine atom to give the CH_2Cl^+ cation, would this ion be more or less stable than the corresponding methyl cation, CH_3^+?

It would be less stable because the chlorine atom would make the carbon atom even more positive and so destabilize it.

Now consider the related anion with the chlorine atom substituting a hydrogen atom, CH_2Cl^-, would this be more or less stable than a methyl anion?

In this case the chloroanion would be more stable than the related hydrocarbon anion because the electronegative chlorine would decrease the charge which is present on the carbon.

Suggest the order of the stability of the following anions: $CHCl_2^-$, CH_2Cl^- and CCl_3^-.

$$CCl_3^- > CHCl_2^- > CH_2Cl^-$$

The greater the number of electronegative atoms that are joined to the carbon which bears the extra electron, the greater the degree of stabilization of the resultant anion. The trihalomethylcarbanion will be seen later as a good leaving group in the haloform reaction.

5.3 Mesomeric Effects

5.3.1 Charged Hydrocarbon Species

The mesomeric effect operates through the π bond system. Draw the dot and cross structure of the cation, $CH_2=CH\text{-}CH_2^+$.

The system in which the carbonium ion is adjacent to a carbon/carbon double bond is called an allylic grouping. The diagram represents one of the canonical structures for this cation.

Draw the stereochemical diagram of the allylic cation indicating both the

p_z/p_z interaction to yield π bond

empty p_z orbital

p_z orbitals which are involved in the double bond and also the empty p_z orbital on the carbon that bears the positive charge.

The p_z orbitals which are involved in the carbon/carbon double bond are parallel to the empty p_z orbital, with all the carbon and hydrogen bonds being in the same plane. This means that the electrons within the double bond may easily be delocalized into the empty orbital.

Write the mechanism for the rearrangement of the electrons within the π bond system so as to give the other canonical structure of this allylic group.

Draw the resultant species if these two canonical structures are time averaged.

Each terminal carbon is now joined to the central carbon by a bond which is half way between a single and double bond. Furthermore, each terminal carbon bears half a unit of positive charge. Thus the charge has very effectively been spread out over two carbons, *i.e.* delocalized. The structure with the intermediate bonds closely represents the actual electronic distribution which is present in the ion, and is this structure is called the resonance structure.

If a further single and double bond is added to this system we obtain the cation with five carbons, $CH_2=CH-CH_2=CH-CH_2^+$. Draw the three canonical structures of this cation.

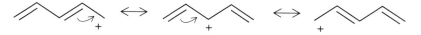

The addition of another $CH_2=CH$ unit is said to form the vinylogue of the original compound.

Write down the mechanism for the mesomeric transmission of the positive charge along the chain.

It is readily seen that the charge is very efficiently transmitted along a chain which possess alternate double and single bonds, *i.e.* a system which is conjugated. In such a system the degree of delocalization is greater and so the amount of stabilization is greater. There is one major restriction in that this rearrangement of electrons must not result in a double bond at a bridge head, that is a junction point between three ring systems. This is Bredt's rule.

Earlier we mentioned the shape of various carbocations. Remind yourself of the shape of the *t*-butyl cation, $(CH_3)C_3^+$, and then draw this.

The cation is trigonal planar. Now redraw this cation using the stereo-chemical notation, indicating the position of the empty p_z orbital and also drawing one of the methyl groups with one of its carbon/hydrogen bonds parallel to the p_z orbital.

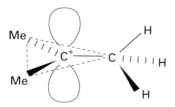

The electrons within the carbon/hydrogen bond are capable of moving into the empty p_z orbital. Draw the mechanism for this and hence the resultant product.

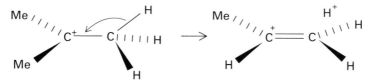

The unsaturated product is 2-methylpropene, and there is a bare proton positioned above it. There are nine possible hydrogens which could be involved in this rearrangement. This mechanism has been suggested as a possible way in which hydrogens which are adjacent to the positive charge may help stabilize it. It is called hyperconjugation and as may be seen from the required geometry it most readily occurs when the carbonium ion adopts a planar conformation. This is another reason why carbonium ions are more stable if they are able to adopt a flat shape.

5.3.2 Charged Non-Hydrocarbon Species

It is often found that a carbon which is bearing a positive charge is bonded to a heteroatom which has a lone pair of electrons, *e.g.* in the R_2C^+-OH system.

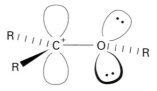

Michael Edenborough

Draw the stereochemical diagram of this cation, indicating the empty p_z orbital and the lone pairs on the oxygen atom.

Like the hyperconjugation that was discussed above the lone pair may move into the empty p_z orbital to form a π bond.

Write the mechanism for this reaction.

Now there are two canonical structures to represent this cation. Draw the resonance structure.

In this diagram the extra bond is half formed, with the charge is shared between the carbon and the oxygen.

A similar stabilization may occur in the amino derivative, R_2C^+-NH_2.

Draw the mechanism which represents this stabilization.

Here the lone pair on the nitrogen is donated into the empty p_z orbital.

So far we have looked at the stabilization of a positive charge using the mesomeric effect, but it is equally possible to stabilize a negative charge. Instead of having a group which has an extra lone pair which may be used to form a double bond, what is needed is a group which has a multiple bond which can be broken heterolytically with the result that the extra charge is placed upon an electronegative element.

The first example is concerned with the stabilization of a negative charge from an oxygen in the carboxylate anion, RCO_2^-. Draw the dot and cross structure of this anion.

If the carbon/oxygen double bond breaks heterolytically and so place the negative charge upon the oxygen, the carboxylate carbon will be left with a positive charge. Draw the mechanism for this initial step.

This species is symmetrical about the charged carbon. Now draw the second part of the mechanism in which the oxygen which originally bore the negative charge donates this to the carboxylate carbon so as to form a bond.

Here the arrangement of the carbon/oxygen single and double bonds has changed over.

Usually these two steps are joined together. Draw a mechanism which shows both parts on one diagram.

Now draw the resonance structure of the carboxylate anion.

If instead of the extra charge coming from the oxygen in the carboxylate anion it comes from a carbon which is attached to the carboxylic acid group, then the same principles would apply.

For the anion, $R_2C^-\text{-}CO_2H$, write a mechanism which would delocalize the charge on to one of the carboxylic acid oxygens.

Here the carbon/oxygen double bond has been broken, which places the negative charge on that oxygen, and a new carbon/carbon double bond has been formed.

Instead of the charge being rearranged on to the other oxygen of the carboxylate anion it may be placed upon the carbon which is joined to the carboxylate group. In this case the carboxylate group will cleaved itself form the rest of the molecule. In order for this to occur it must be possible for the negative charge to be stabilized while on the carbon atom.

Write a mechanism for the cleavage of the $R_3C\text{-}CO_2^-$ anionic species.

The carboxylate anion leaves as the uncharged molecule of carbon dioxide, which usually escapes from the reaction mixture and so may not react again to reform the original compound. If the negative charge may be accommodated elsewhere then the carboxylate may be used as an effective leaving group.

The bond which is broken need not be a double bond, but it may be a triple bond which rearranges to form a double bond. For example the cyano group, as $NCR_2C\text{-}CO_2^-$, in is capable of stabilizing a negative charge. Write a mechanism for this.

In this case a carbon/carbon double bond is directly adjacent to a carbon/ nitrogen double bond with the terminal nitrogen bearing the extra charge.

The stabilization of negative charges, initially formed at a carbon atom, plays a very important part in organic chemistry. One of the commonest groups which performs this function is the carbonyl group. Write down the mechanism whereby a negative charge which is resident on a carbon atom may be delocalized on to a carbonyl group which is adjacent.

If this anion is now protonated on the oxygen a neutral species is formed. Write down the structure of this compound.

This compound is called an enol, and the rearrangement of the carbon/ oxygen double bond in the carbonyl group to the carbon/carbon double bond in the enol compound is called the keto/enol tautomerism. Notice also that a hydrogen atom has changed place, from the carbon to the oxygen. Tautomers are different from canonical structures in that the former represent real compounds which may be isolated, while the latter represent only hypothetical structures which help in the understanding of the electron distribution.

The nitro group, NO_2, also performs the function of stabilizing a negative charge on an adjacent carbon atom. It does this by carrying out a similar rearrangement to the keto/enol tautomerism. Write a mechanism for the delocalization of a negative charge which is initial upon a carbon atom adjacent to a nitro group, R_2C^--NO_2.

Here again a double bond has been broken, this time a nitrogen/oxygen bond, and a new carbon/nitrogen double bond has been formed in its place. This is called the nitro/aci tautomerism, and again each tautomer may be isolated, because each tautomer represents a real compound and not just a hypothetical model.

Inductive and mesomeric effects are the two most common properties which affect the stability of a carbanion or carbonium ion. However, there are a number of other properties which when they are relevant have a very powerful influence. We will now consider these in turn.

5.4 Degree of S Orbital Character

A lone pair of electrons only orbits one nucleus. The nucleus carries a positive charge, while the lone pair is a negative charge which is distanced from the nucleus.

Would the degree of stabilization of the electrons increase or decrease as the average distance between the electrons and the nucleus increases?

As the charge separation increases the stabilization would decrease as the Coulombic interaction decreases.

For any given quantum level, is the s sub-orbital closer to the nucleus than the p sub-orbitals?

The s sub-orbital is closer to the nucleus. Thus place the following hybrid orbitals sp^3, sp and sp^2 in order of increasing average distance from the nucleus.

$$sp < sp^2 < sp^3$$

With increasing p sub-orbital content the hybrid atomic orbital becomes increasing distant from the nucleus. Suggest the order of increasing stability for a lone pair in these hybrid orbitals.

$$sp > sp^2 > sp^3$$

This order reflects the ease with which a proton may be removed from a carbon hybridized in any of the above ways, *i.e.* this reflects the acidity of such a hydrogen.

Now suggest the order of stability of the various dicarbon anions, $C_2H_5^-$, $C_2H_3^-$ and C_2H^-.

$$C_2H^- > C_2H_3^- > C_2H_5^-$$

The ethyne anion is by far the most stable of the three, and may be made under quite mild conditions by the addition of a suitable base such as sodium hydroxide. To remove the proton from a carbon which is hybridized in the sp^3 format requires a very strong base indeed such as butyllithium.

Suggest the order for the stabilities of the corresponding carbonium ions, $C_2H_5^+$, $C_2H_3^+$ and C_2H^+.

$$C_2H_5^+ >> C_2H_3^+ > C_2H^+$$

Only the ethyl carbonium ion is produced under normal conditions. The vinyl carbonium ion, $C_2H_3^+$ is only produced under extreme conditions and if you need to invoke it in a mechanism you are either wrong, or suggesting a rather advanced mechanism.

5.5 D Orbital Involvement

Third row elements such as phosphorus and sulfur have empty d sub-orbitals into which electrons may be placed from an adjacent carbon which bears a negative charge. These elements are not bound by the Octet Rule and may accommodate more than eight electrons in their valence shell. Thus these elements may act as an electron sink.

Suggest a possible mechanism for the stabilization of the carbanion, $RSO_2CR_2^-$, in which sulfur has expanded its octet.

Here a sulfur/oxygen double bond is broken and replaced by a sulfur/carbon double bond.

Phosphorus exhibits similar properties in that it may expand its octet. In the Wittig reaction, the reagent is a phosphorus/carbon compound in which the carbon bears a negative charge next to a phosphorus which is carrying a positive charge. Such an arrangement is called an ylid. Draw the structure for the ylid $Ph_3P^+\text{-}CR_2^-$ indicating the lone pair and the position of the charges.

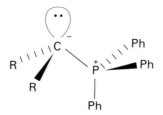

This is only one of the possible canonical structures, but this is the one that is used when involving the ylid in a reaction, because it allows the movement of electrons to be followed easily.

Suggest another canonical structure of this compound which involves the stabilization of the negative charge by the empty d orbitals of phosphorus.

Now the phosphorus has expanded its valency to five and accommodates ten electrons around its nucleus. Later on in this reaction sequence the phosphorus forms an oxide of formula $Ph_3P=O$, which again has a valency of five. It is the propensity of phosphorus to form this oxide that is one of the driving forces behind this reaction.

5.6 Aromatic Character

It has been observed that cyclic conjugated systems that have six electrons in the π bond system are particular stable. In general this applies to systems which have $(4n+2)\pi$ electrons in a continuous closed loop of π orbitals, where n varies from zero upwards in steps of one unit. This property is called aromaticity.

Draw out the two major canonical structures of the cylcohexatriene compound, where n equals one and hence there are six π electrons.

This molecule is benzene. The resonance structure would have six bonds which are half way between a single and double bond. Benzene is often represented by a regular hexagon with a circle inside to indicate the delocalization, but this is not very useful when the electrons in bonds within a benzene ring are being utilized in a reaction pathway. When this occurs it is more useful to indicate each single and double bond separately, because it is then easier to account for the all the electrons.

How many π electrons are there in the cyclopentadiene molecule?

There are four π electrons, however, there is not a continuous circuit of π orbitals around the ring. One of the carbon atoms in the ring is sp^3 hybridized, *i.e.* is a methylene group, CH_2. If a hydrogen cation is removed from this carbon, and then that carbon re-hybridizes to sp^2, which means that there is now a continuous ring of p orbitals.

Draw this anion, $C_5H_5^-$, showing the p orbitals of the double bonds and the newly formed p orbital which is occupied by the lone pair.

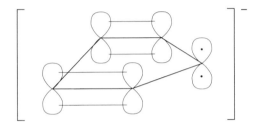

This system now has a continuous ring of five p orbitals which contain a total of six electrons, two from each of the two double bonds and two more from the lone pair. Thus this system is aromatic. Notice that in an open chain version of this molecule a methylene hydrogen would be the least acidic hydrogen in the molecule, because all the other hydrogens are bonded to sp^2 carbons, while this one is bonded to a sp^3 carbon. However, as a result of the extra stability that is accrued to the molecule by obtaining an aromatic character, the normally least acidic hydrogen is the one which is actually removed.

This anion is sometimes drawn as a regular pentagon inside of which is circle, and with a negative sign in the middle of the whole diagram, *e.g.*:

The circle is used to indicate the delocalized nature of the bonding, and in particular the aromatic character.

So far we have looked at two species in which n equals one. There is another species in which n equals one. In this case a cation is formed from the cycloheptatriene. Suggest how this may be formed and the resultant structure.

In the cycloheptatriene there were three double bonds and hence six p orbitals; if a hydride ion is removed from the only methylene unit and that carbon is re-hybridized from sp^3 to sp^2 so as to create an empty p orbital this then completes the circuit around the cycloheptane ring. There is then a continuous series of p orbitals which contain six π electrons, which gives rise to an aromatic system.

If a seven membered ring is fused with a five membered ring to form a 5/7 ring system, and if in the larger ring there are three double bonds while in the smaller ring there are two double bonds, suggest what, if any, movement of electrons may occur so as to form a dipole and the direction of any such dipole that may be formed.

If the electrons in one double bond rearrange so as to leave the seven membered ring electron poor, while correspondingly making the five membered ring electron rich, each ring will become aromatic. This would create a permanent dipole with the negative end on the smaller ring.

So far we have looked at aromatic systems where n equals one. If n equals zero then there are only two π electrons. Suggest a system in which this arrangement may exist.

Start by thinking about the formation of the two aromatic ions illustrated above. In those cases the aim was to obtain six π electrons in a continuous ring of p orbitals. When forming the anion, there were only four electrons initially in two double bonds and the extra two electrons were supplied by the removal of a hydrogen cation from the methylene unit. In the case of the cation, there were already six π electrons, but there was a need for an extra p orbital to complete the ring. This extra orbital had to be empty, and so a hydride anion was removed from the methylene unit.

When n equals zero, only two π electrons are required. It would seem logical to achieve this by removing a hydride anion from a methylene unit which is adjacent to a double bond containing the two π electrons required. Draw the mechanism for the removal of a hydride anion from cyclopropene to give the cyclopropenyl cation.

Here there are two π electrons in three p orbitals which form a continuous ring and so the system is aromatic. Notice that for the case of n equals zero, there is no aromatic system which is anionic in nature, because for this to happen there must be one fewer p orbital than there are π electrons which in this case would require only one p orbital, *i.e.* no ring.

If instead of (4n + 2)π electrons in a continuous ring of p orbitals there are 4n π electrons, then the system is destabilized instead of being stabilized. This is called anti-aromaticity. Suggest an ion which may be formed from cyclopentadiene which would be anti-aromatic in nature.

Here n equals one, and so there are four π electrons in five p orbitals. This results in the destabilization of the ring, and so the formation of this ion is disfavored.

Suggest a neutral species in which n equals two which has anti-aromatic properties.

Now there are eight π electrons in eight p orbitals which are continuous. The system shows anti-aromatic character, *e.g.* it is not flat and the single and double bonds are distinguished by being different lengths.

5.7 Hydrogen Bonding

In the hydrogen chloride molecule there is a permanent dipole. Draw this molecule showing the direction of this dipole.

$$\overset{\delta+}{H} \longrightarrow \overset{\delta-}{Cl}$$

The chlorine has a small negative charge while the hydrogen carries a corresponding positive charge. Suggest which atom has the higher charge density.

Each atom must carry the same charge, but the chlorine atom is much larger than the hydrogen atom. This means that the hydrogen atom has the higher charge density. In fact, because hydrogen is so small, that the charge density is relatively high, so that it is capable of interacting with a negative charge. Suggest a source which is electron rich in the hydrogen chloride molecule.

The lone pairs of the chlorine atom are electron rich and so may have an electrostatic interaction with the positive centre of the polarized hydrogen atom.

This Coulombic interaction is called the hydrogen bond, and is a source of stabilization in a variety of molecules, in particular in those which have an atom bearing a negative charge which is close to a polarized hydrogen atom. Such a configuration is present in *ortho*-hydroxybenzoic acid after the carboxylic acid group has been deprotonated. Draw the resultant anion.

In this case there is an *intra*molecular hydrogen bond, while in the case of the hydrogen chloride molecule the hydrogen bond was *inter*molecular. Both give rise to a favorable thermodynamic interaction.

In dicarboxylic acids which are capable of forming intramolecular hydrogen bonds it is observed that the first proton is removed more easily than would be predicted by just considering the inductive effect of the second carboxylic acid. Suggest the structure which *cis*-butenedioic acid adopts after one deprotonation.

Here a seven membered ring is formed by the deprotonated carboxylate group and the polarized hydrogen of the second carboxylic acid group. Suggest the form of the hydrogen bonding that may exist in the monodeprotonated anion of *trans*-butenedioic acid.

Michael Edenborough

Note that in this case only intermolecular hydrogen bonding is possible, because of the rigid nature of the central carbon/carbon double bond. The difference in the favoring of the formation of the intermolecular and intra-molecular hydrogen bonding is reflected in the difference in the acidities of the first hydrogen to be removed in each case. The *trans* isomer has a dissociation constant of 3.02, while for the *cis* isomer it is 1.92, *i.e.* it is a stronger acid.

Suggest the relative ease of removal of the second carboxylic hydrogen from each of these two species.

In each case it is more difficult to remove the second hydrogen than it was to remove the first. This reflects that it is harder to remove a positive proton from an anion than it is from a neutral species. The second hydrogen in the *trans* isomer is more acidic than the equivalent hydrogen in the *cis* isomer, because it is harder to remove the proton from the seven membered ring than it is from an open form of hydrogen bonding.

5.8 Steric Effects

The concept of intramolecular hydrogen bonding may be taken a little further by looking at some further steric considerations.

Draw out a possible structure of 2,4,6-trinitroaniline which shows the intramolecular bonding.

Now write down another canonical form of this molecule which still has the intramolecular hydrogen bonding present.

In this species the lone pair of the nitrogen is involved in the aromatic ring and so can no longer act as a proton acceptor.

Now consider the related molecule 2,4,6-trinitro-N,N-dimethylaniline. Draw this molecule, and in particular draw out the hydrogen atoms on the methyl groups.

It is apparent when the hydrogen atoms are displayed on the methyl groups that there is not enough room for the N,N-dimethylnitrogen group to lie flat in the same plane as the benzene ring, but instead this group must be orthogonal to the ring. As a consequence the lone pair on the nitrogen would also be orthogonal to the delocalized π system and so there could not interact mesomerically with it.

Now suggest which of the two aniline derivatives is the more basic.

2,4,6-trinitro-N,N-dimethylaniline is about 40,000 times more basic than 2,4,6-trinitroaniline because of the impossibility of the lone pair on the nitrogen being delocalized into the aromatic ring, and so it is free to act as a base and react with a proton.

6

Thermodynamic and Kinetic Effects

6.1 Introduction

The interplay between thermodynamic and kinetic effects is very important and as such it is possible to spend vast amounts of time and effort on it. However, there are only a few basic ideas which are needed to guide you through most of the situations that will arise.

In brief, reversible reactions are governed by thermodynamic considerations, while irreversible reactions are controlled by kinetic factors.

In a general reaction between A and B to give C and D, the reagent A must physically approach the substrate B so that the electrons within each molecule may interact; this results in some bonds being broken and reformed so as to produce the new molecules C and D. This process may be represented on a diagram called a reaction co-ordinate diagram.

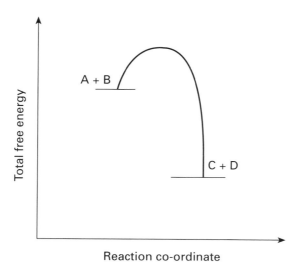

The ordinate, *i.e.* the *y*-axis, represents the total free energy of the system, *i.e.* the exo and endothermic factors of the reaction. The abscissa, *i.e.* the *x*-axis, represents how far the reaction has proceeded from the starting materials, A and B, to give the products, C and D. For a simple one step reaction in which the products are more stable than the starting materials, the curve which represents the reaction pathway is such that at the end of the reaction, *i.e.* on the right hand side of the diagram, finishes lower than the starting point.

The peak of the curve represents the transition state between the starting materials and the products. The height of the peak from the starting materials represents the amount of energy that is needed to start the reaction, *i.e.* the activation energy. The difference between the levels representing the starting materials and the products represents the overall free energy of the reaction, called the Gibbs free energy ΔG. If the entropy factors, that is those factors which are concerned with the disorder within the system, are similar on each side of the reaction, then the free energy corresponds to the enthalpy of the reaction, that is the difference in the energy of the starting materials and the products. If the products have less energy stored in them than the starting materials, then energy will be released as the reaction proceeds. In such a case the reaction is said to be exothermic. If energy is absorbed as the reaction proceeds, then the reaction is said to be endothermic.

For a two step reaction the reaction co-ordinate diagram is as follows.

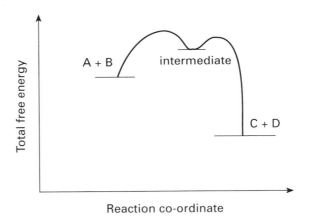

Reaction co-ordinate

In this case there is an energy well which has a peak either side of it. The species that is caught in the energy well is called an intermediate and may be isolated. The presence of the peaks either side of the intermediate indicates that energy is required for either the forward or the reverse reaction to proceed from this intermediate species, *i.e.* there is an activation energy barrier which the intermediate must overcome before it may revert to the starting materials or proceed to the final products.

There is a simple rule of thumb which states that the larger the energy difference between the starting materials and the products then the equilibrium will lie further over to the side which has the lower energy. Also, the greater the difference in energies the faster that equilibrium is reached, because the activation energy barrier tends to be smaller.

6.2 Thermodynamic Considerations

For a reaction to proceed spontaneously the change in Gibbs free energy, ΔG, must be negative. Hence deduce whether this means that the free energy of the products must be lower or higher than that of the starting materials for the reaction to proceed in the forward direction.

For the free energy to be negative, then the free energy of the products must be lower than that of the starting material. Gibbs free energy is composed of two parts, the enthalpy, ΔH and the entropy ΔS, and these terms are related by the equation:

$$\Delta G = \Delta H - T\Delta S$$

The enthalpy refers to the bond energies: more and stronger bonds increase the enthalpy of the system. The entropy refers to the amount of disorder: the greater the disorder of the species present in the system the higher the entropy of that system. If the starting materials and the products have approximately the same amount of disorder, then the reaction is controlled by the changes in the enthalpy.

Write down a general chemical equation for the hydrolysis of alkyl halides.

$$RX + OH^- = ROH + X^-$$

To a first approximation the same sort of species exist in the same relative ratios on each side of this equation, *i.e.* there is one anion and one neutral molecule on each side of the equation. Thus it may be assumed that the entropy difference between the starting materials and the products is small.

If we now look at the alkyl halide in a little more detail, suggest what will be the major differences between the alkyl chloride and the alkyl iodide.

There are two major differences. The first is that in the alkyl chloride the C-X bond is more polarized than in the alkyl iodide; the second is that the C-Cl bond is much stronger than the C-I bond. Assuming initially that the degree of polarization is the dominant effect upon the rate of hydrolysis, then would an alkyl chloride hydrolyse faster than an alkyl iodide, or *vice versa*?

Assuming that the rate of hydrolysis is determined by the degree of polarization of the carbon/halide bond, then the alkyl chloride would hydrolyse faster than the alkyl iodide. This hypothesis may be justified by saying that the more highly polarized the bond is initially, the closer it is to breaking.

Now, however, let us assume that it is the strength of the C-X bond that is the dominant effect upon the rate of hydrolysis of the alkyl halide. In this case what would be the relative rates of hydrolysis?

Now the opposite result is achieved, in that the compound with the weaker bond would hydrolyse faster, *i.e* the alkyl iodide. This result comes from the rule of thumb that was given earlier, namely that the larger the difference between the energies of the starting materials and the products the faster the rate, because the activation energy is lower. In this case the alkyl iodide is less stable than the alkyl chloride because it has the weaker C-X bond. Thus the iodo compound would be higher up the ordinate axis on the reaction co-ordinate diagram. In each case, though, the product is the same, *i.e.* the alcohol. Thus there is larger energy difference for the iodo compound and so that species reacts faster. In practice it is observed that the alkyl iodide hydrolyses faster and thus it may be assumed that the strength of the C-X bond has more influence on the rate than does the degree of polarization of the bond.

This highlights the point that even though a particular hypothesis can be well justified, as was the case when it was suggested that the rate of hydro-lysis was determined by the degree of polarization, this does not mean that either the justification or hypothesis is correct. One must always base a hypothesis on observed facts, and then attempt to find an explanation.

The bond strengths may be used to identify which isomer is more stable. For example, a ketone such as propanone may form an enol under basic conditions. The structures of both isomers are shown below.

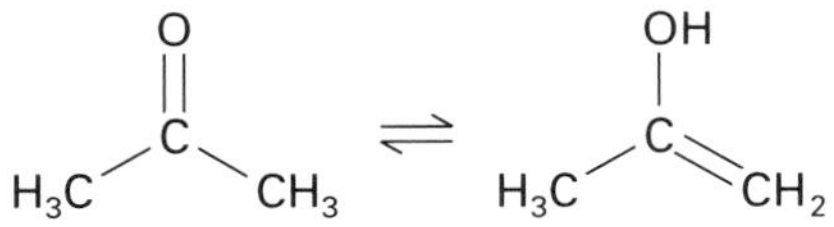

$$
\underset{\text{keto}}{\underset{H_3C}{\overset{O}{\underset{}{\overset{\|}{C}}}}\text{CH}_3} \quad \rightleftharpoons \quad \underset{\text{enol}}{\underset{H_3C}{\overset{OH}{\underset{}{\overset{|}{C}}}}\text{CH}_2}
$$

This particular isomeric relationship is called tautomerism. Now list which bonds which are different between the two compounds.

In the ketone there is a carbon/oxygen double bon,; three C-H bonds from a sp^3 carbon and one carbon/carbon single bond. In the enol tautomer there are two C-H bonds from a sp^2 carbon, a carbon/carbon double bond, an O-H bond and a carbon/oxygen single bond. This may be simplified by assuming that a C-H bond from a sp^3 carbon is about the same a C-H bond from a sp^2 carbon. Now one is looking at the difference between one carbon/oxygen double bond plus one C-H bond for the ketone, and one carbon/carbon double bond plus one O-H bond for the enol tautomer. Every type of bond requires a different amount of energy to separate the component parts. This value is called the bond energy and is measured in kJmol^{-1}. Using the value given below calculate the difference in the bond energies of the keto and enol isomer of the carbonyl compound.

O-H	110 kJmol^{-1}
C-H	100 kJmol^{-1}
C=O	180 kJmol^{-1}
C=C	150 kJmol^{-1}
C-O	90 kJmol^{-1}
C-C	85 kJmol^{-1}

The figures for the ketone are 180 kJmol^{-1} + 85 kJmol^{-1} + 100 kJmol^{-1} which equals 365 kJmol^{-1}; while for the enol tautomer they are 150 kJmol^{-1} +90 kJmol^{-1} +110 kJmol^{-1} which equals 350 kJmol^{-1}. Hence the enol is less stable isomer by about 15 kJmol^{-1}. Note, however, that in the enol tautomer there is also the possibility of hydrogen bonding which could contribute some enthalpic stabilization, but it may also increase the order of the system and so decrease the entropy which would be a destabilizing factor; thus its effect has been ignored!

There is another common tautomeric system, and this involves the nitro/aci compounds, which is illustrated below.

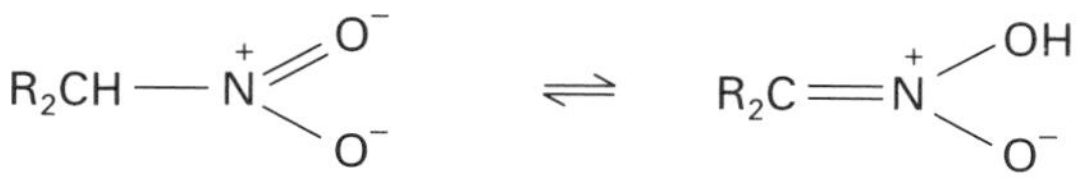

nitro   aci

Given the following values for the relevant bond energies, calculate the relative stability of the nitro/aci tautomers:

C-H	100 kJmol^{-1}
O-H	110 kJmol^{-1}
C-N	75 kJmol^{-1}
C=N	145 kJmol^{-1}
N-O	50 kJmol^{-1}
N=O	150 kJmol^{-1}

The figures for the nitro compound are 75 kJmol^{-1} + 150 kJmol^{-1} + 100 kJmol^{-1} which equals 325 kJmol^{-1}, while for the aci tautomer they are 145 kJmol^{-1} + 50 kJmol^{-1} + 110 kJmol^{-1} which gives 305 kJmol^{-1}. So in this case there is a larger difference in the stability of the two tautomers than in the case of the keto/enol tautomerism. In each of these two tautomeric systems, both isomers actually exist and may be isolated as identifiable compounds. Often, the tautomers are in equilibrium with each other.

Lastly it is interesting to look at the relative bond energies for the single and multiple bonds of carbon with another carbon atom, an oxygen atom and a nitrogen atom. The figures are:

C-C	85 kJmol^{-1}
C=C	150 kJmol^{-1}

$C \equiv C$ 200 kJmol^{-1}
$C\text{-}N$ 75 kJmol^{-1}
$C = N$ 145 kJmol^{-1}
$C \equiv N$ 205 kJmol^{-1}
$C\text{-}O$ 90 kJmol^{-1}
$C = O$ 180 kJmol^{-1}

It maybe seen that, all other things being equal, there is little difference between a carbon atom forming one carbon/oxygen double bond or two carbon/oxygen single bonds. However, for the carbon/nitrogen and the carbon/carbon systems the situation is different. By using the above numbers suggest what are the relative stabilities of the carbon/nitrogen systems.

There is little difference between the double and single bond systems. However, if a triple carbon/nitrogen bond is being broken this will release about 60 kJmol^{-1}, and a further 70 kJmol^{-1} would be released if the reaction proceeded to the carbon/nitrogen single bond stage. Using the rule of thumb mentioned above that the greater the difference in energies between the starting materials and the products the faster the reaction, then the reaction from the double bond to the single bonds should be faster than that from the triple to the double bond. The conclusion is that if a triple carbon/nitrogen bond is being broken then it is unlikely that the reaction will stop at the double bond stage, but that it will continue to the compound containing a single carbon/nitrogen bond.

By using the values for the carbon/carbon system suggest the relative stabilities of those compounds.

Here it is apparent that both the multiple bond systems are less stable than the single bond configuration. This is reflected in the fact that addition reactions are very facile.

Now we shall turn our attention to the other thermodynamic parameter, namely ΔS. Entropy is a measure of the disorder of the system; the greater the disorder the more the system is favored, in other words chaos is the favored state of things, and so introducing order requires energy. For the three phases of matter suggest which is the most disorder and which is the least.

$$\text{Gas} \gg \text{liquid} > \text{solid}$$

This means that if the phase of the products is different from that of the starting materials then there will be a large entropy factor that must be taken into consideration.

For example, if a reaction produces three molecules from only two molecules of starting materials then the disorder of the system is greatly increased.

A different situation is where a charged species is formed from uncharged precursors in a polar solvent. Suggest whether the entropy would increase or decrease in such a situation, and why.

The entropy would decrease because the charged species would order the solvent molecules around it, either with hydrogen bonding if the solvent is protic, *i.e.* had acidic protons, or otherwise by electrostatic interactions.

From the equation that was given at the beginning of this section, it may be seen that temperature has an effect on the entropy factor. As the temperature is raised is the effect of entropy greater or smaller?

The Gibbs free energy is directly related to the product of the temperature and entropy, and hence as the temperature is raised the effect of entropy becomes larger. This may be used to advantage, as a reaction may be controlled by altering the temperature at which it is performed.

In a reaction mixture in which every step is reversible then, so along as sufficient time is allowed, the final ratio of all the possible products will be determined by the relative thermodynamic stabilities of each of the compounds. Such a reaction is said to be under thermodynamic control. However, if for any reason a step is irreversible, or is in essence irreversible because the reverse reaction is very slow with respect to the length of time which is being allowed for the reaction, then the ratio of the possible products is no longer determined by thermodynamic considerations alone, but also by the speed at which each reaction takes place, *i.e.* the kinetic factors. A reaction in which the product distribution is determined by the rate of formation of the various products is said to be under kinetic control. We will now examine these factors.

6.3 Kinetic Considerations

As with thermodynamic factors, it is possible to spend a great deal of time and effort in understanding the kinetic factors which affect a reaction. In particular it is possible to use some very elaborate mathematical models which are very elegant but tend to be a bit daunting on first impression.

However, there are several very useful guidelines which may be used to advantage without any need to have a detailed understanding of the underlying mathematics.

In many reactions it is possible for the products to react in such a manner that the starting materials are reformed. This is the case when a general ester is hydrolysed under acid conditions, *e.g.* ethyl ethanoate:

$$CH_3CO_2CH_2CH_3 + H_2O + H_3O^+ = CH_3CO_2H + CH_3CH_2OH + H_3O^+$$

Once the carboxylic acid and the alcohol are formed they are capable of re-reacting to form the ester. If the *t*-butyl ester is substituted for the ethyl ester, then write down the equation for the hydrolysis of this new ester under acid conditions.

$$CH_3CO_2C(CH_3)_3 + H_2O + H_3O^+ = CH_3CO_2H + C(CH_3)_3OH + H_3O^+$$

This looks like the same problem. However, under certain acidic conditions the *t*-butyl carbonium ion is produced instead of the alcohol. This carbonium ion readily loses a proton. Write down the equation for this deprotonation.

$$^+C(CH_3)_3 = H^+ + (CH_3)_2C=CH_2$$

The product is 2-methylpropylene, which is a gas under normal laboratory conditions. Suggest what will happen to this product so long as the reaction vessel is not sealed.

Assuming that the reaction is being carried out in liquid solvent, then the 2-methylpropylene will bubble off from the mixture. Suggest what will be the consequence of the loss of the 2-methylpropylene.

If the 2-methylpropylene is no longer in the reaction mixture then it cannot under any circumstances react with the other product, *i.e.* the carboxylic acid, so as to reform the starting materials. This means that the reaction will be driven over to form more of the product. Suggest a thermodynamic reason why the equilibrium lies over to the product side.

There is a large increase in the entropy of the system, due to the formation of the gaseous product and this favorably affects the position of the equilibrium towards the product. So there are two good reasons which favor the formation of the other product when a gaseous co-product is formed in a reaction.

This may be seen clearly in the following example. If 1-aminobenzoic acid is treated with sodium nitrite then the diazo intermediate may be formed. This compound may then lose a proton from the carboxylic acid group to form an ion which contains both a positive and a negative charge. The skeletal formulae of this compound is shown below.

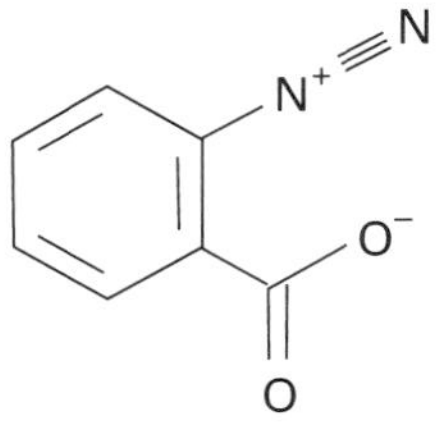

Suggest a mechanism in which a molecule each of nitrogen and carbon dioxide is formed.

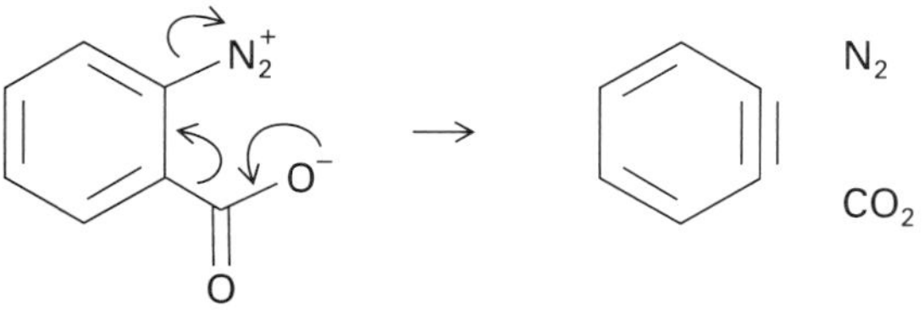

Here the electron rich carboxylate group donates a pair of electrons to the aromatic ring, while the electron poor diazo group reclaims a pair. This leaves three neutral molecules, two of which are gases and bubble off from the reaction mixture, and one hydrocarbon of the formula C_6H_4, in which there is a triple bond imposed on an aromatic ring. This molecule is called benzyne, and is rather unstable and very reactive. The fact that two gaseous molecules were also formed in this reaction helps drives the reaction over to produce this product.

Another useful concept is that certain sizes of ring are more easily formed than others. Draw out a six carbon aliphatic chain in such a manner that the ends are close together. Remember that in an aliphatic carbon the hybridization is sp^3, and so the CCC bond angle is about 109°.

If the hexane chain adopts a shape which is similar to the chair conformation of cyclohexane, then it is observed that the terminal carbons are within a normal single bond length of each other. Thus they are close enough to react to form a bond which may close the ring if that is chemically feasible.

Now look at the aliphatic pentane molecule and write down a conformation which will allow the terminal carbons to be close to each other.

Here if the carbon chain adopts a similar shape to that of cyclopentane then the end carbons would be close enough to react together.

Now look at the aliphatic butane chain and suggest what conformer may be adopted here to minimize the distance between the terminal carbons.

In this case there is no conformation which brings the terminal carbons close together. Suggest what will be the consequence of this for the ease of formation of a four membered ring by the closure of a butyl chain.

As the ends of a four membered chain cannot get close together it means that they cannot react together easily. This means that four membered rings are hard to form by this method.

The tetrahedral sp^3 hybridization is very common in second row elements. However, in the third row elements there is the possibility of using one or more of the d sub-orbitals that are now available. We used the electron repulsion theory in an earlier chapter to predict the shape of molecules of general formula AB_n, where n varied from two to four. If n equals five or six, what do you think will be the shape of such molecules?

When n equals five, the shape adopted is a trigonal bipyramid, while when n equals six it is a square bipyramid, or an octahedron. What is the angle between an axial and an equatorial bond in either geometry?

In both cases the axial/equatorial angle is 90°.
Suggest two third row elements, one of which has a valency of five and the other of which has a valency of six.

Phosphorus may have a valency of five, and sulfur may have a valency of six; when they do so, they adopt a trigonal bipyramidical or an octahedral geometry respectively.

So now we have an element which may have a bond angle which is naturally only 90° and thus would not be under any strain if it formed part of a four membered ring. This is demonstrated in the Wittig reaction in which an intermediate species has a four membered ring containing a phosphorus/oxygen bond along one side. So one way around the problem of forming a four membered ring is to use a third row element in one corner of the ring.

Lets us now look at the formation of three membered rings. Draw down a conformation in which the terminal carbons of a propyl chain are close together.

$$CH_3 \diagdown \diagup CH_3$$

Here, the terminal carbons are within one bond length of each other and so may in theory react to close the ring. What would be the internal bond angle in such a compound?

The internal bond angle would be 60°, which is a lot less than the normal angle of 109° for an sp^3 hybridized carbon atom. What does this suggest about the stability of such a ring?

Even though three membered rings may be fairly easily formed, they do have a tendency to re-open, because of the great strain imposed on the bonds being forced to adopt such a small bond angle. This is revealed in their chemistry; for example under normal conditions is it not possible to hydrogenate a carbon/carbon single bond in an alicyclic compound like cyclohexane, but in cyclopropane the ring opens up to form the aliphatic compound, a propane derivative.

Another general guideline involves the reactivity of hydrogens attached to carbon atoms in environments which have different accessibility. The first step in many organic reactions often involves the abstraction of a proton from the substrate so as to form a carbanion. This of course requires that there is an available hydrogen which may be attacked. Suggest which group of hydrogens in the butane molecule would most likely to be attacked.

On a purely statistical basis there are more hydrogens on the terminal methyl groups, *i.e.* six, than there are on the methylene units which only have four. So if all other things are equal then the hydrogens on the terminal methyl groups would be attacked in preference to those on the methylene carbons. Note also that in this case the effect is further enhanced because the resultant anion would be more stable if the charge was on the terminal carbon than on one of the middle carbons: this will be discussed in more detail in the next chapter.

The idea of accessibility may be taken further than a mere statistical effect. If there are two hydrogens which are equal in all respects except that one of them is partly hidden by some other part of the molecule, suggest what the difference in reactivity of these two hydrogens will be towards a base which is small in size or large in size.

If the base is very small such as the hydroxyl ion, then it will be able to approach and so react with a hydrogen in almost any position, even if it is partly obscured by another part of the molecule. The converse is true if the base being used is very large, *i.e.* a hindered base which has a large steric demand. Such a base will only be able to react with hydrogens that are accessible.Thus a degree of selectivity may be introduced by choosing the size of the base which will be used to abstract the proton.

So far we have only considered a single step reaction sequence. In many organic reactions there are many individual steps between the starting materials and the final products, with many separate intermediates. The principles which we have discussed with respect to a single step apply equally to each individual step of a multi-step reaction pathway. The overall rate of such a pathway is determined by the rate of the slowest step, which is called the rate determining step, rds.

6.4 Catalysis

6.4.1 General Considerations

A catalyst is a component of the reaction mixture which affects the rate of the reaction, while not being consumed itself. It may speed up or slow down the rate of the reaction, however, normally only those catalysts which increase the rate of a reaction are considered to be useful. Even though a catalyst is not consumed in the reaction, it may become involved, and change its oxidation state for example, so long as by the end of the reaction any changes that have occurred to it have been reversed so that there is no net change in its chemical composition. However, the physical state of the catalyst may be permanently changed.

If a catalyst has a different physical state from the reagents which are involved in the reaction which it is catalysing, then it is said to be a hetero-geneous catalyst. An example of this would be the metal catalysts, such as Pt and Pd, which catalyse the hydrogenation of alkenes. If, however, the catalyst has the same physical state, then it is said to be a homogeneous catalyst, *e.g.* the solvated proton in an acid catalysed reaction which is occurring in solution.

Many catalysts are only required in trace amounts in order to affect the rate of the reaction. Some, however, are required in similar quantities to the reagents, in which case they are called stoicheiometric catalyst. This often indicates that the catalyst is forming a complex with one of the reagents and it is this complex which is the attacking species in the reaction. An example of this is the Lewis acid, aluminium trichloride, catalysis of the Friedel-Craft reaction.

We saw earlier that a catalyst affects the rate of a reaction. In thermo-dynamic terms what controls the rate of a reaction?

The rate of a reaction is related to the activation energy barrier which exists between the starting materials and the transition state. On the reaction co-ordinate diagram, this is represented by the difference in height between the starting materials and the peak of the reaction co-ordinate curve. If the activation barrier is large how will this affect the rate of the reaction?

The smaller the activation energy barrier, then the faster the rate of that reaction. Suggest, then how a catalyst may affect the rate of a reaction.

The simplest way a catalyst may affect the rate is by lowering the activa-tion energy barrier, and this will result in a faster rate of reaction. This is usually achieved by the catalyst providing an alternative reaction pathway that involves the catalyst. The activation energy for this alternative route must obviously be smaller than the normal route in order to increase the overall rate of the reaction. This alternative route will presumably pass through a different transition state. The final products of a catalysed reaction are the same as in the uncatalysed reaction. Bearing this in mind, what is the effect which a catalyst has on the equilibrium position of the reaction?

The equilibrium position of any given reaction is determined by the difference in energy between the starting materials and the products. The catalyst does not effect the stability of the starting materials or products involved in the reaction and so cannot have any effect on the equilibrium position of a reaction. This means that a catalyst cannot be used in order to make a reaction proceed that would otherwise be thermodynamically impossible. A catalyst may only alter the rate of a reaction which is possible.

We will now look at a particular type of homogeneous catalyst that is very important in organic chemistry, that is acid/base catalysis.

6.4.2 Acid/Base Catalysis

The simplest example of acid/base catalysis is when the rate of a reaction is found to be directly proportional to the pH. If water is the solvent, then the rate is proportional to the concentration of hydroxonium ions, $[H_3O^+]$. This is the case in the acid catalysed hydrolysis of acetals, $CR_2(OR)_2$. This is called specific acid catalysis, because the rate is affected only by the concentration of the specific acid species H_3O^+. The rate is thus unaffected by the concentration of any other acid, *i.e.* proton donor, which is in the reaction solution, so long as the concentration of H_3O^+ ions remains constant.

Specific acid catalysis is characteristic of reactions in which there is a rapid, reversible protonation of the substrate before the rate determining step.

If the rate of a reaction is found to be dependant not only of the concentration of the hydroxonium ion, but also on the concentration of any other acid species which may be present, then the reaction is said to be general acid catalysed. An example is the acid catalysed hydrolysis of orthoesters, $RC(OR)_3$. In such a case the catalysis is by proton donors in general and is not just limited to hydroxonium ions.

General acid catalysis is characteristic of reactions in which the protonation of the substrate is the rate limiting step.

The same distinctions may be drawn for base catalysed reactions. In the reverse of the aldol condensation reaction, the rate of the reaction is found to be dependant upon the concentration of hydroxide ions only, and so is this is an example of specific base catalysis. This is indicative of rapid reversible deprotonation of the substrate before the rate limiting step.

In general base catalysis, any base may affect the rate of the reaction. Again by analogy with the case of general acid catalysis, this type of reaction is characterized by the fact that there is a slow deprotonation of the substrate which precedes the fast subsequent reactions to the products. An example of this type of reaction is the base catalysed bromination of a ketone.

7

Acid/Base Characteristics

7.1 Introduction

All the previous properties that have been discussed have been inherent in the molecule in question, and have not required the presence of another molecule for those properties to be revealed, namely non-induced charge distribution; shape; the ability to stabilize charge and thermodynamic properties. However, acid/base properties of a molecule do depend on the presence of other molecules in the reaction mixture. For example, nitric acid is usually considered to be an acid, but this is because we normally encounter nitric acid when it is dissolved in water. When nitric acid is dissolved in concentrated sulfuric acid it acts as a base. So to display the acid/base properties of a compound one needs the presence of a molecule with the opposite property.

Many mechanisms in organic chemistry start with an acid/base reaction. This may be just a simple Brönsted-Lowry protonation of a hydroxyl group which results in the activation of a C-OH bond; or else, it may be a Lewis acid/base reaction as, for example, when aluminium trichloride complexes with a haloalkane in the first step of the Friedel-Crafts reaction. In each case the initial intermediate usually reacts further and leads on to the desired product. In inorganic chemistry, the acid/base reaction may be all that is of interest, e.g. the treatment of a carbonate with an acid to liberate carbon dioxide. However, it is unusual in organic chemistry for the acid/base reaction to be an end in itself; it is for this reason that acid/base characteristics are normally considered as a property of the molecule, similar to nucleophilic and electrophilic properties to which they are closely related, rather than as a fundamental reaction type as is the case in inorganic chemistry.

We will now look at what is an acid or a base and then investigate some of the reactions which display these properties.

7.2 Definitions of Acids and Bases

The first thing is to define what we mean by an acid or a base. This is not straightforward as there are several definitions, each of which has its uses. We will look at four principal definitions in turn.

99

7.2.1 Arrhenius

Historically, this is the original definition of what is meant by an acid or a base. It is also the one with which you are most likely to be familiar. An acid is a hydrogen cation donor, while a base is a hydroxyl ion donor.

Identify the acid and base in the following reaction:

$$HCl_{(aq)} + NaOH_{(aq)} = NaCl_{(aq)} + H_2O$$

The hydrogen chloride is the acid because it donates the proton, *i.e.* the hydrogen cation; while the sodium hydroxide is the base because it donates the hydroxyl ion. The above reaction is the classical: acid plus base gives salt plus water. This definition is limited because the base is restricted to those compounds that can produce a hydroxyl ion, and so this effectively limits a base to being a metal hydroxide. The solvent is limited to being water, which may be convenient for the inorganic chemist working in aqueous solutions, it is less so for the organic chemist working in non-aqueous solvents. Thus we will not consider this definition further.

7.2.2 Brönsted-Lowry

Acids and bases are defined as proton donors and acceptors respectively. This now includes a much wider array of compounds as possible bases. Which is the acid and which is the base in the following reaction?

$$HCl_{(g)} + NH_{3(g)} = NH_4^+Cl^-_{(s)}$$

The hydrogen chloride is still considered to be the acid because it donates a proton. The ammonia molecule is now considered to be the base because it accepts a proton in this reaction. Why would the ammonia molecule not be considered a base under the Arrhenius definition?

The ammonia molecule does not produce a hydroxyl ion. If the reaction had been carried out in the aqueous phase would the same have been true?

No, because in the aqueous phase the ammonia molecule would have reacted with a water molecule to form an ammonium ion and a hydroxyl ion. Write an equation for this reaction.

$$NH_{3(aq)} + H_2O = NH_4^+{}_{(aq)} + OH^-{}_{(aq)}$$

If ammonia had been allowed to react with the hydrogen chloride in an aqueous phase then it would have fallen within the Arrhenius definition of an acid/base reaction. Notice that it is the water molecule which is central to the Arrhenius definition and this is one of the main reasons why it is of little use in organic chemistry, because these reactions are usually performed in non-aqueous phase. Similarly, because the Brönsted-Lowry definition does not depend on water, but only upon the presence of two compounds of complementary properties, it is this definition that is of more use to organic chemists.

Now consider the following reaction, performed in an organic solvent, and identify which is the acid and which is the base:

$$RCO_2H + RNH_2 = RCO_2^- + RNH_3^+$$

The carboxylic acid, RCO_2H, is the acid; while the amine, RNH_2, is the base. Now consider the reverse reaction, and identify the acid and the base:

$$RCO_2^- + RNH_3^+ = RCO_2H + RNH_2$$

Now the carboxylate anion, RCO_2^-, is the base; while the amino cation, RNH_3^+ is the acid. Notice that the carboxylic acid after deprotonation formed a base, *i.e.* the carboxylate anion. This is called its conjugate base. Suggest the name for the amino cation, RNH_3^+, which is formed after the protonation of the amine, RNH_2.

The amino cation is called the conjugate acid of the amine. Write down the general equation for a Brönsted-Lowry acid/base reaction, labelling the acid and the base, as well as the related conjugate base and acid.

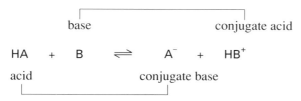

The related conjugate acid and base are always formed in a Brönsted-Lowry acid/base reaction.

On rare occasions the acid/base reaction is the sole purpose of the reaction. For example, what will be the product when the benzyloate anion, $C_6H_5CO_2^-$, which is dissolved in an aqueous solution, is then acidified with hydrochloric acid? Write an equation for this reaction.

$$C_6H_5CO_2^-{}_{(aq)} + HCl_{(aq)} = C_6H_5CO_2H_{(s)} + Cl^-{}_{(aq)}$$

The protonated form of the benzoate anion is benzoic acid and is neutral. Suggest whether it would be more or less soluble in the aqueous solution than its related conjugate base.

The benzoic acid is less soluble than the related anion. In fact it is so insoluble that it precipitates out from the solution as white flaky crystals. This great difference in solubility may be used to advantage in that the free acid may then be extracted into an organic solvent which is in turn immiscible with the aqueous phase. This extraction process allows the acid to be separated from impurities which are soluble in water.

The same technique may be used to form the free amino compound from its conjugate acid, this time by using a base. Write down the equation for the reaction between the hydrochloride of a general amine, $RNH_3^+Cl^-$, and a base.

$$B + RNH_3^+Cl^- = BH^+ + Cl^- + RNH_2$$

Some organic molecules are stored as the related salt because they are more stable and maybe easier to purify in that form. An example is the hydrochloride of glycine, $CH_2(CO_2H)NH_3^+Cl^-$, which is a white crystalline solid. Write down the equation of this compound when it reacts with a base.

$$B + CH_2(CO_2H)NH_3^+Cl^- = BH^+ + Cl^- + CH_2(CO_2H)NH_2$$

This reaction liberates the free amino acid, glycine. If glycine is allowed to stand while in solution it reacts with itself to form a cyclic dehydration product called a hydrazone:

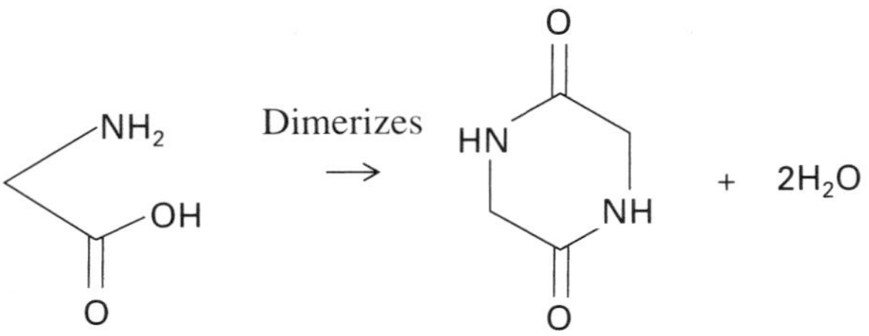

This hydrazone has limited uses. If you want to perform reactions on glycine, the first step is normally to release it from its hydrochloride and use the free glycine as it is freshly formed *in situ*.

So far we have mainly looked at organic molecules, or their ionic derivatives, reacting with an acid or base in an aqueous environment. Before we look at the acid/base reactions of organic molecules in a non-aqueous solvent it is instructive to look more closely at the reaction of a general acid, HA, dissolved in water and its reaction to form the hydroxonium ion, H_3O^+. Write down the equation for the reaction of this general acid with water, labelling the acid and base and the related conjugate base and acid.

$$\underset{\text{acid}}{HA} + \underset{\text{base}}{H_2O} \rightleftharpoons \underset{\text{conjugate base}}{A^-} + \underset{\text{conjugate acid}}{H_3O^+}$$

From this equation an expression for the equilibrium constant, K, may be obtained. Write down this expression.

$$K = ([A^-][H_3O^+])/([HA][H_2O])$$

The concentration of the water is almost constant, so rearranging the equation and defining K_a as equal to $K[H_2O]$, we arrive at the following expression:

$$[H_3O^+] = K_a[HA]/[A^-]$$

Now take the negative logarithm to the base ten of each side:

$$pH = pK_a + \log_{10}([A^-]/[HA])$$

This expression now contains the pH of the solution and also the term pK_a. The pK_a term is often used to indicate the strength of any particular acid. For a strong acid is this value low or high?

The stronger the acid the lower the value of the pK_a term. These values may even become negative for very strong acids. However, very nearly all organic acids have pK_a values from about plus two upwards. In the equation which defined pK_a, water was used as the complementary molecule to the acid so that the latter could display its acidic properties. (The subscript "a" refers to the fact that this equilibrium constant is concerned with acidic properties, and not to the fact that the reaction occurs in an aqueous environment, hence the opposite term which is used to describe the strength of bases has the subscript "b"). However, we could have used any other complementary molecule, such as a general alcohol, ROH. Write down the acid/base equation for the reaction between an alcohol and a general carboxylic acid, RCO_2H.

$$ROH + RCO_2H = ROH_2^+ + RCO_2^-$$

The stronger the acid, RCO_2H, the more of the protonated form of the alcohol, ROH_2^+, will be formed. This will be related to the pK_a value of the acid, but remember that the latter is defined for the aqueous system. This even excludes D_2O from being the solvent, let along a general alcohol, ROH.

A general alcohol is protic, which means that it may be a donor, as well as an acceptor, of hydrogen ions. We have seen it acting as an acceptor of hydrogen ions; now write down the equation for the reaction of an alcohol acting as a proton donor to a general primary amine.

$$ROH + RNH_2 = RO^- + RNH_3^+$$

The degree to which the alcohol is deprotonated is related to the relative strength of the amine as a base, which in an aqueous system is measured by the pK_b value.

Write down the chemical equation of the reaction of a general base, B, reacting with water to form the hydroxyl anion, OH^-, and so deduce the expression for pK_b.

$$B + H_2O = BH^+ + OH^-$$
$$K_b = [BH^+][OH^-]/[B]$$
$$pOH^- = pK_b + \log_{10}([BH^+]/[B])$$

At room temperature and pressure, pOH^- is related to pH by the expression:

$$pOH^- + pH = 14$$

Hence the pH of the solution may be found. Notice, though, that the pK_b of a base, which is a measure of its relative strength as a base, is only defined with water as the solvent, and so even though these values may be of interest in organic solvents they are not strictly applicable.

We have now looked at reactions which involved a protic solvent such as an alcohol. However, there are many organic solvents which are not protic, such as benzene or hexane. In these solvents there are no hydrogens which are joined to heteroatoms like oxygen or nitrogen, but only ones that are involved in carbon/hydrogen bonds. These bonds do not break easily and under normal conditions do not cleave to give hydrogen ions.

There are some solvents which are not protic, but under extreme conditions may lose a hydrogen ion, an example is trichloromethane, of which the trivial name is chloroform. Suggest what would be the result of losing a hydrogen ion from this molecule.

$$CHCl_3 + B = BH^+ + CCl_3^-$$

This anion is stabilized by having three -I groups attached to the carbon which bears the negative charge. This species may lose a chloride ion to form the dichloromethane carbene which itself may react further.

Another solvent which under extreme conditions may lose a proton is tetrahydrofuran, THF. Suggest which hydrogen is lost.

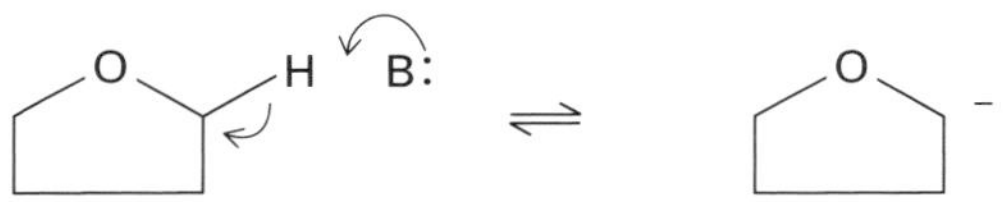

The hydrogen which is lost is the one which is closest to the oxygen. This anion may react further, by rearranging the distribution of the lone electron pair. Suggest a mechanism for such a rearrangement.

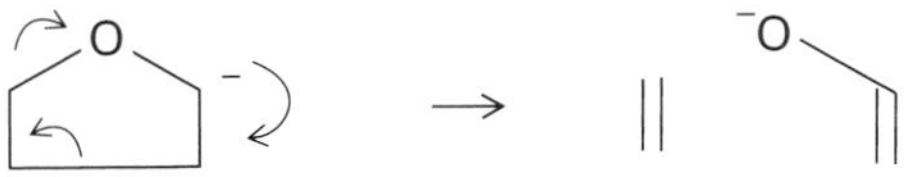

This reaction is driven by the fact that once the ethene is formed it bubbles off from the reaction mixture and so the ethanal enolate anion is unable to perform the reverse reaction. This decomposition has practical consequences. It means that THF cannot be used as the solvent when a very strong base is to be employed, unless precautions are taken to avoid the above reaction. The commonest precaution is to perform the reaction at low temperature, *e.g.* -78°, when the deprotonation of the THF is very slow.

The acid/base property of a molecule is a thermodynamic property and as such it is affected by temperature. This may be illustrated by the autolysis of water, in which two molecules of water react with each other, one acting as a base while the other acts as an acid. Write down the autolytic reaction of water.

$$2H_2O = H_3O^+ + OH^-$$

What is the pH of a sample of pure water at standard temperature and pressure, STP?

The pH of pure water at STP is 7. This reaction is endothermic, so what would be the effect on the pH of this sample when the temperature is increased?

According to Le Chatalier's Principle for an endothermic reaction, as the temperature rises, the reaction will proceed further to the right as written, which in this case, means that more hydrogen ions will be produced and thus the pH will fall. Note, however, that at each temperature the concentration of hydrogen cations still remains equal to the concentration of hydroxyl anions, even though the solution has become more acidic, as indicated by the drop in the pH value for the solution. This highlights the point that pH is only a measure of the hydrogen ion concentration, and that there is nothing particularly special about 7 being the normal value which is quoted for a neutral solution.

The essence of thermodynamic reactions is that they are reversible. If a very strong base is allowed to react with a molecule which has a removable proton, and the resultant conjugate base is much weaker, then the reaction will be in effect irreversible, and so kinetic factors may become important. What factors will affect which proton will be removed when the acid/base reaction is controlled by kinetic considerations?

There will be two principal driving forces which will determine which proton is removed. The first is which hydrogen is most weakly bonded; and the second is which hydrogen is accessible to the base. The first is a reflection of the which hydrogen is the most acidic in thermodynamic terms. The second reflects the steric demand which may be required by the base which is being used.

If the base that is to be used is large, and hence has a large steric demand, would it be more likely to remove a primary or tertiary proton, assuming that they each had the same pK_a?

The sterically hindered base would be more likely to remove the primary proton, because it would not be able to reach the crowded tertiary proton. When $CH_3COCH_2CO_2Et$ is treated with a base, which proton is removed?

This proton is removed regardless of whether the base employed is NaOEt or lithium diisopropylamine, LDA, because that proton is the most acidic and is also readily accessible. If the reagent is modified slightly, to become, $CH_3COCHRCO_2Et$, so that now the most acidic proton is attached to a tertiary carbon, there is a difference in how this molecule will react with NaOEt or LDA. Suggest what will be this difference.

The OEt⁻ anion may still gain access to the tertiary proton, which is the most acidic, and so remove it. However, LDA has a much larger steric demand and as such cannot approach this tertiary proton, thus it attacks the next most acidic proton which is readily available, which in this case is one of the terminal methyl protons. Draw the two resultant anions from the removal of the two different protons.

Notice that in both cases the resultant anions are stabilized by a carbonyl group, in one case by only one, while in the other by two. The essence of acid/base reactions which are controlled by kinetic considerations is that the reaction is irreversible. For each example given above, write down the conjugate acid of the reacting base which is formed after the deprotonation of the initial reagent.

$$EtO^- + H^+ = EtOH$$
$$Pr_2N^- + H^+ = Pr_2NH$$

Both these conjugate acids are very weak acids. The weaker the resultant conjugate acid the less likely that it will react to form the original reagent, and so the reaction becomes in essence irreversible. Charge stabilization of the anion which results after deprotonation is still important, but to a lesser extent than when the acid/base reaction is reversible. If butyllithium is used as a base, then very unreactive hydrogens may be removed, which result in unstabilized anions. However the conjugate acid is butane, which is a very, very weak acid, and secondly it is a gas and so leaves the reaction vessel, the deprotonation is irreversible.

7.2.3 Lewis

Acids and bases are defined as electron pair acceptors and donors respectively. This now includes a much wider array of compounds. An example is the reaction between boron trichloride and ammonia to give the adduct, H_3NBCl_3. Suggest which is the acid and which is the base in this example.

The ammonia is the base, while the boron trichloride is the acid as it accepts the lone pair from the ammonia. This definition is of great use to the organic chemist in, for example, the Friedel-Crafts reaction. Here the reagent, an alkyl halide, reacts with an aromatic compound such as toluene, in the presence of a catalyst such as aluminium trichloride. There must be approximately the same amount of the catalyst as there is alkyl halide, and not just trace amounts of the catalyst. The reason becomes apparent when the detailed mechanism is revealed. Suggest what may be the reaction between the alkyl halide and the aluminium trichloride.

$$R\ddot{X} \quad AlCl_3 \quad \rightarrow \quad R - \overset{+}{X} \rightarrow \overset{-}{A}lCl_3$$

Here the alkyl halide acts as a base, while the aluminium trichloride acts as the electron pair acceptor. The aluminium trichloride is called a stiocheiometric catalyst.

The adduct formed between the alkyl halide and the aluminium trichloride may further react by fragmenting. Suggest how this may happen.

$$R - \overset{+}{X} \rightarrow \overset{-}{A}lCl_3 \quad \rightarrow \quad R^+ \ + \quad [AlCl_3X]^-$$

This is a Lewis acid/base reaction in reverse, in that the adduct is fragmenting to form an acid and a base, however, they are different from those that originally formed the adduct. The carbonium ion may rearrange and so react *via* a different carbon atom than that to which the halide was originally attached. This rearrangement is called the Wagner-Meerwein rearrangement and is a complicating factor in the Friedel-Crafts reaction. Both of these reactions will be explored in more detail in Part II of this book.

7.2.4 Cady-Elsey

This definition is really included for completeness and out of interest, because it has little application in organic chemistry as it is concerned with ionic solvents. However, a few important organic species may be synthesized in the superacid solvents to which this definition has most application.

In this definition an acid is defined as a species which raises the concentration of the cation of the solvent, while a base raises the concentration of the anion of the solvent. This definition is of use when dealing with autolytic solvents such as liquid ammonia. Write down the autolytic reaction of ammonia, and so identify the solvent anion and cation.

$$2NH_3 = NH_2^- + NH_4^+$$

The anion is an amide ion, while the cation is an ammonium ion. So suggest what would be an acid in this solvent.

Ammonium chloride would be an acid as it would raise the concentration of the cation, namely the ammonium ion. Now suggest what would act as a base in this solvent.

Sodium amide would act as a base as it would raise the concentration of the anion, namely the amide ion. This definition may be used for water as well. Suggest what is the autolytic equation for water, and then what would be a potential acid and base using this definition.

$$2H_2O = OH^- + H_3O^+$$

Here the anion is the hydroxyl ion, and so sodium hydroxide would act as a base. Hydrochloric acid would act as an acid as it would raise the concentration of hydroxonium ions. This, you will recall, is similar to the original definition with which we started, namely the Arrhenius definition.

The main use of this definition is for solvents which are capable of forming both anions and cations by autolysis. There are other examples which do not depend upon the transfer of a proton to effect this ionization of the solvent, such as liquid antimony pentafluoride. Suggest the autolytic equation for this solvent.

$$SbF_5 = SbF_4^+ + SbF_6^-$$

Here the anion is formed by the addition of a fluoride ion, while correspondingly the cation is formed by the loss of the fluoride ion. Suggest in what manner hydrogen fluoride will act in this solvent.

$$HF + SbF_5 = H^+ + SbF_6^-$$

In this solvent hydrogen fluoride acts as a base because it raises the concentration of the anion of the solvent. This is rather different to the reaction of HF in water, where is acts as an acid.

As stated at the beginning of this section, this definition is of little use to the organic chemist in the normal course of events. However, there are rare occasions when it is of some limited application. For example, the Birch reduction is performed in liquid ammonia. Sodium metal is dissolved in liquid ammonia to give solvated electrons, which may be represented, rather simply, by NH_3/e^-. This species may decompose to give NH_2^- and a hydrogen atom. This nascent hydrogen may then react in a rather different manner to that in which the solvated electron would have done and so introduce an unwanted side reaction.

Also liquid antimony pentafluoride has been used as a solvent to study the properties of carbonium ions, because such ions are relatively stable in that solvent.

7.3 Electrophilic and Nucleophilic Properties

A species which is attracted to electrons is called an electrophile. It is usually electron deficient in that it has an empty orbital. Suggest a molecule in which one element has an empty orbital.

Such molecules as boron trihalide or aluminium trihalide would have an empty orbital on the central atom, because in each case there are only three electron pairs around it. Electrophiles may have a full complement of electrons, if they also bear a full positive charge. Suggest an example which falls into this category.

Such species would include the ammonium cation or the hydroxonium cation. These species will attract electrons Coulombically. Some species are both electron deficient and positively charged and so, not surprisingly, are very electrophilic. Suggest an example of such a species.

The commonest example would be the carbonium cation. There is another category in which the electrophilic species has a full compliment of electrons and does not bear a full positive charge yet still attracts electrons. Suggest an example which fits into this group.

An example would be a hydrogen atom which is joined to an element which has a high electronegativity such as chlorine, nitrogen or oxygen. In such a case the electronegative element draws electrons away from the hydrogen and so, even though the hydrogen does not bear a full positive charge, nor has a completely empty orbital, it is electron deficient and so may attract other electrons. Suggest a situation in which such an electron deficient hydrogen does attract electrons.

An example is when a hydrogen is involved in a hydrogen bond, such as that which exists between two water molecules. The opposite property to electrophilicity is called nucleophilicity. This is when a species is attracted to areas which are electron deficient. This property may be displayed in a number of different ways. One of the commonest ways is for the nucleophile to have an excess of electrons, which may be either in the form of a lone pair on a neutral species, or it may be as a full negative charge on an anion. Suggest an example of each type.

The ammonia molecule has a lone pair, but is electrically neutral; while the alkoxide anion has a lone pair and a full negative charge. The excess of electrons need not arise from having a lone pair, but may be found in a multiple covalent bond. Suggest an example which would fall into this category.

A molecule like ethene, which has a carbon/carbon double bond, has a centre of high electron density in the π orbital. Nucleophiles generally have a high electron density, and so would they be attracted to a positive or negative centre?

An area of high electron density would be attracted to positive centres. Ammonia, which has a lone pair and so a high density of electrons, is attracted towards a positive centre. Draw the reaction between the ammonia molecule and a general positive centre, A^+.

$$H_3N: \quad A^+ \quad \rightarrow \quad H_3\overset{+}{N} \rightarrow A$$

This is the interaction of a nucleophile with an electrophile. However, this type of reaction is already familiar. Replacing the general electrophile, A^+, with a proton highlights this similarity. Write down the reaction of the ammonia with the proton, labelling each species as is appropriate with the terms nucleophile, electrophile, acid and base.

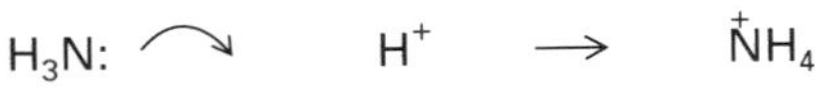

$$H_3N: \quad H^+ \quad \rightarrow \quad \overset{+}{N}H_4$$

nucleophile   electrophile   adduct
base   acid

The nucleophile, the ammonia molecule, is also acting as a base, while the electrophile, the proton, is also acting as an acid. So a base may also be thought of as a nucleophile, because both are electron rich species and seek positive centres. The above reaction between the ammonia and the proton may be classified as a Brönsted-Lowry acid/base reaction. It may also be classed as a Lewis acid/base reaction, depending on whether one views the ammonia as a proton acceptor or as a donor of a lone pair of electrons.

Thus there is a relationship between nucleophiles and electrophiles on the one hand, and acids and bases on the other. We will now develop this idea a little further.

The hydroxyl anion may classified as either a base or a nucleophile. Suggest reactions in which it acts as one or the other.

as a base: $OH^- + H^+ = H_2O$

as a nucleophile: $HO^- \quad \overset{+}{C}R_3 \quad \rightarrow \quad HO \text{——} CR_3$

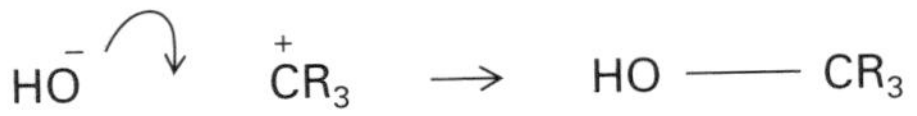

The hydroxyl anion may act as a base when it reacts with a proton to form a water molecule. This is a typical acid/base reaction under the Brönsted-Lowry definition. However, it may also act as a nucleophile when it is attracted towards a positive carbon centre. There is, however, an important difference between the positive centre of the proton and that of the carbonium ion. The difference lies in their relative charge densities: has the proton or the carbonium ion the greater charge density?

The hydrogen cation has a much higher charge density than the carbonium ion. A charged centre with a high charge density generates a very high potential gradient around itself. Thus would a charged centre with a high charge density be better at polarizing another charged centre than one with low charge density?

A centre with a high charge density would be better at polarizing another charge centre. A centre with a high charge density is called hard, while one with a low charge density is called soft. Suggest a charged species which has a low negative charge density and thus would be classified as soft. Also suggest a neutral species which has low density of excess electrons.

The iodide anion has a low charge density, as does the carbon/carbon double bond. Both of these would be classified as soft nucleophiles. Species with the opposite property are the electrophiles and acids. Suggest a charged electrophilic species which has a high charge density.

A common example would be the proton. Now suggest a species which is an electrophile, but with a low charge density.

The carbonium ion has a low charge density, and is an electrophile. Another electrophile which has a low charge density is the bromine molecule. Here the covalent bond between the bromine atoms may be polarized by being in close proximity to a charged centre. Any dipole which is induced would only be small and hence the corresponding charge density must be low.

Nucleophiles with a high charge density, such as hydroxyl ions tend to react more readily with electrophiles with a high charge density such as protons. Thus hard nucleophiles tend to react with hard electrophiles. This reaction of a hydroxyl ion with a proton is also a typical Brönsted-Lowry acid/base reaction.

Soft nucleophiles tend to react more readily with soft electrophiles. Suggest an example of this type of reaction.

For example the reaction between the soft nucleophile, bromine, and a soft electrophile such as an olefin, is shown in the following reaction.

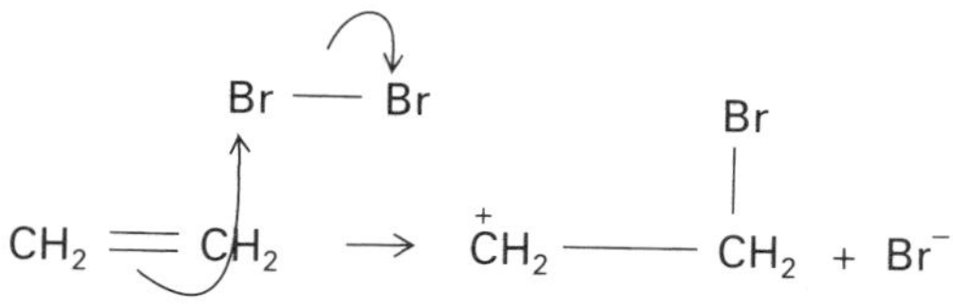

$$CH_2 = CH_2 \quad + \quad Br - Br \quad \rightarrow \quad \overset{+}{C}H_2 - CH_2 - Br \quad + \quad Br^-$$

This general rule of thumb is of assistance when a reagent like the hydroxyl anion has the possibility of reacting with either a hard electrophilic centre like a proton or a soft one like a carbonium ion. Usually the hydroxyl ion, which is a hard nucleophile, will react with a hard electrophile like an acidic hydrogen. So commonly, a hydroxyl ion will abstract a hydrogen and not react with carbon centres.

7.4 External Factors

There are a number of other factors which will affect how a species reacts in a particular set of conditions.

One of the most easily appreciated is when an acid and its conjugate base bear a different charge, for example when ethanoic acid, $MeCO_2H$, loses a proton to give the ethanoate anion. Write down the equation for this reaction, clearly showing the change in the charge of the organic component.

$$MeCO_2H = MeCO_2^- + H^+$$

The neutral carboxylic acid becomes the carboxylate anion which bears a single negative charge. This charged species will interact with the solvent molecules in a different manner from that of the uncharged species. If the ethanoic acid was originally dissolved in ethanol, CH_3CH_2OH, suggest how the solvent/solute interaction changes as the ethanoic acid loses its proton.

$$\underset{O-H}{\overset{O\cdots H-O-Et}{\diagdown\!\!\diagup}} \quad \overset{-H^+}{\longrightarrow} \quad \underset{O^-\cdots H-OEt}{\overset{O}{\diagdown\!\!\diagup}}$$

The uncharged ethanoic acid interacts by means of a weak hydrogen bond between the carbonyl oxygen and the hydrogen of the hydroxyl group of the ethanol. However, the carboxylate anion interacts more strongly because of the full charge now resident on the carboxylate group and so attracts the hydrogen on the hydroxyl group more strongly.

This demonstrates that there is a differential solvation of the conjugate acid and base. This differential solvation will of course affect the relative solubilities of the conjugate acid and base. This may be illustrated by using as an example the acid/base reaction of benzoic acid. Sodium benzoate, $NaC_6H_5CO_2$, is an ionic species. Suggest whether it would be soluble in water.

Sodium benzoate is readily soluble in water. When the benzoate anion is acidified with hydrochloric acid, its conjugate acid is formed. Write down the equation for this reaction.

$$NaC_6H_5CO_2 + HCl = NaCl + C_6H_5CO_2H$$

The conjugate acid is neutral. Suggest whether this species would be more or less soluble in water than the conjugate base, and give reasons for any difference in solubility proposed.

The acid is less soluble in water because there is less scope for hydrogen bonding in the free acid than there is in the charged base. Also there is a favorable Coulombic interaction between the solvent and charged solute, while there is no such Coulombic interaction for the neutral species.

The precipitation of the free uncharged acid from its conjugate base may be used to purify the benzoic acid. Molecules with terminal triple carbon/carbon bonds may be purified in a similar manner. If the neutral species has its terminal alkynyl proton removed by the reaction with a base suggest the formula of the resulting conjugate base.

$$R\text{-}C \equiv C^-$$

This alkynyl anion may then be reacted with a metal cation, such as Cu^+ or Ag^+ to form a compound which has a low solubility and so precipitates from the reaction solution. Thus this is a potential method for purifying a terminal alkyne. However, these precipitates tend to be explosive and so this method is not used regularly.

The difference between a conjugate base and the acid is a full unit of charge. This often has the effect of altering which resonance structure is favored. In aniline the lone pair on the nitrogen is delocalized into the aromatic ring and hence this compound is not very basic in nature. Once it is protonated, the delocalization into the ring is no longer possible.

In 2,6 dimethyl aniline the amino group is free to rotate as in aniline itself, and so the lone pair is capable of delocalization into the aromatic ring and so this compound is not very basic in nature. In the N,N-dimethyl derivative the dialkyl group may only position itself orthogonally to the aromatic ring and thus the lone pair on the nitrogen is no longer capable of being incorporated into the aromatic ring. It is now capable of acting as a base and reacting with a proton.

Thus the *ortho* methyl groups prevent the mesomeric delocalization of the lone pair, and so leads to a great difference in the basicity and nucleophilicity of the nitrogen group.

Further, changing a solvent may have a great effect on the course of a reaction by changing the balance of solvation of the different solutes. For example, the change from water to ethanol will shift a reaction from substitution to elimination, because there is a greater charge separation in the substitution reaction which is favored by the more polar nature of the water.

That brings to a end, not only the last section on thermodynamic and kinetic effects but also, the last introductory chapter on the basic principles of mechanistic organic chemistry. Now that we have studied the basic tools of the trade we will apply them to help us understand the main types of reaction mechanism.

Part II:
Mechanisms

8

Introduction

In inorganic chemistry, even though there are a large number of reactions which each element may undergo, it is possible to divide all the reactions into four general reaction types, namely redox; acid/base; precipitation and complexation. Furthermore, complexation reactions may be considered as being the first stage along the path towards precipitation. Complexation reactions may also be seen as a type of acid/base reaction, particularly when one uses the Lewis definition for an acid and base. From this very brief analysis it is clear that the vast range of inorganic reactions may be ordered into a small number of fundamental reaction types, and this ready division greatly assist in the understanding of these reactions.

In organic chemistry there is an even larger number of reactions which are possible, many of these at first sight seem to be unique, and so seemingly unrelated to any other example. However, as is the case in inorganic chemistry, it is possible to organize this multitude of reactions into a small number of fundamental reaction types. These are substitution, addition, elimination, rearrangement and redox reactions.

In substitution reactions one group is replaced by another. There is a subdivision of substitution reactions, that is called displacement reactions, in which the group being replaced is a hydrogen atom as opposed to some heteroatom or other moiety. In many addition reactions a small molecule is broken into approximately equal parts, and these parts then add across a multiple bond, which is commonly a carbon/carbon or carbon/oxygen double bond. In an elimination reaction the reverse is true. Thus a compound is formed which has more multiple bonds than did the starting material. There are some reactions where the overall substitution of a group is achieved by an addition reaction which is then followed by an elimination reaction, or *vice versa*. These reactions will be considered separately in this book under the heading of sequential addition/elimination reactions. In rearrangement reactions the backbone of the molecule is broken and reformed in a different configuration. Often this involves the breakage of a carbon/ carbon bond, but on some occasions it may involve the cleavage of a carbon/oxygen or carbon/nitrogen bond. Many redox reactions may be considered conveniently under the first four headings; but for some others, it is more convenient to consider them separately.

117

Within each major division, a secondary classification may be employed which divides along the lines of the mechanistic characteristics, *i.e.* by the nature of the species which attacks the starting material, *e.g.* nucleophilic, electrophilic or free radical.

Nucleophilic reagents are electron rich, usually because they have a lone pair or possess a multiple bond, and so they have two electrons with which to form a new bond with the species which is attacked. Nucleophiles may be neutral or bear a negative charge. They tend to attack centres of low electron density, such as empty orbitals or positive charges.

Electrophilic reagents are the opposite of nucleophilic ones. They are often positively charged, or at least are electron poor by virtue of having an empty orbital, *i.e.* an atomic orbital which accommodates no electrons.

Free radicals are molecules which have an odd number of electrons and so possess an unpaired electron. Unlike the nucleophilic and electrophilic species which are often charged, free radicals are usually neutral, although on occasions charged radical species may exist.

There is another type of reaction division in addition to the nucleophilic, electrophilic and radicals divisions already considered, and this is the pericyclic type of reaction. The earlier divisions can be characterized by the fact that they involve species which usually only involve the movement of one or two electrons in any given reaction step. However, pericyclic reactions usually involve more than one pair of electrons, which is may be assumed react simultaneously in a concerted manner.

However logical the above divisions may appear, there are other possible divisions and many textbooks divide the presentation of the reactions by reference to the dominant carbon functional moiety, *e.g.* aliphatic, aromatic or carbonyl. This approach is of value if one wishes to highlight the differences between the ways in that compounds that contain these moieties react with different reagents. However, underlying the differences, there are many fundamental similarities in the manner in which these different moieties react, and it is these similarities which this book will highlight. Once these parallels have been understood, then the differences may be appreciated more fully.

9

Nucleophilic Substitution Reactions

9.1 Introduction

Substitution reactions form the mechanistic basis for one of the most important groups of synthetic reactions that is available to the organic chemist. In its simplest form a functional group, X, is replaced by a different functional group, Y. This overall reaction may be represented by the following equation, in which R represents a general carbon moiety.

$$RX + Y = RY + X$$

Y may be a nucleophile, an electrophile or a radical. These terms may be used to distinguish the different mechanistic pathways which are characterize by the involvement of these respective reagents.

The substitution of one functional group for another is called a functional group interchange, FGI. This is particularly so when multi-step synthetic procedures are being discussed. If X is a hydrogen atom in the original molecule and this atom is exchanged, this type of substitution reaction is called a displacement reaction.

Often in a functional group interchange the groups concerned contain only heteroatoms, *e.g.* a hydroxyl group is substituted for a bromo group. However, it is possible for a heteroatomic functional group to be replaced by one containing carbon atoms, *e.g.* a cyanide or even an alkyl group. In this way the carbon chain of the original molecule can be extended, and so a simple precursor may be elaborated into a more complex molecule. Such reactions are of fundamental importance in organic synthesis.

The commonest type of substitution reaction involves the attack of a nucleophile, Nuc⁻, and the concomitant expulsion of a different nucleophile in order to ensure the conservation of charge. Attacks by electrophiles and radicals will be discussed in subsequent chapters.

9.2 Substitution at a Saturated Carbon

9.2.1 Introduction

In a nucleophilic reaction at a saturated carbon centre the incoming nucleophile acts as a soft nucleophile, that is it attacks the soft electrophilic centre of a δ+ charged carbon atom. In a later chapter we will study a class of reactions in which the incoming nucleophilic agent acts as a hard nucleophile, that is as a Brönsted-Lowry base and removes a proton from the reacting carbon species. This latter type of reaction results in an overall elimination reaction occurring. In practice no reagent is either purely a soft or hard nucleophile and so a mixture of substitution and elimination reactions may occur in any one reaction mixture. However, various factors may favor one route over another, and in this chapter and the next two, we will concentrate upon substitution reactions before we look at elimination reactions in later chapters.

At a carbon centre which is involved in four covalent single bonds, how many electrons are there around the carbon atom? Demonstrate your answer by focusing on the carbon atom in a molecule of methane, CH_4, and drawing a dot and cross structure for it.

$$H \overset{\overset{\textstyle H}{\times \ \bullet}}{\underset{\underset{\textstyle H}{\bullet \ \times}}{\overset{\bullet}{\underset{\times}{C}}}} \overset{\times}{\underset{\bullet}{H}}$$

The answer is, of course, eight: a full octet of electrons is accommodated around the carbon atom in this molecule. What is the maximum number of electrons that a carbon atom may normally accommodate in its outermost, that is the valence, orbital?

As carbon is a second row element the maximum number of electrons which it may normally accommodate in its valence orbital is eight.

Now let us turn our attention to the electronic characteristics of the incoming nucleophile. Give two examples of typical nucleophiles.

The incoming nucleophile is electron rich, and so your examples may have included such anionic species as OH^-, halide ion, H^- or CN^-; or may have included such neutral species as NH_3 or H_2O. Note that in all cases, whether the species in question was anionic or neutral, there was always a lone pair of electrons. Nucleophiles are never positive species, because the positive charge on a cation indicates that it is electron poor.

In a simple nucleophilic substitution reaction, a carbon atom that already has eight electrons is approached by an electron rich nucleophile. This would lead to an excessive number of electrons on the carbon atom. This simple exercise of electron counting highlights the central problem with substitution reactions: how does a carbon atom which already has its full complement of electrons manage to substitute one group for another without exceeding the octet of electrons?

It is possible to identify one mechanistic pathway which cannot be correct: that is the simple addition of the incoming nucleophile to give an adduct, that then subsequently falls apart to give the desired products. This cannot be a permissible pathway, because in forming the adduct more than eight electrons would reside on the carbon atom. If you are in any doubt about this, count the number of electrons around the central carbon atom in the following hypothetical reaction.

$$CH_4 + H^- = CH_5^-$$

The methane carbon had eight electrons to start with and then with the addition of the hydride ion, two more are added to give ten in all. Hence this mechanism is not possible under normal circumstances.

There are two solutions to this problem that are found to occur in practice. Try to suggest what they may be.

The first solution proposes that the group that is to be substituted leaves the carbon species taking two electrons with it. In such a case the remaining carbon species now has only six electrons and so may readily accept the electrons which the incoming nucleophile offers. This mechanism is called unimolecular substitution and may be written in general terms as follows.

$$1. \quad CH_3X = CH_3^+ + X^-$$
$$2. \quad CH_3^+ + Y^- = CH_3Y$$

It occurs in two discrete steps, with a carbonium ion intermediate.

In the second mechanism, as the new group approaches, simultaneously the old group leaves, and so at any one time the central carbon atom sees only eight electrons. This reaction may be represented in the following manner.

$$CH_3X + Y^- = CH_3Y + X^-$$

This reaction occurs in only one step, and is called bimolecular substitution.

Each of these two types of reaction, unimolecular and bimolecular will now be discussed in turn.

9.2.2 Unimolecular Substitution

Originally the difference between unimolecular and bimolecular substitution reactions was deduced from kinetic studies on a wide range of reagents. It was observed that for some reactions the overall rate of substitution depended only upon the concentration of the species undergoing substitution and that the rate was independent of the concentration of the nucleophile. These reactions are called unimolecular nucleophilic substitution reactions and given the label S_N1.

In this mechanism, it is proposed that the first step consists of the hetero-lytic fission of the bond which joins the carbon atom and the group which will be substituted. In the molecule $C(CH_3)_3Br$, identify the polarized bond, and indicate in which direction it is polarised.

Only the C-Br bond is polarized to any significant extent, and it is polar-ised towards the bromine atom. Hence it is likely that this is the bond which will be broken most easily. Write down the first step of the SN1 reaction in which the carbon/heteroatomic bond is broken, using curly arrows to indicate the movement of electrons. Identify the products.

The neutral starting compound has been split into two parts, the first is an electron deficient carbonium ion, and the second an electron rich leaving group, Br⁻. The sum of the charges of the products equals zero.

The kinetic data from this type of reaction indicated that only the con-centration of the starting material affected the overall rate of substitution. From this may be deduced that only the starting material is required in the rate determining step, which is the slowest step of the pathway. This explains why the reaction is called unimolecular, because only one molecule is required in the rate determining step. Whatever happens thereafter must occur more quickly that this initial step, so the subsequent formation of the product by the attack of the incoming nucleophile on the carbonium ion must be fast.

There are a number of factors which may affect the rate of this type of reaction. Some of these factors are internal to the molecule itself, while others are external, and so related to the environment in which the reaction is performed. The most obvious internal factor is the bond strength of C-X. Suggest whether a strong or a weak heteroatomic bond will increase the rate of a SN1 reaction and why.

A weak C-X bond will increase the rate, because it will break more easily and thus form the carbonium ion more easily. Another internal factor that is related to the leaving group is the stability of the resulting species after cleavage. Suggest how the stability of the leaving group may affect the rate, and give some examples of good and bad leaving groups.

Good leaving groups are those that form soft nucleophiles such as the bromide or iodide anions, and those that form weak Brönsted-Lowry conjugate bases, *e.g.* the anions of strong acids such as tosylate (*p*-$MeC_6H_4SO_3^-$), or neutral molecules like water and amines. Soft nucleophiles are good leaving groups because they are easily polarized, and so the heterolytic fission of the C-X bond may take place more readily. Further it is advantageous if the leaving group is readily solvated, because this means that there is some favorable energy of solvation which is available to compensate for the energy which was required to break the C-X bond initially.

One of the best leaving groups is the diazo group, N_2^+. By using this group it is possible to form the unfavored phenyl cation, $C_6H_5^+$ from $C_6H_5N_2^+$. Suggest the mechanism for this reaction and the reason why the diazo group is such a good leaving group.

$$C_6H_5 \!\!-\!\! \overset{+}{N} \equiv N \longrightarrow C_6H_5^+ + N_2$$

The diazo group leaves as a nitrogen molecule, which is a very thermodynamically stable compound. The nitrogen molecule is also a gaseous product and so there is a large increase in the degree of entropy for the forward reaction. The phenyl cation cannot be readily formed except by using this decomposition, which is the reason why the diazonium salts of benzene derivatives are so important in the synthetic preparation of aromatic compounds.

Turning our attention from the leaving group to the carbon moiety, it is observed that the structure of the resulting carbonium ion has a large effect upon the ease of substitution. Consider the relative stability of the carbonium ions, CH_3^+, $C_2H_5^+$ and CMe_3^+, and then suggest which related parent species would undergo substitution fastest by the SN1 mechanism, and why?

$$CMe_3^+ > C_2H_5^+ > CH_3^+$$

The tertiary carbonium ion is more stable than the secondary, which in turn is more stable than the primary. The consequence of this is that tertiary substituted compounds undergo SN1 substitution much faster than primary ones. This is because as the tertiary carbonium ion is more stable it is preceded by a smaller energy of activation, and so its rate of formation is faster. Notice also that the greater the degree of substitution in the starting material, then the greater the amount of release of steric strain that will be experienced in going from a sp^3 configuration to a sp^2 configuration.

Two further examples may be used to illustrate the dependance of the rate of SN1 substitution on the stability of the resultant carbonium ion intermediate. Firstly, suggest what are the relative rates of substitution between EtBr and $PhCH_2Br$; and secondly, the relative rates of substitution between CMe_3Br and 1-bromonorborane, that latter has a bicyclic structure with the bromine bonded at the bridgehead carbon.

$$PhCH_2Br \; > \; EtBr$$
$$CMe_3Br \; > \; \text{1-bromonorborane}$$

In the first case, even though both bromides are primary, for the phenyl derivative the resultant carbonium ion will be delocalized into the benzene ring and so will be greatly stabilized. In the second case even though both the compounds are tertiary, the norbanyl derivative is almost inert to SN1 substitution because the required carbonium ion is unable to obtain the planar conformation which is needed for it to be stabilized, Thus it is a thermodynamically unfavored intermediate and so the preceding activation energy is high.

Further factors which affect the stability of the carbonium ion intermediate will also affect the rate of SN1 substitution. In particular the effect of inductive or mesomeric groups that influence the electronic distribution around the charged carbon have a great effect. This may be seen in the successive deuteration of the α-carbons. Thus suggest whether $C(CH_3)_3Br$ or $C(CH_3)_2CD_3Br$ will be substituted faster and why?

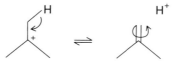

The SN1 mechanism proposes a carbonium ion as the intermediate, and this cation may be stabilized by $+I$ or $+M$ groups. The alkyl groups are $+I$, and so help to stabilize the positive charge. However, they may also be involved in hyperconjugation. This is a special type of mesomeric effect in which the α-hydrogens donate the electron pair in their hydrogen/carbon σ bond to help stabilize the positive charge. The involvement of these hydrogen/carbon bonds may be deduced from the observation that when the hydrogen atoms are replaced by deuterium atoms the overall rate of substitution is reduced. This because deuterium atoms are less effective at stabilizing the positive charge. This effect is an example of the secondary isotope effect, *i.e.* an isotopic substitution has an effect on the rate of the reaction even though that isotope is not directly involved in the bond which is being broken.

Apart from these internal properties which affect the rate of SN1 substitution, there are several external factors which may have an effect. Firstly, it is preferable to have a weak incoming nucleophile. This then will not interfere with the heterolytic fission of the starting material by attacking that molecule before the bond is fully broken. There are, however, other important factors concerning the solvent. Suggest whether a solvent of high or low polarity would favor the SN1 reaction.

The result of the first step in a SN1 reaction is to produce a pair of ions. If these ions can be solvated, then this will favor the forward reaction because of the favorable energy of solvation. Furthermore, as both a cation and an anion are produced it is best to have a solvent which is capable of solvating both species of ion.

If the ionic concentration of the reaction mixture is increased by the addition of other ions, then it is found that the overall rate of substitution is increased. This is called the salt effect. However, if the anion of the leaving group is added to the reaction mixture, then, in contrast, it is found that the rate decreases. Suggest a reason for this observation.

This latter effect is an example of Le Chatelier principle, and is also called the common ion or mass-law effect in this particular case. The first step of the S_N1 reaction is to produce the leaving group, and if there is already a large concentration of this present, the forward reaction is disfavored, as it would produce more of that species.

The interplay of all these different factors affects the overall rate for the substitution reaction. Given any two compounds it is usually possible to predict which will react the faster.

Apart from the speed with which a reaction occurs it is also of interest to follow the stereochemical consequences of a reaction. This may be studied by observing the result of a group departing form a chiral carbon centre. What is the structure of the carbonium ion which results from the chiral precursor, CHMeEtBr, and hence what is the stereochemical consequence of any substitution reaction which involves this carbonium ion as an intermediate?

Like other carbonium ions it is planar. The incoming nucleophile may then attack from either side of the carbonium ion with equal ease, and so equal amounts of each enantiomer of the product will be formed. Thus the stereochemical consequence is that there will a complete loss of the steric information that was contained in the original compound, *i.e.* there will be complete racemization.

Suggest what would be the stereochemical consequences if the intermediate carbonium ion only existed for a very short time before it was converted to the product.

The shorter the half life of the carbonium ion then the less the degree of racemization, and the greater the amount of inversion of configuration that occurs. Perfect racemization will only result when there is an exactly equal opportunity for attack on each side of the carbonium ion.

Illustrate this point of the protection of one face by the leaving group.

While the leaving group occupies the space around the carbonium ion it in effect blocks the attack by the incoming nucleophile from that side. Thus during that period the nucleophile must attack from the other side, and this will lead to inversion at the chiral centre. This protection only operates for the short period of time while the leaving group is departing and so only has an effect on the resultant stereochemistry if the carbonium ion is very short lived.

If one takes this one step further, and the incoming nucleophile arrives at the same time as the leaving group departs, one arrives at a situation in which there is no carbonium ion intermediate, *i.e.* the substitution reaction proceeds in one step and so it has now become a SN2 reaction.

9.2.3 Bimolecular Substitution

The other major type of nucleophilic substitution reaction is bimolecular, and is given the label SN2. In this case the rate of the reaction is observed to be dependent upon both the concentration of the species being substituted, *i.e.* the substrate, and also the incoming nucleophile. The reaction is thus a second order reaction: that is, the sum of the indices of the concentration of the reagents in the rate equation equals two. The order of a reaction is an experimentally observed number and need not be an integer.

The mechanism for the SN2 reaction occurs in only one step, in contrast to the two steps involved in the SN1 reaction. The incoming nucleophile arrives at the same time as the leaving group departs, *i.e.* there is simultaneous bond cleavage and bond formation. Any reaction step that involves two molecules is said to have a molecularity of two. The molecularity is a theoretical concept and depends upon the details of the mechanistic step that is being proposed. By its very nature, it must be a whole number as it represents the number of molecules involved in that step.

The rate of a SN2 reaction may be affected by many different factors which are similar to those which affected the rate of a SN1 reaction. These points may be illustrated by examining in detail the approach and subsequent departure of the incoming and leaving group respectively. First, draw a diagram to illustrate the initial approach of a nucleophile, Nuc⁻, to a general substrate, CR_3X, where X will be the eventual leaving group. Consider the direction from which the incoming nucleophile will approach.

It seems reasonable to suppose that the incoming nucleophile will approach the substrate along a line that extends away from the C-X bond that is to be broken, *i.e.* the attack is from the backside of the leaving group so that the incoming nucleophile will not interfere with the leaving group's eventual departure as the anion, X⁻.

Now consider the steric factors that will operate to hinder or facilitate this approach. Illustrate your answer by comparing the attach upon *n*-BuBr and 1-bromo-2,2-dimethylpropane (*neo*-pentylbromide).

n-BuBr   *neo*-pentylbromide

In both cases the bromine is the leaving group, and it is attached to a primary carbon atom. In the case of the *neo*-pentylbromide there is the large *t*-butyl group which crowds the line of approach of the incoming nucleophile, hindering its approach, and so slowing down the rate of SN2 substitution.

Now consider the electronic interactions, and the changes in electron distribution, that are occurring as the nucleophile approaches the substrate.

Nuc ·········→ δ^+ δ^- — LG   LG = leaving group

The C-X bond is polarized towards the leaving group, and so the carbon atom bears a δ+ charge. The incoming nucleophile is usually negatively charged, but always electron rich, and so there is a Coulombic attraction between the these two species. At first sight it would be reasonable to suppose that the rate of SN2 substitution increased with the degree of polarization of the C-X bond. This would mean that fluorine compounds would be substituted faster than iodine compounds, under otherwise similar conditions. The opposite is found to be the case.

In order to understand why this is so, we must consider the next stage in the approach of the incoming nucleophile. Draw a diagram to illustrate the mid-point of the transition state, and consider the changes that have occurred in the electronic distribution of the bonds which are being broken and formed.

(−) Nuc ------ ------ LG (−)

In forming the transition state, the incoming nucleophile needs to form a partial bond with the carbon atom which is to undergo substitution. This carbon must not exceed its octet of electrons, and so the bond to the leaving group must be simultaneously broken. Thus there is a movement of electrons within this bond towards the leaving group. A C-F bond is very strong and also is not easily polarized, while a C-I bond is both weak and easily polarized. In forming the transition state it is preferable for the bond to the leaving group to be easily polarized, because this will facilitate the subsequent heterolytic fission of the C-X bond.

The same considerations apply to the incoming nucleophile; more polarizable nucleophiles are better attacking species. Bearing this point in mind suggest why the overall rate of substitution of an alkyl fluoride by hydroxide anions is increased by the addition of iodide ions, and so suggest the what intermediate compound may be formed.

$$R\text{-}F + I^- = R\text{-}I + F^-$$
$$R\text{-}I + OH^- = ROH + I^-$$

The C-F bond is both strong and not easily polarised, and the OH⁻ ion, even though electron rich and capable of forming strong C-O bonds, is not easily polarized. Hence there is a reluctance in either species to undergo the required movement of electrons within the bonds to form the transition state. Iodide ions, however, are easily polarized and so may readily attack a saturated $\delta+$ carbon, and form the transition state, which may then yield the iodo compound, R-I, as an intermediate. This in turn may now be more readily attacked by the OH⁻ anion as the C-I bond is easily polarizable. The use of an iodide ion in this manner is called nucleophilic catalysis.

As the incoming nucleophile approaches the substrate the electron density on the carbon to be substituted increases. Suggest what substituent groups on this carbon would reduce this build up of negative charge.

-I and -M groups would help reduce the potential build up of negative charge on that carbon which is to be substituted. Furthermore, -I groups would increase the positive character on that carbon in the substrate and so enhance the initial Coulombic attraction between it and the incoming nucleophile.

The effect of -I groups also partly accounts for the reason why SN2 reactions are favored at primary as opposed to tertiary carbon sites, because alkyl groups are slightly +I in nature. There are two other reasons which favor primary over tertiary carbons in SN2 reactions: suggest what they may be.

The first reason is that sterically there is less hindrance of the approaching nucleophile. The second reason is that primary carbons only reluctantly form carbonium ions, and so the rate of any competing SN1 reaction is greatly reduced, which means that the presence of any SN2 reaction that is competing will be noticed. This latter point also explains why it is preferable for SN2 reactions to have poor leaving groups because such groups will be unlikely to cleave from the substrate of their volition, but rather wait until they are expelled by the incoming nucleophile.

The last stage of the SN2 reaction in which the leaving group departs is the mirror of the first stage in which in the incoming nucleophile approaches.

The factors which have been considered so far have concerned the inherent properties of the molecules involved in the substitution reaction. The nature of the solvent also has an effect on the rate of a SN2 reaction as it did with a SN1 reaction. Suggest whether a solvent of high polarity would favor or disfavor the SN2 pathway.

In the transition state the charge which was originally concentrated on the attacking nucleophile and the carbon/leaving group bond is now spread out over the two partially formed bonds. So there is a decrease in the charge density in going from the starting reagents to the transition state. Thus if the solvent is polar, there would now be a smaller energy of solvation. The result is that the more polar the solvent the more disfavored is the SN2 reaction.

There is a further property of the solvent which will affect the rate of SN2 substitution, apart from just its polarity. In the reaction of the azide anion, N_3^-, on iodomethane, the rate of substitution is increased by a factor of 10,000 in changing the solvent from methanol to N,N-dimethylmethanamide (dimethyl formamide, DMF), even though the two solvents have approximately the same polarity. Suggest why this is so.

Methanol is a hydroxylic solvent, which means that it is capable of hydrogen bonding with its solutes, and so forms a solvent shell around the potent nucleophile. DMF, however, cannot hydrogen bond with the azide ion, but can only interact with it Coulombically. In effect this means that the azide ion is now free of a solvent shell and so may more readily attack the substrate.

So far we have considered the rate of SN2 substitution and factors which affect it. In so doing we have, however, already noted one of the most important points that affects the stereochemical consequences of this type of substitution reaction, namely that the incoming nucleophile approaches the substrate from the opposite side to which the leaving group departs. Suggest what will be the stereochemical consequence of this.

In a reaction which proceeds purely by the SN2 pathway there ought to be complete inversion at the carbon site that undergoes substitution, and hence if that was a chiral centre beforehand, it will afterwards have the opposite enantiomeric configuration. This is called the Walden inversion. Notice that it is an absolute requirement for the substituted carbon to have its configuration inverted. Hence predict the susceptibility of a bridge head carbon such as exists in 1-bromonorbornane to undergo SN2 substitution.

At such a bridgehead carbon neither backside attack nor inversion of the carbon centre is possible, so such a carbon is unable to be substituted in this manner. You will recall that in the last section the inertness of such a carbon

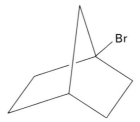

centre to SN1 substitution was noted, because of the impossibility of forming a flat carbonium ion. Thus it will be appreciated that such a carbon will not readily undergo either SN1 or SN2 substitution and so is inert.

Notice also that in cyclopropyl derivatives SN2 substitution does not occur even though there is room for the approach of an incoming nucleophile, because, the carbon being attacked cannot invert due to the restraint of the three membered ring. Inversion would require the substituted carbon to go from one sp^3 configuration, with a normal bond angle of 109°, to another sp^3 configuration, via a trigonal bipyramid, which normally requires the bond angles for the groups in the trigonal plane to be 120°. The introduction of this extra bond strain is prohibitory to inversion occurring.

There are occasions when the initial substrate is not sufficiently prone to attack by an incoming nucleophile and it needs activating in some manner. The commonest way is to make the potential leaving group a weaker base by protonation so as to form the conjugate acid of the substrate. For example, a simple open chain ether is normally resistant to cleavage, and so usually requires protonation before it will cleave. This may be achieved by reacting an ether with hydrogen iodide. Write down the pathway for this cleavage and identify the factors that have enabled this reaction to proceed.

$$HI + ROR = R_2OH^+ + I^-$$
$$R_2OH^+ + I^- = RI + ROH$$

The HI molecule protonates the ether oxygen. This increases the polarization on the carbon which is to be substituted, and also improves the leaving group from the hard alkoxide ion to the soft alcohol molecule. Furthermore, the HI molecule provides a very good nucleophile in the form of an iodide ion which readily performs the SN2 reaction which effects the cleavage of the ether. A substitution reaction which proceeds via the conjugate acid of the substrate is labelled SN1cA or SN2cA, or simply A1 or A2, depending upon whether the subsequent reaction follows a bimolecular or unimolecular pathway.

All of the reactions which we have studied so far have taken place between two molecules, *i.e.* there were intermolecular reactions. We will now turn our attention to the situation where one part of a molecule interacts with another part, *i.e.* there is an intramolecular reaction.

9.2.4 Intramolecular Substitution

We have seen that primary and secondary haloalkanes when attacked by a hydroxyl anion normally undergo inversion at the carbon centre at which substitution occurs, because they proceed via a SN2 pathway. In the case of 1-chloro-2-hydroxyalkanes it is observed that after hydrolysis with hydroxide anions to give the 1,2-diol there has been retention at the carbon site that bore the chlorine atom. Suggest what may be the first step in the reaction mixture, bearing in mind that hydroxyl anions are hard nucleophiles.

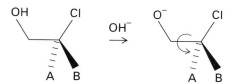

In this case the hard hydroxyl ion has removed the hard proton on the hydroxyl group to form an alkoxide ion in a simple Brönsted-Lowry acid/base reaction. Suggest what is the next step in this reaction, remembering that overall there is retention of the stereochemistry at the carbon that bears the chlorine atom.

The alkoxide ion performs an internal attack upon the C-Cl bond. This has proceeded in a similar manner to a normal SN2 reaction, and so there has been an attack from the back side of the C-Cl bond, with the required inversion of configuration, to produce the epoxide. Suggest what may follow in order to produce the 1,2-diol with overall retention at the carbon centre.

There is now an attack upon the epoxide by a hydroxyl anion which inverts again the carbon centre and this produces the monoanion of the diol, which then is protonated from the solvent to form the 1,2-diol. Notice how the two sequential inversions lead to overall retention of configuration.

This is an example of the neighbouring group effect, in which a near-by group which has an available lone pair performs an internal SN2 reaction on the centre which is to be substituted. In so doing it forms a cyclic compound which is then opened up by the original nucleophile to form the expected product, but with overall retention of configuration.

α-bromocarboxylic acids undergo substitution of the bromine atom by hydroxyl anions with retention of configuration, but only when the concentration of the hydroxyl anions is low. Suggest what the mechanism is for the substitution reaction at low and high concentration of hydroxyl anions.

At low concentration the reaction proceeds via an α-lactone, *i.e.* the carboxylate anion which is generated by the hydroxyl anion acts as a neighbouring group and reacts internally. This lactone is then attacked by another hydroxyl ion that opens up the ring to form the α-hydroxylcarboxylate anion, with overall retention of configuration at the α-carbon. At higher concentration of hydroxyl ions the reaction proceeds directly by a normal intermolecular SN2 reaction with overall inversion at the α-carbon.

The neighbouring group effect is not limited to the internal reaction of anions, but may also be effected by the lone pair on a sulfur or nitrogen atom. In a later chapter we will see a bromine atom acting as a neighbouring group, when it forms a cyclic bromonium ion intermediate, during the addition of a bromine molecule to an alkene.

The examples given so far have involved the formation of a three membered ring. This effect is not limited to cases in which a three membered ring may be form. Suggest which other ring sizes may be observed, and why.

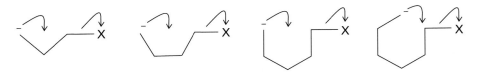

Apart from three membered rings, the two other common sizes are five and six. Four membered rings are not observed in this type of mechanism.

Three, five and six membered rings may be easily formed, because with such ring sizes it is possible to bring the ends within close proximity of each other and so allow them to interact. In the case of a four membered ring, because the normal bond angle of a sp^3 hybridized carbon is 109°, this means that the ends of a four carbon chain are a long way apart and so the molecule must be put under strain in order for the ends to become close enough to interact. Thus there is a large activation energy required for the formation of four membered rings in this way, and hence the rate of reaction is slow.

Explain why the intramolecular reaction occurs in preference to an inter-molecular attack.

When two molecules react to form one molecule, there is a great reduction in the entropy of the system. There is not such a large decrease in entropy when a molecule reacts with itself, for there is only a loss of some free rotation within the molecule. Thus intramolecular reactions are favored en-tropically over intermolecular reactions. Thus if the other factors are similar, intramolecular reactions are more facile than intermolecular reactions. Hence substitution reactions that occur via a neighbouring group mechanism are often faster than ones that occur by a simple intermolecular SN2 pathway. The neighbouring group is said to be lending anchimeric assistance.

When an alcohol is treated with thionyl chloride, SOCl$_2$, the rate equation is found to be dependent upon the concentration of both the substrate and the thionyl chloride. Yet it is observed that there is retention at the hydroxyl carbon. The rate of this reaction is found to increase with increasing polarity of the solvent. By considering which atom bears the greatest positive charge, suggest what may be the first step of this reaction.

The sulfur atom is the most positive atom, and is prone to attack by the hydroxyl group to form a chlorsulfite intermediate. Notice that the C-O bond has not been broken in this step and so there can have been no change in the configuration at this centre. This intermediate now fragments. Suggest how this may occur.

The C-O bond breaks to form the chlorosulfite anion. This, then further decomposes to form sulfur dioxide and a chloride ion. This is the slow step of the reaction.

Suggest what occurs next and explain why there is retention of configuration.

The sulfur dioxide escapes, then the chloride ion attacks the carbonium ion to form the substituted compound. The fragmentation of the chlorosulfite to give the sulfur dioxide and chloride ion occurs within a shell of solvent molecules. The sulfur dioxide escapes as a gaseous product, while the anion is still held in place by the solvent shell. Thus the nucleophile is formed on the same side as the original C-O bond; it then attacks the carbonium resulting in retention of configuration. The attack by the chloride anion occurs very quickly after the sulfur dioxide is lost and so there is no time for the solvent cage to collapse and release the anion. This reaction type is called internal nucleophilic substitution, Sni. Note that the chlorosulfite intermediate may be attacked from the reverse side if there is another nucleophile present, and in this case the reaction would proceed with inversion of the carbon centre.

In this section we have seen the formation of three membered ring intermediates when a neighbouring group participates in a reaction. We will now look at the cleavage of a three membered ring present in a substrate and the formation of one in the final product.

9.2.5 Formation and Cleavage of Three Membered Rings

One of the commonest three membered ring systems is the three membered cyclic ether, otherwise known as an epoxide. The first thing to appreciate is that a three membered ring is under strain, because of the compression from the normal bond angle of 109° to about 60°. This strain is released when the cyclic compound forms the open chain compound. Unlike the cleavage of open chain ethers which is quite difficult, the opening up of the epoxide ring is easy and may be achieved by water at neutral pH, even though it is rather slow. The reaction is faster if either acid or base catalysis is employed. Write down the three possible hydrolysis reactions which may occur depending upon the pH of the solution. Pay particular care to the protonation state of the substrate or intermediate and also the nature of the attacking nucleophile.

The formation of the *vicinal* diol, *i.e.* the 1,2 diol, may thus be achieved under a wide range of conditions.

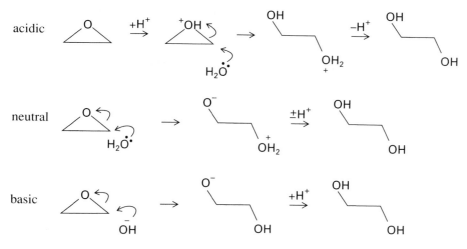

We saw earlier that a cyclopropane ring will not undergo a SN2 reaction because of the impossibility of the carbon inverting its configuration. The carbon atoms which constitute the ring are secondary and thus capable of forming carbonium ions. Such an ion though, would be even more strained than the neutral molecule because the carbonium ion normally adopts a sp^2 geometry, which has a bond angle of 120°. Suggest what would happen to a cyclopropyl cation if it were formed.

One of the bonds in the three membered ring may break and so form an allylic carbonium ion which is stabilized due to delocalization. This ion may then react further to form the final substituted product.

This reaction may proceed in the reverse direction. For example suggest the pathway whereby the allylic compound, 3-bromo-2-methylpropene, is converted to the cyclopropyl cation.

In this case there is a cyclic tertiary carbonium ion which helps to stabilize an otherwise strained system.

The cyclopropylmethyl system which has been formed by this reaction also undergoes some interesting reactions of its own. If the primary carbonium ion is formed, this may then rearrange to form the cyclobutyl carbonium ion. Suggest how this may occur.

It is also possible to form the cyclopropylmethyl system when a 4-halobutene system is attacked by a nucleophile. Suggest how this conversion may occur.

The 4-halobutene system is a homoallylic compound, which means that the substituent is one methylene unit, $\leftarrow CH_2 \rightarrow$, removed from the allylic position, which in turn is one methylene unit removed from the alkenyl carbons. Instead of the incoming nucleophile directly attacking the carbon which bears the halogen atom, it attacks the unsaturated carbon, and the π bond acts as an internal nucleophile to displace the halogen, and at the same time forms a cyclopropyl ring.

In this last example, a SN2 reaction occurred in which the initial attack was not on the substituted carbon, but instead it was at a conjugated unsaturated carbon. The π bond of the unsaturated carbon then performed the displacement reaction. This type of reaction is called a conjugated attack, and is indicated by a prime sign after the basic form of the reaction, so in this case SN2'. These conjugated reactions will be studied in more detail in the chapter on rearrangements, because they are an example of an allylic rearrangement.

As may be readily appreciated from the forgoing examples, substitution reactions on cyclopropyl systems and their derivatives may take some unexpected routes and lead to a variety of interesting products. Even though it may be possible to draw a mechanism which is reasonable for a given reaction, which pathway will actually be followed, or more realistically, what the spread of products will be, may be only determined by performing the experiment and identifying the products. The value of the mechanistic approach is that it will give some indication of those reactions which may be worth pursuing, because there is at least a plausible pathway by which the desired products may be formed.

9.3 Substitution at an Unsaturated Carbon

9.3.1 SN1 and SN2 Mechanisms

So far we have looked at substitution reactions in which the leaving group was attached to a saturated carbon. We will now turn our attention to unsa-

turated carbons and examine how this modifies the mechanistic pathways
that we have already studied.

When an alkene derivative, $CH_2=CHX$, forms a carbonium ion, consider
whether this fragmentation would be more or less difficult than the similar
reaction in a saturated derivative.

We already know from the chapter on the acidity of protons that in a sp^2
hybridized bond the electrons are held closer to the carbon than in a sp^3
hybridized bond. So it is harder to break the C-X as it is, firstly, a stronger
bond, and secondly, less polarized towards the leaving group.

It is found that an alkenyl carbonium ion is slightly less stable than the
corresponding alkyl carbonium ion, so a secondary alkenyl carbonium ion is
less stable than a secondary alkyl carbonium ion, but both are more stable
than the primary alkyl carbonium ion.

Bearing all these factors in mind, it is apparent that alkenyl carbons tend
not to perform the SN1 reaction unless conditions are so arranged as to stabilize the resultant carbonium ion intermediate. Suggest two possible ways in
which this could be done.

It is possible to force an alkenyl carbon to undergo a SN1 reaction by
having a very good leaving group, such as triflate, $\leftarrow OSO_2CF_3$, or by stabilizing the positive charge with an α-aryl group.

Suggest what will be the shape of the intermediate alkenyl carbonium ion,
and then suggest what will be the stereochemical consequences of a SN1
reaction on a vinyl carbon.

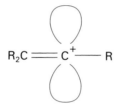

When an alkenyl carbonium ion is formed it is linear, and so the stereochemistry of the final product after the addition of the electrophile will be
randomized.

We will now consider the possibility of the SN2 reaction mechanism in
unsaturated systems. Firstly, draw a diagram of bromoethene, and showing
the π orbitals of the double bond. Then consider the arrangement of the
electrons in space and how they may interact with an incoming nucleophile.

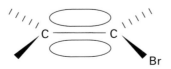

It is appears that in the case of an alkene, there is a large electron cloud which is close to the line of approach of any potential nucleophile. This will exert a disfavorable Coulombic interaction and so inhibit the approach of the nucleophile. Furthermore, the already high electron density on the carbon undergoing substitution will not readily accept the increase in electron density which accompanies the SN2 reaction.

Lastly, consider the possibility of inversion at such an unsaturated carbon atom.

The transition state at a saturated carbon is a trigonal bipyramid, in which the three groups which are not involved in the substitution reaction orientate themselves in the trigonal plane, while the approaching and leaving groups interact with the perpendicular p orbital. The hypothetical transition state at an unsaturated carbon atom would be octahedral, with the approaching and leaving groups interacting along a p orbital, that is orthogonal to the double bond π orbital, while the two other groups are not involved in bond cleavage or formation. It is hypothetically possible for this arrangement to exist, however, it is observed that in a sp^2 hybridized carbon inversion of the configuration does not take place.

Consider the case of a hypothetical SN2 reaction on an alkynyl halide, and predict whether such a molecule would react by this pathway.

An alkynyl halide cannot undergo a SN2 reaction at all, because firstly, that would require the incoming nucleophile to approach through the carbon at the other end of the triple bond; and secondly, there is no mechanism whereby the carbon could invert its configuration.

Note also that in the case of an alkynyl halide, SN1 substitution is even more disfavored than in the case of an alkenyl halide because it would result in a positive charge on a sp hybridized carbon, after having broken a bond which is stronger and less polarized than in the case of the alkenyl halide.

9.3.2 Tetrahedral Mechanism

We have seen above that in an alkenyl carbon the SN1 is not particularly favored and that the SN2 mechanism is highly unlikely.

Instead of performing the one step bimolecular SN2 reaction, alkenes react via two closely related bimolecular pathways. The first of these is called the tetrahedral mechanism and proceeds via a negatively charged intermediate. This mechanism is sometimes called the addition-elimination reaction which is given the label Adn-E. This label is unfortunate because the other pathway is also called the addition-elimination mechanism and proceeds via a readily detectable neutral intermediate. This latter mechanism will be considered in the chapter on sequential addition/elimination reactions. In this book we will call the mechanism which proceeds via an anionic intermediate the tetrahedral mechanism, and the other mechanism, the addition-elimination mechanism in order to reduce the confusion.

Consider the general alkene, ZCH=CHX, and let it be attacked by a nucleophile, Y⁻. Draw the mechanism for this attack, and suggest how the anionic intermediate may be stabilized.

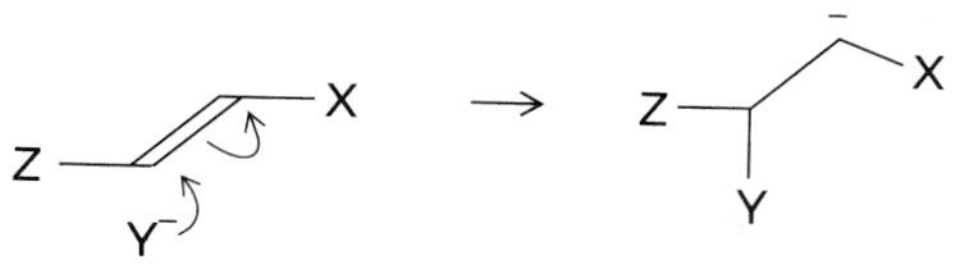

The α-carbon now bears a negative charge and this may be stabilized by either strongly -I groups, or better still -M groups such as carbonyl or cyano. Suggest how this anionic intermediate may now react.

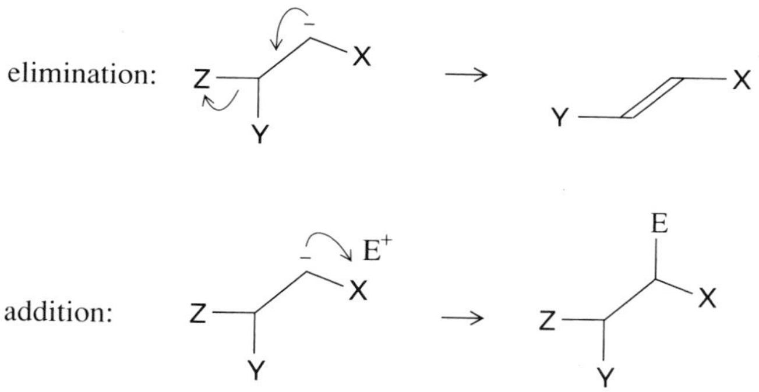

elimination:

addition:

The intermediate may either eliminate the X⁻ group, which would result in overall substitution, or it could add an electrophile, which would result in overall addition to the alkene bond. These two reactions usually compete.

9.3.3 Ester Hydrolysis

In the two preceding subsections we examined the substitution reactions that occur at an unsaturated carbon joined to another unsaturated carbon to form an alkene. When, however, a carbon atom is doubly bonded to an oxygen atom it forms a carbonyl. Such compounds undergo a wide variety of reactions. In particular when one of the substituents on the carbonyl carbon is an electron withdrawing group, substitution reactions via the tetrahedral mechanism occur very easily, with concomitant substitution of the electron withdrawing group for the incoming nucleophile. In the chapter on sequential addition/elimination reactions we will look at those reactions of the carbonyl group in which the carbonyl oxygen is itself substituted. These reactions again proceed via the tetrahedral mechanism, but often there is an intermediate which may be isolated.

Returning to substitution reactions of the electron withdrawing group on the carbonyl carbon, write down the structure of an ester, RCO_2R, and identify the electron withdrawing group.

The electron withdrawing group is the alkoxide moiety, and the carbonyl oxygen.

Write down the overall reaction for the hydrolysis of an ester compound.

$$RCO_2R' + H_2O = RCO_2H + R'OH$$

The ester yields the carboxylic acid and the alcohol. This reaction proceeds under either acid or base conditions. The hydrolysis reaction may be either unimolecular or bimolecular. Furthermore, it is possible for either the bond between the carbonyl carbon and the alkoxide oxygen to break (acyl cleavage); or for the bond between the alkoxide oxygen and the alkoxide carbon to break (alkyl cleavage). This gives rise to eight different possible reactions, which are distinguished by the labels, A or B for acid or base conditions; ac or al for acyl or alkyl cleavage; and, 1 or 2 for unimolecular or bimolecular pathway. The hydrolysis of an ester under basic conditions is called saponification, because it was the process whereby soap was originally made.

The two most important pathways which occur in practice are the AAC2 and the BAC2, both of which proceed via the tetrahedral mechanism. Write down these two pathways, paying careful attention to the nature of the attacking species and the protonation and deprotonation steps.

These are the normal pathways for acid or base hydrolysis of an ester. Notice that if the alcohol was chiral then there would be retention of configuration and no rearrangement of the carbon chain.

The other six possible pathways only occur under special circumstances. We will look at the unimolecular reactions first, *i.e.* AAC1, AAL1, BAC1 and BAL1. In each case these reactions proceed in a manner which is related to the SN1 reaction.

The AAC1 reaction bears a resemblance to the SN1cA reaction in that in the first two steps, firstly, the carbonyl oxygen and then the alkoxide oxygen are protonated. Write down these two steps and then suggest what the remaining steps are.

The cleavage of the acyl bond and the attack by the water molecule are both slow steps. This pathway is followed when R is very bulky and so hinders approach by the nucleophile, and when there is a strong acid in an ionizing solvent.

In the AAL1 mechanism, after the initial protonation of the carbonyl oxygen, the alkyl bond breaks to yield a carbonium ion. Write down this pathway and suggest how it proceeds. Suggest the characteristic of R′ which would favor this mechanism.

R′ must be able to form a stable carbonium ion, and so a tertiary carbon would be an example. This is also the requirement in the related BAL1 reaction pathway that occurs only under very mild basic conditions or sometimes even neutral conditions. Write down this pathway. Only the final deprotonation of the conjugate acid of the alcohol requires a base in this case.

In all cases that involve alkyl cleavage, the proposed reaction proceeds via a SN1 or SN2 type substitution on the alkoxide carbon with the acyloxy, OCOR, group, or its conjugate acid, as the leaving group.

The last possible unimolecular pathway, BAC1, is not observed in practice. Write down this pathway and then suggest a reason why this reaction mechanism does not occur.

BAC 1

$$R-\overset{O}{\underset{}{C}}-OR' \quad \overset{slow}{\rightarrow} \quad R-\overset{+}{C}=O \quad + R'O^-$$

$$\overset{OH^-}{\rightarrow} \quad R-\overset{O}{\underset{OH}{C}} \quad + R'O^- \quad \rightarrow \quad R-\overset{O}{\underset{O^-}{C}} \quad + R'OH$$

The pathway requires that an alkoxide anion leaves to form the carbonium ion, however, this anion is not a good enough leaving group to do that because the C-O bond is not easily polarized.

There are two remaining possible pathways. Both of these proceed via a bimolecular mechanism. The acid catalysed version, AAL2, is similar to the SN2cA pathway. This mechanism, like the one before, is not observed experimentally. Write down a route for this reaction, and also suggest why it is not observed.

AAL 2

$$R-\overset{O}{\underset{}{C}}-OR' \quad \overset{H^+}{\rightarrow} \quad R-\overset{OH}{\underset{}{C}}-OR' \quad \rightleftharpoons \quad R-\overset{O}{\underset{H}{C}}-\overset{+}{O}R'$$

$$\overset{H_2O}{\rightleftharpoons} \quad R-\overset{O}{\underset{}{C}}-OH \quad + R'\overset{+}{O}H_2 \quad \overset{-H^+}{\rightleftharpoons} \quad R'OH$$

This reaction would require a water molecule to be a nucleophile in a SN2 reaction, which it is not capable of doing.

The last reaction type, BAL2, requires an hydroxyl anion to attack an alkyl carbon in preference to the carbonyl carbon. Write down this mechanism, and deduce what will be the stereochemical consequences on the configuration of the alcohol carbon.

BAL 2

$$R-\overset{O}{\underset{}{C}}-OR' \quad \overset{OH^-}{\rightarrow} \quad R-\overset{O}{\underset{}{C}}-O^- \quad + R'OH$$

The alcohol carbon will be inverted, and so this pathway is easy to distinguish as it is the only route in which this is the result. The BAL2 mechanism is very rare. One of the few examples is the hydrolysis of β-lactones, under neutral or very mild basic conditions.

In summary the AAC2 mechanism is the commonest under acid conditions, with AAL1 occurring when R' is able to form a stable carbonium ion; while under basic conditions BAC2 operates in almost all cases. The other pathways only occur, if at all, in limited circumstances.

9.4 Trigonal Bipyramidal Substitution

This last section deals with substitution which occurs in the organic salts of sulfonyl salts, RSO_2X. There are two possible mechanisms whereby this substitution reaction could proceed. The first pathway proceeds via an anionic intermediate which would have a trigonal bipyramidal geometry, and is very similar to the tetrahedral mechanism that occurs in the hydrolysis of carboxylic acid derivatives at the carbonyl group. Write down the structure of this intermediate.

In this case, the sulfur has five groups covalently bonded to it, and so has ten electrons in its valence shell. This is permissible, because sulfur is a third row element and so may accommodate up to twelve electrons in that shell, *i.e.* it is not limited by the octet rule of the second row elements.

The other possible mechanism resembles the SN2 pathway which is found at a saturated carbon centre. This would involve a trigonal bipyramidal transition state. This is the pathway which the majority of the kinetic and stereochemical studies support. For example, in the conversion of an aromatic sulfonyl chloride with an aromatic Grignard reagent to yield the sulfone, what would be the stereochemical consequences of this latter SN2 type mechanism, and how could it be tested?

This mechanism would result in an inversion at the sulphur centre, and it may be tested by stereospecifically labelling the oxygens, one ^{16}O and the other ^{18}O.

When nucleophilic substitution is being considered in the special case of an sulfonic ester, RSO_2OR', and R' is an alkyl group, then cleavage of the alkoxide bond is far more likely than cleavage of the S-O bond. This is because the sulfonate group, RSO_2O^-, is such a good leaving group as it is the conjugate base of a strong acid. However, when R' is an aryl group, then S-O cleavage is more likely because of the generally low reactivity of aryl substrates towards nucleophilic substitution.

10

Electrophilic Substitution Reactions

10.1 Introduction

So far we have looked at substitution reactions in which the attacking species has been a nucleophile. We will now look at reactions in which the attacking species has the opposite electronic properties.

In an electrophilic substitution the attacking species is an electrophile, *i.e.* an electron deficient species. Therefore the leaving group must also be electron deficient, *i.e.* an electrofuge, in order to maintain the overall charge of the substrate. Write down the equation for the completed reaction of the electrophilic substitution of RX, by an electrophile, E^+.

$$RX + E^+ = RE + X^+$$

It may be deduced that the facility of this pathway will depend upon the stability of the departing group as an electrofuge.

In a nucleophilic substitution reaction, it was found that either the leaving group left before the incoming nucleophile arrived, or their arrival and departure were simultaneous. These different pathways were required so that the quota of electrons around the carbon undergoing substitution did not exceed eight. In the case of electrophilic substitution the incoming electrophile is electron deficient and so may attack the substrate to form an adduct intermediate, from which an electrofuge leaves in order to generate the substituted product. Write down the general equation for this two step process.

$$RX + E^+ = [RXE]^+ = RE + X^+$$

This type of sequence may be characterized as being an addition/elimination process. The reverse of such a sequence is also possible. In this latter case, the reaction would proceed via a two step process which is comparable to that which was found observed in the SN1 reaction. Write down the general equation for the two stages of such a pathway for the substitution of RX by E^+.

$$\text{step 1:} \quad RX = R^- + X^+$$
$$\text{step 2:} \quad R^- + E^+ = RE$$

Such a heterolytic cleavage gives rise to an intermediate carbanion, and the stability of that species is important in determining the rate of this pathway.

In this chapter we will first study the electrophilic substitution reactions that occur in aromatic systems, because they are well defined, and then we will look at those that occur in aliphatic systems.

10.2 Aromatic Substitution

10.2.1 Arenium Ion Mechanism

In aliphatic systems there is an $\delta+$ carbon centre which is readily accessible to an attack by an incoming nucleophile and so nucleophilic substitution predominates. In aromatic systems, it is immediately apparent that the electronic characteristics of the substrate are different, because there is the delocalized cloud of electrons which exists above and below the aromatic ring. This cloud of high electron density naturally disfavors the approach of an incoming negative species, but correspondingly attracts positive ones.

The attacking species in such cases is an electrophile, being either a positive ion or the positive end of a dipole. Similarly the leaving group must lack an electron pair.

In nucleophilic substitution the best leaving groups were those that could most readily carry an unshared pair of electrons, *i.e.* weak Lewis bases. Suggest what will be the best leaving groups in an electrophilic substitution.

The best leaving groups in an electrophilic substitution reaction will be weak Lewis acids, such as the proton.

In aromatic chemistry, there is one electrophilic pathway which predominates. This is called the arenium ion mechanism. It occurs in three steps. The first is the formation of the attacking electrophile; the second is the attack by this electrophile on the aromatic ring; and finally, the departure of an electrofuge, which is usually an proton with the assistance of a base. Write down the general equations for this sequence of reactions.

$$\text{step 1:} \quad Z\text{-}X = Z^- + X^+$$
$$\text{step 2:} \quad PhY + X^+ = PhYX^+$$
$$\text{step 3:} \quad PhYX^+ = PhX + Y^+$$

The production of the attacking electrophile may occur in many different ways, and aromatic chemistry is often studied by reference to the nature of this species. This initial step is not usually rate limiting.

The remaining two steps proceed along very similar lines for most electrophilic aromatic substitution reactions. The attack by the electrophile is usually the rate limiting step. The cationic intermediate is called a Wheland intermediate, or σ-complex or an arenium ion, and can sometimes be isolated.

The stability of the Wheland intermediate will obviously have a large effect upon the overall rate of this reaction pathway. First, write down the two principal canonical forms of the benzene ring, PhH, and then write down the resonance structure of benzene.

canonical structure   resonance structure

It is this extensive ability to form delocalized structures which accounts for the increased stability of the cyclohexatri-enyl system over the hexatri-enyl system.

Now write down the mechanistic step which shows the attack of an electrophile on a benzene ring and so forms the Wheland intermediate.

Now write down the three possible canonical structures for the Wheland intermediate, and so suggest a structure for the resonance form.

Even though there is still extensive delocalization of the positive charge that has been introduced by the electrophile, it is limited to that which would have been available to a linear, non-cyclic system. The extra delocalizing possibilities that were available to the benzene system because it was cyclic are no longer possible. This means that the cyclohexadi-enyl cation is significantly less stable than the initial benzene ring, because it has lost the aromatic stabilization of the cyclic sextet of electrons.

The last step in this pathway is the loss of an electrofuge. If, as in this example, a proton is lost with the help of a base, then overall a substitution reaction has occurred. When a hydrogen atom is substituted then the reaction is called a displacement reaction. This is the commonest type of substitution reaction which occurs in aromatic chemistry. Write down this step and suggest what is the thermodynamic driving force.

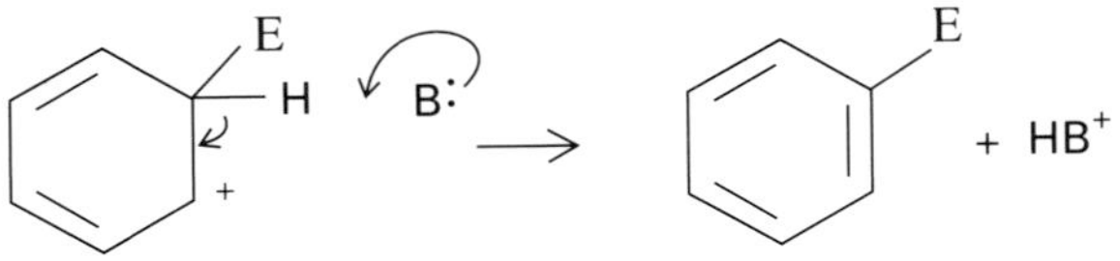

The restoration of the aromatic sextet with the accompanying energy of delocalization provides the driving force for this step of the reaction.

It has been suggested that there is a step prior to the formation of the Wheland intermediate, which involves the complexation of the attacking electrophile with the π-system of the aromatic ring to form a π-complex, which then converts into the σ-complex. The structure of such an intermediate may be represented as:

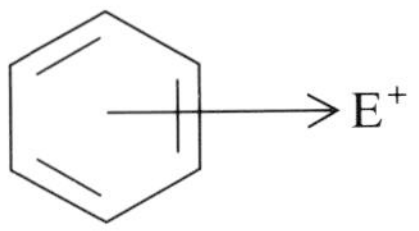

The involvement of this extra step is very difficult to prove in practice, and we will not consider it further, because for the present purposes, it is sufficient to postulate that there is only one step between the substrate and the Wheland intermediate.

One of the commonest simple electrophilic substitution reactions of a benzene is nitration. In this reaction the attacking electrophile is the nitronium ion, NO_2^+, which may be formed by the action of concentrated sulphuric acid on concentrated nitric acid. Bearing in mind that sulphuric acid is a stronger aid than nitric acid, write down the equation for the formation of this cation.

$$H_2SO_4 + HNO_3 = HSO_4^- + H_2NO_3^+$$
$$H_2NO_3^+ = H_2O + NO_2^+$$
$$H_2O + H_2SO_4 = H_3O^+ \ HSO_4^-$$

Note that two molecules of sulphuric acid are consumed, the first to produce the protonated nitric acid which decomposes to produce the nitronium ion; and the second, to protonate the water molecule which is also produced in the decomposition.

Write down the second step of this reaction, which is the attack of the nitronium ion on the benzene ring, paying particular attention to the position of the positive charge on the intermediate.

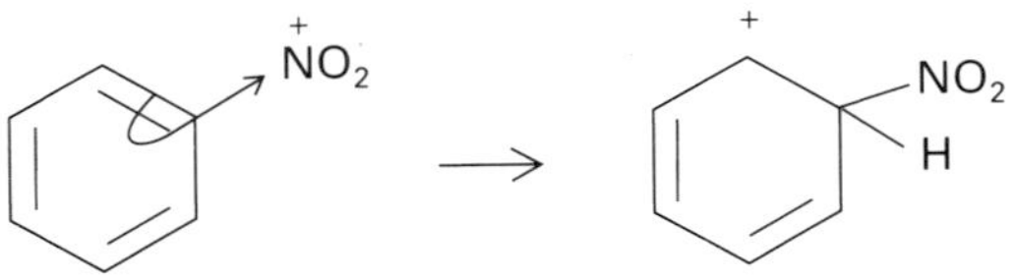

Notice that the positive charge initially resides on the carbon adjacent to the one which is about to undergo substitution. The charge is of course delocalized by the conjugated double bond system.

Write down the last step in this reaction. Suggest what base participate in this step.

The hydrogen sulphite anion acts as a base to help remove the proton so as to reform the aromatic ring, which results in the nitrobenzene product. The nitronium ion may be formed in a number of different ways, one of which is directly from N_2O_5 in chloroform.

Another common aromatic electrophilic substitution reaction is chlorination. In this case the electrophile is produced by the action of chlorine on the Lewis acid, $AlCl_3$. Suggest how these two molecules may react together, and so suggest what is the nature of the electrophile.

The $AlCl_3$, acting as a Lewis acid, complexes with the chlorine molecule, and this polarizes the chlorine/chlorine bond so that the strongly $\delta+$ end may act as the electrophile in the attack on the benzene ring.

Write down the next step of this reaction.

The formation of the Wheland intermediate results in the heterolytic cleavage of the chlorine/chlorine bond to form the $AlCl_4^-$ complex, which later acts as the base to help remove the proton that is to be displaced.

Benzene rings may also be alkylated, thereby adding a carbon side chain to the ring. This is the Friedel-Crafts reaction, and is commonly performed by reacting an alkyl halide with the aromatic compound in the presence of a Lewis acid such as $AlCl_3$. Suggest how the electrophile is formed.

The carbonium ion may also be formed from alkenes or alcohols. The carbonium ion which is formed in this manner is particular prone to rearrangement reactions, that are called Wagner-Meervien rearrangements, and these severely limit the synthetic utility of this reaction to form simple alkyl substituted aromatic compounds. The tendency to rearrange may be reduced if the acyl derivative is used instead. This modification called the Friedel-Crafts acylation reaction, and it has the further advantage that normally only monoacylation occurs, instead of the polyalkyation which happens using the simple Friedel-Crafts reaction.

A chloromethyl group may be added by the reaction of methanal and HCl with a $ZnCl_2$ catalyst. The overall reaction is called chloromethylation, but the first step is hydroxyalkylation due to the reaction of the aldehyde with the benzene ring, followed by the acid catalysed substitution reaction of the hydroxyl group by the chlorine. Write down the steps in this reaction sequence and identify the role which is played by the Lewis acid.

$$HCl + ZnCl_2 = H^+ + ZnCl_3^-$$
$$H^+ + CH_2O = CH_2OH^+$$
$$CH_2OH^+ + C_6H_6 = [C_6H_6CH_2OH]^+$$
$$[C_6H_6CH_2OH]^+ = C_6H_5CH_2OH$$
$$C_6H_5CH_2OH + Cl^- = C_6H_5CH_2Cl + OH^-$$

The $ZnCl_2$ raises the acidity of the medium, and so causes an increase in the concentration of the CH_2OH^+ ions.

There are many examples of different electrophiles being added to an aromatic system. In each case the synthetic utility lies in the ease of producing the electrophile and the usefulness of the final product.

The substitution of pure benzene by an electrophile will result in the formation of a monosubstituted product which is capable of undergoing further substitution reactions. The reactivity of this monosubstituted product compared with the original benzene, and also the position in the aromatic ring where the second substitution reaction occurs, is of great importance when designing the strategy for the synthetic routes of aromatic compounds.

10.2.2 Orientation and Reactivity in Monosubstituted Benzene Rings

The rate of a subsequent substitution reaction may be either faster or slower than benzene, *i.e.* the ring may be activated or deactivated. The position of substitution may be at any one of three places. If the position of the first substituent is given the number "1", then the second substitution may occur at position 2, 3 or 4. These positions are often referred to by their old labels: *ortho*, *meta* and *para* respectively.

It is important to note that aromatic electrophilic substitution reactions are under kinetic and not thermodynamic control. This is because most of the reactions are irreversible, and the remainder are usually stopped before equilibrium is reached. In a kinetically controlled reaction the product spread is determined not by the thermodynamic stabilities of the products, but by the activation energy barrier which controls the rate determining step. In a two step reaction, it is a reasonable assumption that the transition state of the rate determining step is close in energy to that of the intermediate, which in this case is the Wheland intermediate, and so by invoking the Hammond postulate, one may assume that they have similar geometries.

The result of this analysis is that by looking at what factors affect the stability of the Wheland intermediate, we will be able to deduce which products will be favored kinetically.

If a particular Wheland intermediate is stabilized over its related intermediates, would that favor or disfavor the product which it precedes?

A stable intermediate would normally have a lower energy of activation preceding it, and so would be formed more rapidly than the related intermediates. Once formed the next step is fast and so a stable intermediate would favor the product which it preceded.

We will study the effect of inductive and mesomeric groups separately. First, for the general benzene ring, substituted with a $+I$ group, there are three possible Wheland intermediate which may be formed by the addition of a general electrophile, Y^+. Write down the *ortho*, *meta* and *para* substituted Wheland intermediates, and for each write out the three canonical structures. Then predict the effect upon the stability of the intermediate of a group which had a positive or a negative inductive effect.

In all cases the ring bears a positive charge. Thus a $+I$ groups will stabilize all the intermediates to some extent, and a $-I$ group will destabilize all of them to some extent. A further refinement may be made. Inductive properties are field effects and so their influence rapidly diminishes beyond the atom to which they are directly attached. In the *ortho* and *para* Wheland intermediates, one of the possible canonical structures places a positive charge upon the carbon atom which is joined to the inductive group. Thus the inductive stabilization of a $+I$ group will be greatest for the *ortho* and *para* substituted Wheland intermediates, and hence the disubstituted products with these orientations will be kinetically favored. The reverse is true for $-I$ groups, they will destabilize all positions, but the *ortho* and *para* positions more than the *meta* position. Now if we turn out attention to the case of a mesomeric group, Z, which is capable of donating a lone pair of electrons into the ring. As before there are three possible positions of attack for the incoming electrophile. There are the nine canonical structures that we considered for the inductive group, but in addition there is now an extra canonical structure for both the *ortho* and *para* substituted Wheland intermediates. Write down these two extra canonical forms, and suggest what effect this will have on the production distribution.

In the case of the *ortho* and *para* substituted Wheland intermediates, this extra canonical form greatly increases the stability of that intermediate, and so greatly favors the production of the related products.

From this analysis it is possible to divide the various substituents into four different groups. The first contains those groups which possess an unshared pair of electrons that are capable of exerting a $+M$ effect on the ring. This would include such groups as $\leftarrow OR$, $\leftarrow O^-$ and $\leftarrow NR_2$. These groups activate the ring, *i.e.* substitution occurs faster in these derivatives than in pure benzene. They direct the incoming electrophile to the *ortho*/*para* positions.

The second group contains the halogens, which also direct the incoming electrophile to the *ortho*/*para* positions, but exert a such a strong $-I$ effect that they withdraw sufficient electron density from the benzene ring that they deactivate it.

The third group comprises of those moieties which lack a lone pair of electrons and exert a $-I$ effect, *e.g.* $\leftarrow NR_3^+$, $\leftarrow NO_2$, $\leftarrow CN$ and carbonyl derivatives. This group deactivates the ring and direct the incoming substituent to the *meta* position, because that is the least disfavored position.

The fourth, and last group, has no lone pair of electrons and is $+I$ in effect. It consists of such moieties as alkyl and aryl groups, as well as negatively charged entities as carboxylate. This group activates the ring and directs to the *ortho/para* positions. It is also possible for the alkyl groups to activate the ring by hyperconjugation.

10.2.3 Ortho/Para Ratio

In the previous discussion we did not distinguish between the attack of the incoming electrophile at the *ortho* or *para* position. Firstly, from a purely statistical reckoning, what would be the expected distribution between these two sites?

As there are two *ortho* protons for every one *para* proton, then from a statistical point of view the ratio between them ought to be 2:1.

It may be calculated that the charge density at the *ortho/meta/para* positions in the protonated benzene cation are in the ratio of 0.25:0.10:0.30. Using these figures, predict whether the amount of *para* substitution would be more than the 33% predicted on purely statistical grounds or less.

In a Wheland intermediate a disproportionate amount of the positive charge would reside at the *para* position. Hence if there was a substituent there which could stabilize that charge, it would have greater effect than a substituent at the *ortho* position, thus the amount of *para* substitution would be increased over that predicted on purely statistical grounds.

It is observed that *para*-quinonoid structures are more stable than those with *ortho*-quinonoid structures. This suggests that para substitution is favored over *ortho* substitution generally.

For groups which exert an inductive effect, it should be remembered that this effect diminishes quickly with distance, and so such groups will have a far greater influence at the *ortho* position rather than the *para* position. In the case of the halogens, predict whether an iodine or a fluorine substituted benzene would have the greater amount of *ortho* substitution.

Nitration of halobenzenes yields 12% of the *ortho* product for flurobenzene, and 41% for the iodobenzene.

Apart from the purely electronic effects, there is the steric interaction which may occur between the incoming electrophile and the substituent already present. In the nitration of toluene and *t*-butyl benzene, suggest which compound experienced the higher degree of *ortho* substitution.

Toluene gives 58% *ortho* and 37% *para*, while the *t*-butyl derivative gives 16% *ortho* and 73 % *para*. The *t*-butyl group physically blocks the entry of the incoming nitronium ion and so disfavors substitution at the *ortho* position.

The steric blocking of the *ortho* position may be achieved in another manner. Cyclohexaamylose is a compound which exists whose structure is an open cylinder. Molecules such as anisole, C_6H_5OMe, may be dissolved in a solution which contains this compound, and then the anisole molecules enter the central core of the cyclohexaamylose. In such a solution, the ratio of *ortho* to *para* chlorinated product increased from 1:1.48 to 1:21.6. Suggest a reason for the increased in selectivity.

The association of the cyclohexaamylose and anisole molecules is similar to the situation found in an inclusion compound. The result is that the *ortho* positions are blocked by the sheath of the cyclohexaamylose molecule; while the *para* position is exposed at the bottom of the cylinder, and so can still be attacked by the chlorine electrophile.

Having started with the statistical distribution which favored the *ortho* position 2:1 to the *para* position, we have seen a number of factors which favor the *para* position over the *ortho* position. There is, however, an important factor which favors substitution at the *ortho* position. For example the nitration of methyl phenylethyl ether with N_2O_5 gives an usually high proportion of *ortho* substituted product. Suggest an explanation for this experimental observation.

The complexation of the reagent with the lone pair of the ether oxygen, which then holds the attacking electrophile in close proximity to the *ortho* position accounts for this high *ortho* yield.

10.2.4 Multiple Substitutions

The substitution of a pure benzene ring by an electrophile to give a monosubstituted aromatic system proceeds by the arenium ion mechanism which involves the Wheland intermediate. By studying the effects of inductive and mesomeric groups upon the Wheland intermediate, we were able to predict the orientation of substitution which the next incoming electrophile would adopt. The simple electronic considerations did not distinguish between the *ortho* and *para* positions on the aromatic ring, and so the analysis was further refined. Now we will look what at position the third incoming electrophile will attack, when there are already two substituents on the ring.

In the simplest case the effects of the two groups reinforce each other. Suggest where the incoming electrophile will attack in a 1,3-dialkylbenzene, and in *p*-chlorobenzoic acid.

In the first case, the incoming electrophile will attack at the 4 position, because this is *ortho* to one alkyl group and *para* to the other. In the second case, the incoming electrophile will attack at the 3 position, because this *meta* to the carboxyl group and *ortho* to the chloro group.

When the groups already present oppose each other, it is more difficult to predict the position of attack of the third incoming electrophile. However, as few guidelines are possible. First, activating groups have a larger influence than deactivating groups on the position of the incoming electrophile, with mesomeric groups exerting a greater influence than inductive groups. Hence predict the position of attack in *o*-methylphenol.

The hydroxyl group is more strongly activating than the methyl group as it is a +M group. Thus the incoming electrophile will go *ortho* and *para* to the hydroxyl group and the point of substitution will not be governed by the methyl group.

Another guideline is that the incoming group is unlikely to enter between two groups which are *meta* to each other. The reason for this is steric, and so increases in importance as the size of the groups increases.

The last guideline is that when there is a *meta* directing group, which is *meta* to an *ortho/para* directing group, the incoming group attacks *ortho* to the *meta* group in preference to the *para* position. This is called the *ortho* effect. Hence predict the position of chlorination of *m*-chloronitrobenzene.

The incoming chlorine tends to attack *para* to the chlorine already present and *ortho* to the nitro group.

10.3 Aliphatic Substitution

The mechanisms which occur in aliphatic electrophilic substitution reactions are less well defined than those occurring in aliphatic nucleophilic substitution and aromatic electrophilic reactions. There is still, however, the usual division between unimolecular and bimolecular pathways: the former consisting of only the SE1 mechanism, while the latter consists of the SE2 (front), SE2 (back) and the SEi mechanism.

The leaving group in all these cases is an electron deficient species, *i.e.* an electrofuge. One of the commonest types of electrofuges in aliphatic systems is the metal ion, as it is easily capable of bearing a positive charge. Thus these types of mechanisms are often encountered in organometallic pathways. Most of the mechanistic studies have been performed on organomercury compounds. Generally, organometallic chemistry is normally only studied in outline at this level, so these mechanisms will be only examined briefly. It is still useful to study the pathways because, even though this may be considered an advanced topic, a lot of progress and understanding may be achieved by applying the basic tools utilized in early chapters.

In the unimolecular mechanism, the electrophilic substitution takes place in two steps which parallel those which occur in the SN1 mechanism. Write down these two steps for a general electrophilic substitution reaction between RX and E^+, and indicate which is the slow step.

Step 1: $R\text{-}X = R^- + X^+$ (slow)
Step 2: $R^- + E^+ = R\text{-}E$

The first step follows the expected first order kinetics. Suggest what type of substituents on the alkyl carbon would favor this reaction pathway, and then suggest what would be the effect of increasing the polarity of the solvent on this pathway.

The intermediate carbanion bears a negative charge and so -M and -I groups would help to stabilize it. An increase in polarity of the solvent would favor this route, because of the increase in charge separation which occurs on the formation of the two charged intermediates.

The factors which affect the rate of the reaction are easily predictable. The stereochemical consequences, however, of this pathway are slightly more difficult to predict. First, suggest what would be the structure of a simple carbanion intermediate, and then consider the effect a -M group, for example a carbonyl group that was conjugated to the negative charge, would have on this structure.

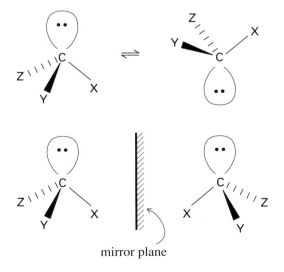

The simple carbanion, would be expected to adopt a tetrahedral conformation. When a -M group is conjugated to the negative charge, then the charge could be delocalized on to that group, with the result that the anion would adopt a planar conformation. The tetrahedral structure of the simple carbanion is similar to that which nitrogen adopts in a tertiary amine. Consider whether tertiary amines may be resolved into their enantiomers, and then by analogy suggest what would be the stereochemical consequences of the S$_E$1 mechanism on the configuration at the substituted centre in two different situations. Firstly, the situation with a simple carbanion intermediate; and secondly, the situation with a carbanion which is stabilized by a -M group.

mirror plane

A tertiary amine cannot be resolved into its enantiomers at normal temperatures, because there is rapid interconversion between the two forms. This rapid interconversion is called the umbrella effect. Hence, neither the simple carbanion nor the stabilized carbanion will retain their configuration, and thus there will be racemization at a unsaturated carbon centre. This result, though, depends for its analysis upon the anionic intermediate existing long enough, and being able, to undergo the umbrella effect or to have the negative charge delocalized onto the -M group. In practice the outcome is highly dependent upon the solvent which may stabilize the charge and interact with the intermediate via the solvation shell, in which case retention or even inversion may be observed.

In the case of alkenyl carbanions, that may maintain their configuration, retention is observed. Write down the reaction pathway for the electrophilic substitution of *trans* 2-bromo-but-2-ene when it is treated with lithium and then carbon dioxide.

The lithium forms the carbanion which maintains its configuration and this anion then reacts with carbon dioxide to form the carboxylic acid, called angelic acid, in about 70% yield. The *cis* isomer, tiglic acid, is only produced in about 5% yield.

We will now turn our attention to the bimolecular pathways. In the case of the bimolecular nucleophilic attach the incoming nucleophile contains an extra pair of electrons. In order to accommodate this pair of electrons, the leaving group has to depart at the same time. The lowest energy arrangement which allows this to occur has the incoming nucleophile arriving on the opposite side to that on which the leaving group departs, *i.e.* the so called back side attack.

In the case of an incoming electrophile there is no such build up of electrons, and so it is possible for the attack to occur on both the front and the back of the substrate. Write down the pathway for these two possibilities.

S$_E$2 (front)   S$_E$2 (back)

These two reactions are called S$_E$2 (front) and S$_E$2 (back). There is a third possibility. When the attacking species enters from the front, a part of it may assist in the removal of the leaving group by forming a bond with it at the same time as the new carbon/electrophile bond is being formed. The bond formation and bond cleavage is approximately simultaneous. Write down the representation of this pathway.

$$
\begin{array}{c}
E \longrightarrow Z \\
\diagdown\!\!\diagup \quad \diagdown \\
\diagdown C \!\!-\!\! X \\
\end{array}
\quad \longrightarrow \quad
\begin{array}{c}
\diagdown \\
\diagdown C \!\!-\!\! E \;+\; X-Z \\
\end{array}
$$

This mechanism is called the internal electrophilic substitution and is usually given the label S$_E$i, however, it sometimes goes under the label of S$_F$2 or S$_E$2 (cyclic).

The S$_E$2 (back) mechanism is different from the other two, because it results in inversion of the configuration of the carbon undergoing substitution. This has been observed, but more usually retention is observed which would indicate that either the S$_E$2 (front) or the S$_N$i mechanism is operating. Suggest how it may be possible to distinguish between these two.

In the S$_E$2 (front) mechanism there would be a greater degree of charge separation in the transition state than in the S$_E$i mechanism. Thus one would expect the rate of the former to be affected by both the polarity of the solvent and the addition of ions to the reaction mixture. The results of such experiments have been used to support the suggestion that one pathway is being favored over the other. However, the results from the same experiments have also been used to support the suggestion that further refinements are needed in order to understanding the pathway of electrophilic substitution reactions. Thus is has been suggested that instead of the simultaneous formation and cleavage of bonds that occurs in S$_E$i, there may be a step wise pathway, which is given the label S$_E$2 (co-ord), or S$_E$C.

The detailed study of electrophilic substitution reactions is beyond the ambit of this book. However, it is instructive to be aware that the a number of different mechanisms may be suggested form an analysis of the experimental data, for it reminds us that all the mechanisms which we are studying are merely hypothesis that so far fit the available observations. If fresh data was forthcoming, which could not be explained by the established mechanisms, then modifications would need to be made to those suggested mechanisms, and so the subject would develop.

11

Radical Substitution

11.1 Introduction

So far we have looked at substitution reactions where the attacking species was either a nucleophile or an electrophile, *i.e.* a charged species. There is a third possibility, which is that the attacking species may be a radical, and so has an unpaired electron and is usually electrically neutral. Even though it is possible to have a radical that bears either a positive or negative charge, in this chapter we will only deal with those radical species which do not bear an electrical charge.

One of the most important features of free radical chemistry is that the reactions are not affected by the normal variations in reaction conditions such as a change in the polarity of the solvent, or the acid/base character-istics of the reagents, except insofar as these changes will favor or disfavor competing ionic reactions. This is because such factors are only relevant when dealing with species which interact Coulombically.

11.2 Photochlorination of Methane

This reaction may be used to illustrate the mechanistic steps which occur in a radical substitution reaction. Its synthetic value is strictly limited, because, as we will see, there is little control over the product distribution. Furthermore, the reaction tends to proceed explosively!

The first step is to form the free radicals which will react further. In contrast to the heterolytic fission which is required in the reactions of polar intermediates, radicals are formed by homolytic fission of a covalent bond. This fission may be induced by light, heat, another radical, or even by a redox reaction. In the case of photochlorination the fission is achieved by shining UV light at the reaction mixture. Write down the mechanism for the simple homolytic fission of a chlorine molecule.

$$Cl \overbrace{} Cl \quad \rightarrow \quad Cl\bullet \quad Cl\bullet$$

161

The chlorine/chlorine single covalent bond is broken symmetrically to yield two radicals. In this first step the overall number of radicals has increased. This is characteristic of an initiation step. Free radicals are by their very nature, very reactive species and so they tend to react quite quickly with any other species that happens to be in the vicinity, which, when the concentration of free radicals is low, will usually be a neutral molecule. The product of such a reaction is another radical species. Many free radical reactions are characterised by the lack of free radical/free radical combination, and instead display a large number of reactions in which the initial free radical reacts with some other molecule in the reaction mixture.

A very common second step in radical substitution reactions is for the initial free radical to react with a hydrogen atom of the organic reagent, and after abstracting this hydrogen atom, to produce a carbon radical. Write down the balanced equation for the reaction of a chlorine radical with a methane molecule, and then write down the mechanism.

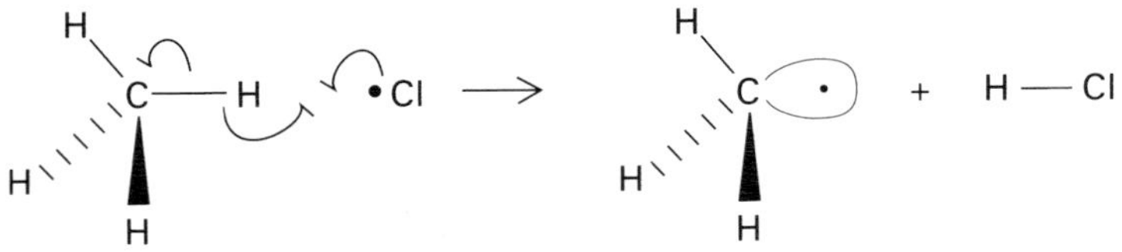

$$CH_4 + Cl\bullet = CH_3\bullet + HCl$$

It is possible to write down an equation in which the initial chlorine radical attacks the methane molecule and, instead of forming HCl and the methyl radical, forms chloromethane directly and a hydrogen radical. In practice only the first route is observed and the reaction does not proceed by this alternative route. Suggest why this is the case.

In order to form the hydrogen radical and the chloromethane directly, the attacking radical would need to interact with the electrons of the central carbon atom. These electrons are masked to a certain extent by the groups that surround the central carbon atom. The observed pathway requires only that the attacking radical interacts with the electrons of the hydrogen atom which is on the surface of the molecule. There is no steric hindrance which inhibits this attack and so it is more easily achieved.

The formation of the methyl radical maintains the number of radical species. This is characteristic of a propagation step. There are often many different propagation steps in any given reaction, whereas there is usually only one initiation step in which the original free radical is formed.

Now a hydrogen chloride molecule and a methyl free radical have been formed. Notice that the carbon radical was not formed originally, but after the reaction with the more reactive chlorine radical that in turn had been formed by the photoinduced homolytic fission. The hydrogen chloride takes no further part in the reaction. The methyl radical may react further with, for example, a chlorine molecule. Write down the equation and mechanism for this step.

$$CH_3\bullet + Cl\text{-}Cl = CH_3\text{-}Cl + Cl\bullet$$

The number of free radicals remains constant in this reaction, and so this is another example of a propagation step. Here one of the products, the chlorine radical, has been used before in the previous propagation step and so may be fed back into the pool and used to continue the reaction, *i.e.* propagate the reaction. The other product is chloromethane which is the monosubstituted product. Write down a balanced equation for the production of this product from the starting materials.

$$CH_4 + Cl\text{-}Cl = CH_3Cl + HCl$$

For every molecule of chlorine consumed, only one chlorine atom ends up in the organic product, while the other combines with the hydrogen to form hydrogen chloride. Note also a hydrogen has been substituted by a chlorine atom in the original methane molecule. You will recall that this type of substitution is called a displacement reaction. Such a reaction is unusual in simple aliphatic compounds, because the hydrogen is normally very difficult to activate *per se*. We have already seen in the previous chapter that a displacement reaction is common in aromatic systems.

After the reaction between chlorine and methane has proceeded for a short time there exists in the reaction mixture chloromethane as well as just methane, and so it is possible for chlorine radicals to attack the monochlorinated species instead of the methane. Write down the equation and mechanism for such an attack.

$$CH_3Cl + Cl\bullet = CH_2Cl\bullet + HCl$$
$$CH_2Cl\bullet + Cl\text{-}Cl = CH_2Cl_2 + Cl\bullet$$

The chlorine radical has abstracted a hydrogen atom from the chloromethane to form the chloromethyl radical which may go on to react as did the methyl radical above, and so form the disubstituted product. This in turn may react with more chlorine radicals and so on. It is this lack of control

over which product is formed that makes this reaction not very useful synthetically.

There is a third type of reaction which may occur in a typical free radical
mechanism. This is exemplified by the reaction of one radical with another.
Write down some possible examples of this third type of reaction.

$$2Cl\bullet = Cl\text{-}Cl$$
$$2CH_3\bullet = H_3C\text{-}CH_3$$
$$CH_3\bullet + Cl\bullet = CH_3Cl$$

All of these reactions share the characteristic that the number of free
radicals is reduced, and thus they are called termination steps, because they
lead to the termination of the chain of propagation of the reaction.

Note that in a few of the examples given above, that the product of the
termination reaction was either a starting material or one of the products
formed by some other route, and so the existence of these termination reactions would not show up. However, the ethane molecule is not formed by
any of the other routes and the presence of this side product in the reaction
mixture indicates the likelihood that these termination reactions do in fact
occur. The presence of unusual side products is very useful in attempting to
elucidate the course which a mechanism follows.

A free radical reaction need only have a initiation step and a termination
step. However, that would be most unusual, because normally one or more
propagation steps interpose. The reason is because, as radicals are very
reactive, the radical intermediates tend to react with the first species with
which they come into contact, and as usually the concentration of radicals is
low, then the most likely species which a radical will meet will be a non-
radical. Hence the presence of the propagation steps. There are four types of
propagation steps, which may be represented as following:

1. $R\bullet + M = R\text{-}M\bullet$
2. $R\bullet + R'H = RH + R'\bullet$
3. $R\text{-}R'\bullet = R + R'\bullet$
4. $R\bullet = R'\bullet$

In the first, the radical adds to a neutral molecule to form a larger new
radical; in the second, one radical attacks a molecule and abstracts a
hydrogen atom so as to form a new radical; in the third, there is a fragmentation reaction, which is similar to the reverse of the first type; and in the
fourth, there is a rearrangement of the radical into another radical, *e.g.*
allylic rearrangement. Propagation steps like type two are given the label
SH2, for bimolecular homolytic substitution reactions. Initiation reactions
are called SH1 reactions.

Fragmentation reactions are important in mass spectroscopy, where a
cationic radical $M^+\bullet$, cleaves to form a daughter molecule and radical. The
existence of these daughter cationic radicals is important in deducing the
structure of the original radical and hence the compound which is being
investigated.

The reaction of one radical to form another and then another, and so on, by propagation steps is called a chain reaction. If the radicals are very reactive would there more or less propagation steps in the chain reaction?

The more reactive the radicals the more propagation steps occur before the chain reaction is brought to an end by a termination reaction. As with propagation reactions, there are often many possible termination reactions, some of which have been seen above, produce intermediates or starting materials of the reaction. The multiplicity of possible reactions make radical reactions very difficult to study kinetically. One possible termination reaction consists of disproportionation. Suggest the mechanism for this and the resultant products of the disproportionation termination reaction between two ethyl radicals.

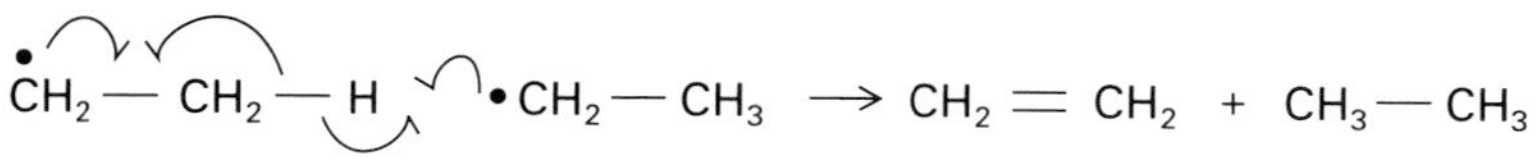

$$\overset{\bullet}{CH_2} - CH_2 - H \quad {}^{\bullet}CH_2 - CH_3 \quad \rightarrow \quad CH_2 = CH_2 \quad + \quad CH_3 - CH_3$$

Another possible termination reaction is the simple dimerization of the radical. Such possible reactions add to the complexity which already exists due to the wide number of propagation steps that are available.

It was mentioned at the beginning of this section that many radical reactions are started by the addition of compounds which promote the formation of radicals. These compounds are called initiators and often contain a weak bond that is easily broken, *e.g.* peroxides. Similarly, there are some compounds which scavenge free radicals and these are called inhibitors. Such compounds include molecular oxygen, nitric acid and benzoquinone. The addition of either initiators or inhibitors will greatly affect the rate of the radical reaction.

Now that we have studied the main type of reactions which occur in a typical radical reaction, using the photochlorination of methane as an example, we will now study the reactivity and structure of radicals in more general terms.

11.3 Reactivity and Structure

In the photochlorination reaction the chlorine radical abstracted a hydrogen atom and so produced a carbon radical. This combination of intermediates was produced and not the possible alternative combination of a hydrogen radical and a chloromethane molecule, because of the steric requirements for the attack by the incoming radical on the substrate.

In the case of methane there was only one type of hydrogen which the chlorine radical could attack, but in a larger alkane there is often a choice. One of the principle factors which plays a role in determining which hydrogen will be abstracted is the stability of the resultant carbon radical.

Suggest what will be the order of stability of primary, secondary and tertiary carbon radicals.

tertiary > secondary > primary
Carbon radicals are like carbonium ions in that they are electron deficient species, *i.e.* the carbon atom does not bear a full octet of electrons. Hence it is not unreasonable to assume that radicals will follow the carbonium ion sequence for stability, and this is found to be the case. It is also observed that the order of preference for the abstraction of a hydrogen atom follows the order of decreasing bond disassociation energy.

In the case of a general alkene, $RCH_2CH=CH_2$, suggest which hydrogen atom will be abstracted by an initiator, In$^{\bullet}$, and give reasons for your choice.

In this case the strongest C-H bond is the vinyl carbon/hydrogen bond and this bond is practically never broken. It is the allylic hydrogen which is abstracted as this results in a very favorable radical, which is capable of delocalization. The allylic carbon/hydrogen bond is also rather weak. It is not surprising that this sort of radical undergoes allylic rearrangements very easily.

In an arylalkyl, such as for example ethyl benzene, which hydrogen will be attacked and why?

In nearly all cases where there is an alkyl group which is joined to an aromatic ring, the abstracted hydrogen will be on the α-carbon of the alkyl side chain, because such a bond is both weak and also allows for delocalization around the ring of the radical which is formed. The aromatic ring is very rarely attacked to form the $C_6H_5^{\bullet}$ radical. Suggest why this is so.

The C-H bond is very strong, as it is in essence a vinylic carbon; also there is no possible stabilization of the resultant radical. When a radical attacks an aromatic system, there is an tendency to form an adduct. Suggest what the structure of such an adduct would be and also suggest why this is the favored route.

The adduct is similar to that which is formed during the electrophilic aromatic substitution reaction. Even though there is the loss of the aromatic delocalization, the resultant radical is itself delocalized to some extent. Such aromatic radicals may dimerize, disproportionate, result in overall substitution after the abstraction of a hydrogen atom, or even under go further addition to result in overall addition. If the last option is followed then often addition takes place at all three double bonds to give the persubstituted product, $C_6H_6X_6$. It is also found that substitution by either an electron donating or withdrawing group activates the ring, and that in all cases substitution at the *ortho* position is favored unless there is steric hindrance due to the original group, in which case the incoming radical will attack at the *para* position.

The shape of the intermediate carbon radicals is important in determining the stereochemical consequences of radical substitution. For the simple alkyl radical, $CR_3^\bullet$, suggest the various conformations which is could adopt, and suggest what would be the stereochemical consequences of these shapes.

tetrahedral

pyramidal, or umbrella, inversion

trigonial

Such a substituted alkyl radical could adopt a tetrahedral arrangement in which the bonding is sp^3 hybridized, or it could adopt a sp^2 arrangement which would result in a trigonal planar shape. In the former case, it would not be unreasonable to suppose that the radical would undergo umbrella inversion, rather like a tertiary amine, and so the configuration of the original molecule would be scrambled upon becoming a radical. The trigonal planar arrangement would also lead to a similar scrambling of the configuration. The exact shape of the carbon radical depends upon the substituents which is are attached to the radical carbon. In the case of the methyl radical, it is almost planar, while in trifluoromethyl radical it is essentially based upon a tetrahedron.

Suggest what will be the practical consequence of the tolerance of a radical species to its conformation, in particular when it is attempted to form a potential radical at a bridgehead.

It is quite possible to form a radical at a bridgehead carbon, and sometimes this is the only way to perform any chemistry at such a position, for we have already seen that a bridgehead carbon is inert to either the SN1 or SN2 mechanism.

It has already been mentioned that radicals are usually very reactive species. Different carbon radicals have different stabilities, which would lead one to presume that there would be some selectivity in the reaction, in fact what is often observed is that there is no selectivity and that all possible products are produced. This, of course reduces the synthetic utility of these reactions.

However, if the reactivity of the radical is reduced sufficiently, then some degree of selectivity may be introduced. An example of this is that the ratio of chlorination at primary, secondary and tertiary carbon sites is 1:4.4:6.7; while for bromination the ratio is 1:80:1600.

Selectivity may be achieved by altering the solvent in certain cases. Thus chlorination of 2,3-dimethylbutane in aliphatic solvents yielded the primary to tertiary product in the ratio of 3:2; while in an aromatic solvent the ratio was 1:9. Suggest an explanation for this result.

The chlorine radical forms a π complex with the aromatic solvent and thus reduces its activity and so its selectivity increases.

Selectivity may also be achieved by having a very bulky radical which is unable to abstract certain protons. A similar point is observed with the use of sterically hindered bases to remove only primary hydrogens instead of the most acidic protons in the molecule. An example of a sterically hindered radical is that which is formed from N-chloro-di-*t*-butylamine which abstracts primary hydrogens 1.7 times faster than tertiary ones.

11.4 Neighbouring Group Assistance

We have seen that photolytic halogenation usually leads to a wide variety of products with little or no selectivity. However, the bromination of alkyl bromides gives about 90% substitution on the adjacent carbon. Furthermore, if the adjacent carbon is chiral then it retains its configuration. Suggest a mechanism which may account for these experimental observations.

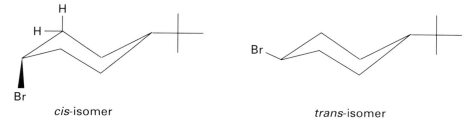

By invoking the assistance of the neighbouring bromine and forming a bridged free radical intermediate, the retention of configuration and the selectivity of substitution is explained.

This bridged free radical has obvious similarities to the involvement of a neighbouring group in SN2 reactions. In those situations the rate of the reaction was often increased because of the favorable entropic factor. Normally in radical reactions if the rate is increased then the selectively goes down. In the case of *cis*-4-bromo-*t*-butylcyclohexane both the rate and the selectivity of substitution by a bromine radical over the *trans* isomer is increased. Bearing in mind that the *t*-butyl group locks the conformation of the cyclohexane ring in such a manner that the *t*-butyl group is in an equatorial position, suggest the mechanism that is in operation.

cis-isomer   *trans*-isomer

The *t*-butyl group must always be in the equatorial position and so this determines the position of the smaller bromine group. In the *cis* isomer the bromine group is in the axial position and so is capable of acting as a neighbouring group in the radical substitution reaction. This promotes substitution on the adjacent carbon, and also increases the rate of reaction. In the *trans* isomer the bromine is in the equatorial position and so cannot assist the substitution reaction in this way.

11.5 Further Examples of Radical Reactions

The first radicals which are formed in the initiation step are usually derived from molecules which have a weak covalent bond in them, which may easily be broken by light or heat. This was the case in the first example in which the chlorine bond was broken by the action of light. However, light induced homolysis may be used to break some strong bonds which would not readily cleave except at high temperatures. An example is the formation of alkyl radicals by the light induced cleavage of azoalkanes, R-N=N-R. Write down the mechanism for this cleavage.

$$R-N=N-R \xrightarrow{h\nu} R\bullet \quad N\equiv N \quad \bullet R$$

Heat is often used to form radicals that may be derived by the fission of weak bonds. A very common example of this is the decomposition of metal alkyls, *e.g.* $PbEt_4$, which readily gives ethyl radicals on heating. This type of compound was often used to prevent pre-ignition, or "knocking", in petrol, which is where the fuel is about to explode rather than burn smoothly. The anti-knocking property derives from the fact that the ethyl radicals may react with the excess petrol radicals and so terminate the chain reaction before it explodes. The problem with using thermal energy alone is that it is unselective, and so thermolysis often yields a variety of products.

A very useful synthetic radical reaction is the bromination of an allylic carbon using N-bromosuccinimide, NBS. This reaction exemplifies another manner in which radical reactions may be initiated. This reaction will not start unless there is a trace of a radical initiator, In•. This may be the result of a small impurity, or more usually a small amount of a peroxide, which readily decomposes to form a radical. This initial radical, In•, then reacts with traces of bromine or HBr molecules to form bromine atoms. These then abstract the allylic hydrogen from the alkane, R-H. Write down the steps which have been considered so far.

Step 1a: In• + Br_2 = In-Br + Br•
Step 1b: In• + HBr = In-H + Br•
Step 2:   Br• + R-H = HBr + R•

So far the NBS has not been involved in the reaction. We know that HBr will react by adding across a double bond, but in this case that would be an unwanted side reaction. This is where the NBS comes into play. It reacts with the HBr to produce Br_2, which then reacts with the allylic radical to form the desired product. For the present, the only point about the structure of NBS you need to know is that it contains a weak nitrogen/bromine bond. Write down these steps.

The NBS thus acts so as to produce a very low steady concentration of bromine molecules. The very low concentration accounts for the fact that there is very little addition of bromine across the double bond. The addition reaction is also reduced because the HBr is consumed by the NBS. This allylic bromination is called the Wohl-Ziegler bromination.

Earlier in this section it was mentioned that PbEt$_4$ was used in order ensure the smooth combustion of petrol by preventing "knocking". Lead compounds are common reagents in radical reactions. An interesting example is the use of Pb(OAc)$_4$ to cyclize alcohols that have a δ-hydrogen to form tetrahydrofurans. The four and six membered cyclic ethers are not formed (oxetanes and tetrahydropyrans respectively). Initially an organolead compound is formed between the alcohol and the Pb(OAc)$_4$, which then cleaves either thermally or photolytically to give an oxygen heteroradical, which in turn abstracts the desired hydrogen. Write down the mechanism for these steps.

Notice that the desired cyclic ether cannot be formed directly by the oxygen radical attacking the δ-carbon, because that would mean the oxygen radical was attacking a quaternary carbon centre in preference to the uni-valent hydrogen that is far more exposed. At this stage of the reaction there is a carbon radical and a lead radical. There is evidence that a carbonium ion is involved in this reaction, because cationic rearrangements may take place at the δ-carbon. Such rearrangements do not occur at radical centres. The carbon cation may be formed by combining the lead and carbon radical and then heterolytically cleaving that same bond. Ring closure then takes place as would be expected. Write down these mechanistic steps.

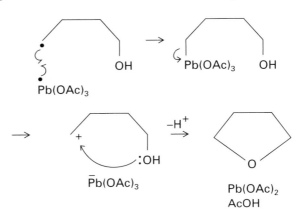

Lead compounds are often involved in radical reactions, because of the weakness of the carbon/lead bond, and the fact that it has two accessible oxidation states. In the case of lead, though, the oxidation states are plus two and plus four, which involves the movement of two electrons. Radical reactions involve the movement of a single electron. Thus one would expect transition metals which have adjacent oxidation states to facilitate radical reactions.

Not surprisingly this is found to be the case! There are many reactions in which the Cu(I)/Cu(II) couple plays a central role. For example in the Sandmeyer reaction an aromatic diazonium salt is converted to the corresponding chloride by the catalytic action of Cu(I)Cl. Suggest what will be the first step, which involves the diazonium ion, ArN_2^+, Cu(I)Cl and a further chloride ion.

$$Ar\text{-}N_2^+X^- + CuX = Ar\bullet + N_2 + CuX_2$$

In this first step the diazonium ion is reduced to form the phenyl radical. This step is helped by the fact that the nitrogen is such a good leaving group. In the next step Cu(I)Cl is regenerated and so it is a true catalyst for this reaction. Write down this reaction step.

$$Ar\bullet + CuX_2 = ArX + CuX$$

If this reaction performed using copper and HCl it is called the Gatterman reaction.

Another reaction in which the Cu(I)/Cu(II) couple plays a central role is the Eglinton reaction. This time though, a Cu(II) salt is placed in the reaction mixture to start with. In this reaction, terminal alkynes, $R-C \equiv C-H$, are coupled in the presence of a Cu(II) salt and a base such as pyridine. The reaction is performed in a vessel which is open to the air. Suggest a route for this reaction.

$$R-C \equiv C-H + pyr = R-C \equiv C^- + pyrH^+$$
$$R-C \equiv C^- + Cu^{2+} = R-C \equiv C\bullet + Cu^+$$
$$2R-C \equiv C\bullet = R-C \equiv C-C \equiv C-R$$
$$4Cu^+ + O_2 = 4Cu^{2+} + 2O^{2-}$$

Notice that the Cu(I) is oxidized back to the Cu(II) by atmospheric oxygen. Sometimes this reaction is done with a balloon of oxygen to help this process. There are many named variations of alkynic coupling, such as the Glaser reaction and the Cadiot-Chodkiewicz reaction.

Another source of single electrons is the anode of an electrolytic cell. In the Kolbe reaction an alkylcarboxylate salt, RCO_2^-, is decarboxylated and the resulting alkyl radicals joined to form the dialkyl product, R-R. Suggest the pathway which this reaction follows.

This is a good method of synthesizing symmetrical alkanes. The Kolbe reaction is usually performed using the potassium or sodium salt of the carboxylic acid. If the silver salt is reacted with bromine, then decarboxylation occurs again, but this time the alkyl bromide is formed. This reaction is called the Hunsdiecker reaction. The first step involves the formation of an acyl hypohalite, RCO_2X. Suggest how this then proceeds to the final product.

If iodine is used instead of bromine, then the exact ratio of the reagents determines which products are formed. From a starting mixture of 1:1, the alkyl halide is produced as above. However, if twice as much salt is used to the iodine then an ester results, RCO_2R. Suggest how this may be formed.

The reaction proceeds as per the Hunsdiecker reaction to yield the R-Br, which then reacts with the silver salt to give the ester. This version of the Hunsdiecker reaction is called the Simonini reaction.

A very important reaction of radicals is the formation of hydroperoxides, *i.e.* RH + O_2 = R-O-O-H. When this occurs by the slow atmospheric oxidation this is called autooxidation. Slow in this context means without combustion. The hydroperoxides that are formed often undergo further reactions, oxidizing other parts of the molecule or different molecules altogether. The reason why this is so important is because this process is responsible for the hardening of paints, the perishing of rubber, the curing of varnishes and fats becoming rancid. The reaction proceeds much faster in light, and usually requires a trace of an initiator, In•, because molecular oxygen is not reactive enough to abstract a hydrogen atom directly. Suggest what the route would be in the presence of such an initiator.

$$In\bullet \ + \ O_2 \ = \ In\text{-}O\text{-}O\bullet$$
$$In\text{-}O\text{-}O\bullet \ + \ RH \ = \ In\text{-}O\text{-}O\text{-}H \ + \ R\bullet$$
$$R\bullet \ + \ O_2 \ = \ R\text{-}O\text{-}O^{\equiv}$$
$$R\text{-}O\text{-}O\bullet \ + \ RH \ = \ R\text{-}O\text{-}O\text{-}H \ + \ R\bullet$$

The initially formed In-O-O• is reactive enough to abstract the hydrogen from any weak C-H bond, such as tertiary, allylic or benzylic. The autooxidation of ethers occurs at the α-carbon. However, the resultant compounds have a tendency to spontaneously explode. This is why ethers should not be stored in sunlight, and preferably should be stored in dark bottles.

Radical reactions may be used to introduce functionality at a remote position from the original functional group. We have already seen an example of this in the formation of cyclic ethers. In that case the ether linkage was introduced at the δ-carbon from the original hydroxyl group. This was an example of an intramolecular radical reaction. Another example is the Hofmann-Löffler reaction in which a N-alkyl-N-haloamine is heated under acidic conditions and gives the N-alkyl-pyrrolidine or N-alkyl-piperidine product. Suggest a route which this reaction may follow.

Protonation followed by homolytic cleavage of the N-Cl bond, produces a nitrogen radical which is capable of abstracting a hydrogen further down the alkyl chain. This carbon radical then combines with the chlorine radical to form the rearranged haloamine, which then closes the ring via a nucleophilic substitution reaction.

A related reaction is the Barton reaction. In this case a methyl group that is δ to a hydroxyl group is converted into an aldehyde functionality. The hydroxyl group is first converted into a nitrite ester by the action of nitrosyl chloride, NOCl, after exposure to light the δ-methyl group undergoes nitrosation. Write down the pathway for these steps.

Suggest how this may now be converted to the desired aldehyde product.

The nitroso compound tautomerizes to give the oxime which in turn may be hydrolysed to give the desired product. This reaction requires a six membered cyclic transition state for the hydrogen abstraction and so is quite specific in where the aldehyde group will be formed.

That concludes this chapter on radical substitution reactions. Radical reactions are vastly underestimated in value, especially considering how important they appear to be in nature. For example the double ring closure which results in the formation of the penicillin structure is thought to proceed by a radical mechanism.

This chapter also brings to a close the study of substitution reactions. We will now take a look at addition reactions.

12

Addition Reactions to Carbon/Carbon Multiple Bonds

12.1 Introduction

Addition reactions represent one of the simplest type of organic mechanisms. They usually involve the addition of a small molecule to a substrate that has a multiple bond, and the resultant product is called an adduct. If the resultant adduct has no more multiple bonds between carbon atoms, then it is said to be saturated. Correspondingly, the original molecule which contained multiple bonds between carbon atoms is said to be unsaturated. There are a number of different ways is which this general reaction type may occur, thus to start with we will just outline these various permutations and introduce some of the terms which we will use in the remainder of the chapter.

Write down a general equation for the reaction of a small symmetrical bimolecular species, A_2, with the symmetrical molecule $R_2C=CR_2$, and indicate which species is saturated, unsaturated and the adduct.

$$R_2C=CR_2 + A_2 = R_2AC\text{-}CAR_2$$

The alkane, $R_2C=CR_2$, is unsaturated, while the adduct, $R_2AC\text{-}CAR_2$, is saturated. This type of reaction is very common, $e.g.$ the hydrogenation of an unsaturated carbon/carbon double bond to give the saturated carbon/carbon single bond. Write down the overall equation for the hydrogenation of ethene, and identify the adduct.

$$H_2C=CH_2 + H_2 = H_3C\text{-}CH_3$$

The addition product is ethane. This type of reaction is fairly elementary and hence you are likely to have come across it before. This is not always the case, and as the examples become more complex and less commonplace, the value of a mechanistic approach increases because you will be able to work out a likely reaction sequence and hence the likely products.

The addition of a small molecule such as hydrogen to an unsaturated molecule may occur in one step, *i.e.* the hydrogen molecule approaches the alkene and they react together via a transition state to form the final product.

Assuming that this is the case, it is not unreasonable to suggest that there is a cyclic intermediate. A consequence of this would be that both hydrogen atoms would be added from the same side. This is called *syn*-addition.

Write down the overall reaction equation for the addition of a bromine molecule to ethene.

$$H_2C = CH_2 + Br_2 = H_2BrC\text{-}CH_2Br$$

The bromine/bromine bond is much weaker than the hydrogen/hydrogen and so it is possible that in this case the bromine molecule breaks into two parts. If this were to happen, then the addition would occur in two steps, with a single bromine species adding one at a time. In such a situation it would be possible for the second bromine atom to add to the intermediate from either the same side as the first bromine species, *i.e. syn*-addition, or from the other side, in which case it is called *anti*-addition.

If the addition occurred stepwise, then there are three possible type of species which may attack the alkane. Suggest what these three species are.

$$E^+, \text{ Nuc}^- \text{ and } R\bullet$$

The attacking species may be an electrophile, nucleophile or a free radical.

If the attacking species is not symmetrical like a dihydrogen molecule or a bromine molecule, but instead is asymmetrical, and if it is attacking an asymmetrical alkene, then the addition may take place in either of two orientations. Illustrate these two orientation with the addition of AB to $X_2C = CY_2$.

$$X_2AC\text{-}CBY_2 \text{ or } X_2BC\text{-}CAY_2$$

If it was possible to arrange the reaction conditions such that one orientation could be favored over the other, then this could prove to be a useful synthetic tool.

If there is a reagent which has more than one multiple bond that is capable of undergoing an addition reaction, or even if there is a multiple bond which is capable of reacting with more than one molecule, then that reagent is said to be polyunsaturated. Suggest an example of each type of polyunsaturated molecule.

Ethyne, $H\text{-}C \equiv C\text{-}H$, and 1,3-butadiene, $H_2C = CH\text{-}CH = CH_2$. In each case, the molecule could under the appropriate conditions react with two molecules of hydrogen. Write down all the possible adducts which result from the addition of just one molecule of hydrogen to a) ethyne and b) 1,3-butadiene.

In the case of ethyne there is only one possible product, *i.e.* ethene, $H_2C=CH_2$, *i.e.* the partial hydrogenated product. However, in the case of the conjugated system of 1,3-butadiene there are two possible products, *i.e.* 1-butene, $H_2C=CH-CH_2-CH_3$ or 2-butene, $H_3C-CH=CH-CH_3$, depending on whether the hydrogen molecule has been added across a double bond as depicted in the original molecule or has added across the terminal carbons and there has been a later rearrangement between the other carbon atoms of the residual unsaturation. The former type is called 1,2-addition, while the latter is called 1,4-addition or conjugated addition.

Having introduced an number of different terms and variations on the general theme of addition reactions, we will now look in greater detail at these possibilities. To start with we will study the simplest type of addition reaction.

12.2 Cyclic Addition

12.2.1 Heterogeneous Catalytic Hydrogenation

In the one step addition of two molecules, it is reasonable to suggest that such an addition occurs via a cyclic transition stage, with new bonds being formed simultaneously with old bonds being broken. It would also be reasonable to suggest that both parts of the molecule which are going to add to the unsaturated molecule approach the double bond from the same side. This would be especially true of the addition of small molecules. This is called *syn*-addition.

The commonest example is the catalytic hydrogenation of an alkane. Write down a balanced equation for the hydrogenation of ethene.

$$H_2C=CH_2 + H_2 = H_3C-CH_3$$

The reaction is commonly heterogeneously catalysed by metals such as nickel, platinum or palladium. The alkane and the hydrogen are adsorbed onto the surface of the metal, which results in the weakening of the bonds, and so the species react with each other more easily. Some of the hydrogen will also be absorbed into the metal lattice, but this is of little use in effecting the reduction of the alkene.

Draw a mechanism which illustrates the movement of the electrons.

The *syn* nature of this addition may be more fully appreciated when a cycloalkene is hydrogenated, such as 1,2-dimethylcyclopentene. Suggest what

the geometry of the resultant alkane will be after hydrogenation with a
nickel catalyst.

A further example may be provided with the partial hydrogenation of
ethyne. Do not confuse partial hydrogenation with partly hydrogenated. The
former means that all of the starting material has been hydrogenated to
produce a intermediate, which is capable of further hydrogenation, but no
starting material is left. The latter term means that only some of the starting
material has been hydrogenated, and so some remains. Lindlar catalyst is
needed to effect this partial hydrogenation, which is palladium on calcium
carbonate, partly "poisoned" with lead (II) acetate. Suggest what the stereochemistry will be of the product from the partial hydrogenation of but-2-yne
using this catalyst.

The precise mechanism of heterogeneous hydrogenation is a matter of
debate, however, the mechanism as proposed above agrees with the experimental observations.

12.2.2 Other Syn-Addition Reactions

Another example of hydrogenation is when diimide, H-N=N-H, reacts
with an alkene to form the alkane. There are two geometric isomers of

diimide. Draw out both isomers, and then write down the mechanism for the reduction of an alkene using the appropriate isomer.

In this reaction only the Z form of the diimide will react, because only this isomer can form the six membered ring transition state.

In this case a molecule of hydrogen has been added across the alkene, but it is possible to add many other species across carbon/carbon double bonds, either directly, as in the case of the hydrogen molecule, or indirectly as in our next example. The overall result of treating an alkene with osmium tetroxide is to add a molecule of hydrogen peroxide, H_2O_2. Write down a balanced equation of this overall reaction, bearing in mind that the initial intermediate is hydrolysed.

$$H_2C=CH_2 + OsO_4 + 2H_2O = CH_2OH\text{-}CH_2OH + H_2OsO_4$$

In this case the osmic acid that is formed is usually reoxidized to the osmium tetroxide by the use of hydrogen peroxide, which means that only a small amount of the toxic and expensive osmium compound is used.

Suggest a mechanism for the reaction between the osmium tetroxide and the alkene.

Again a concerted cyclic mechanism may be drawn with results in the cyclic osmium ester. If this compound is isolated and then hydrolysed using ^{18}O labelled water, no 18O label is found in the resulting 1,2-diol. Suggest a mechanism which explains this result.

The results of the labelling experiment indicate that both oxygens in the 1,2-diol originate from the osmium tetroxide and that the Os-O bonds are broken on hydrolysis. Suggest what is the stereochemical consequence for the disposition of the two hydroxyl groups.

As both of the oxygens originate from the osmium tetroxide, which reacted via a cyclic five membered ring, it means that they must be on the same side as each other, *i.e.* there was *syn*-addition.

So in summary this synthetic route gives rise to overall *syn*-addition of hydrogen peroxide.

12.2.3 Pericyclic Reactions

Suggest the result of the addition of 1,3-butadiene across an alkene such as ethene.

Here the two molecules have reacted together to as cyclic adduct. This mechanism underlies the Diels-Alder reaction. Note that the 1,3-butadiene molecule may rotate around the central carbon/carbon bond. This rotation is not as easy as would be the case with a carbon/carbon single bond in ethane for example. Suggest a reason for the slightly hindered nature of the rotation about this central bond in 1,3-butadiene.

It is possible to draw a resonance structure of this molecule that involves a separation of charges, which in turn confers some double bond character-istics on the central part of the molecule. Thus there are two identifiable conformers of 1,3-butadiene which result from this small energy barrier to free rotation. Suggest what are these two conformers.

cisoid   *transoid*

These two conformers are called the *cisoid* and *transoid* conformers respectively. Only the *transoid* conformer will react in the Diels-Alder reaction indicated above. Suggest whether cyclopentadiene would react more or less rapidly than the open chain 1,3-butadiene, and suggest a reason for any difference which may be observed.

The cyclopentadiene would react more rapidly because for the open chain diene, the *transoid* structure is the more stable due the minimization of steric hindrance. Thus there is only a small concentration of the required *cisoid* conformer present in the reaction mixture. In the case of cyclopentadiene, all of the molecules are in a *cisoid* conformer.

Draw out the three geometric isomers of 1,4-diphenylbutadiene and then suggest which one will under go the Diels-Alder reaction with a suitable reagent.

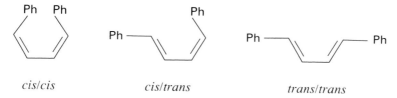

cis/cis   *cis/trans*   *trans/trans*

Only the *trans, trans* isomer will react, because in the other two isomers there is too much steric hindrance for the diene to adopt a planar conformation.

It is observed that the rate of the Diels-Alder reaction is increased by electron donating substituents on the diene, and by electron withdrawing groups on the alkane, or dienophile. This would indicate that the diene is acting as the nucleophile while the dienophile is acting as the electrophile. Suggest the mechanism for the reaction between cyclopentadiene and maleic anhydride, indicating the electronic characteristics of each reagent. This reaction may proceed with two distinct orientations, suggest the geometry of both.

The two products are called the *exo* and *endo* products respectively. The *endo* product is the favored over the *exo* product under all normal conditions. It has been suggested that the reason why the *endo* product is preferred is that in the transition state it is possible for there to be a favorable overlap of secondary orbitals, *i.e.* orbitals which are not primarily involved in the carbon/carbon bonds that are being made or broken. Notice that the result of such a geometry being adopted during the transition state means

exo

endo

that the addition to both the diene and the dienophile is *syn*, and that the stereochemistry of four adjacent carbon atoms is thereby controlled. This control over the stereochemistry of four carbon centres is one of the reasons why the Diels-Alder reaction is so important in synthetic chemistry.

It is usual to perform the Diels-Alder reaction with the type of reagents indicated above, however, it is possible to perform the reaction with reagents that have the electronic characteristics of the diene and dienophile reversed. Bearing this in mind deduce the mechanism for the reaction between per-chlorocyclopentadiene and cyclopentene, and so suggest whether the per-chloro compound would react more or less rapidly with tetracyanoethylene and maleic anhydride.

The perchloro compound reacts the fastest with cyclopentene, and only slowly with maleic anhydride, and not at all with tetracyanoethylene. The latter is usually a very reactive reagent in Diels-Alder reactions which operate under the normal electronic bias.

In the hydrogenation reactions, two atoms of hydrogen were added across a double bond, *i.e.* the addition of H_2, while in the Diels-Alder reaction, two carbon/carbon bonds were formed, *i.e.* the addition of R_2. There is a reaction which is intermediate between these two other reactions, in that one new hydrogen/carbon bond is formed and one new carbon/carbon bond is formed, *i.e.* the addition of RH. This reaction is called the ene synthesis. Suggest the product of the addition between an alkane such as maleic anhydride and a reagent such as 3-phenylpropene.

It is a common feature of both the Diels-Alder reaction and the ene synthesis that the reaction is not affected by the presence of radicals or by changes in the polarity of the solvent, but only by the action of heat and light. The mechanisms suggested above have all been single step mechanisms that involved the concerted movement of electrons utilizing cyclic transition states, in which bond formation and bond breakage occur approximately simultaneously.

Suggest a possible reaction which may occur when two molecules of ethene are forced to react together.

It is found that it is very difficult to perform this reaction. It is suggested that the reason for this reluctance is that this type of reaction is controlled by the symmetry, or phases, of the respective molecular orbitals involved, and in the case of the proposed dimerization of ethene, the phases of the relevant orbitals are not compatible with ring closure. These types of reactions are part of a class called pericyclic reactions. There are three principle divisions, namely electrocyclic, cycloaddition and sigmatropic. The detailed study of symmetry controlled reactions is outside the scope of this book, and whenever a pericyclic reaction appears in this text it will just be assumed that it is concerted without further consideration of the phase of the molecular orbitals.

We will now turn our attention to the two step addition to an alkane.

12.3 Electrophilic Addition

12.3.1 The Addition of a Symmetrical Molecule

The addition of a small symmetrical molecule such as bromine to a symmetrical substrate such as ethene is the simplest type of two step addition reaction. We will look at this reaction in some detail in order to highlight the factors involved.

First, write down an equation for the overall reaction, and name the reagents and the product.

$$H_2C=CH_2 + Br_2 = H_2BrC\text{-}CH_2Br$$

Ethene reacts with a bromine molecule to yield 1,2-dibromoethane. The first important point to notice is that the 1,2-dibromo product was formed. Write down the other possible positional isomer of this compound.

1,1-dibromoethane, $H_3C\text{-}CHBr_2$. In this molecule both of the bromines are attached to the same carbon and not joined to adjacent ones. The presence or absence of isomers often gives very important clues to the route which has been followed by the reagents from the starting materials to the eventual products. This is true not only of structural isomers, but also of stereo-isomers. The 1,1-dibromo compound is called the *geminal* adduct, while the 1,2-dibromo compound is called the *vicinal* adduct. These terms are often abbreviated to *gem* and *vic* respectively.

Draw out the molecular structure of ethene, and in particular show the electron orbitals which are associated with the carbon/carbon double bond. Also describe the geometry of this molecule.

Ethene is a flat molecule, *i.e.* all the six atoms are in a single plane; each carbon is in a trigonal configuration; and the π electrons are arranged above and below the plane in which the atomic nuclei are situated. Given this arrangement, deduce what will be the charge distribution around the molecule, and hence suggest whether this molecule will be electrophilic or nucleophilic.

The π electrons, which are above and below the plane of the atoms, will be centres of negative charge, and so this molecule is nucleophilic.

Turning our attention now to the bromine, suggest what will happen when this molecule approaches the ethene molecule.

As the bromine molecule approaches the negative centre of the π electrons, this centre induces a temporary dipole in the electron cloud of the bromine. This is easily done, because the outer electrons on the bromine molecule are a long way from the positive centre of the nuclei and so are held less firmly than if they were more central. In other words the bromine molecule is a soft electrophile as its electron cloud may be easily distorted. Note that all temporarily induced dipolar interactions are attractive in nature, and so facilitate the further approach of the reaction species.

Suggest what may happen as the bromine molecule is drawn closer by the attractive induced dipolar force to the ethene molecule.

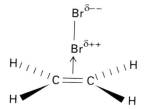

Eventually there comes a stage when, instead of there just being an electrostatic interaction between the two molecules, the two reagents become so close and the interaction becomes so strong, that the electrons rearrange themselves from their old molecular orbitals into new orbitals around different molecular species.

Suggest a likely movement of electrons, and indicate this by drawing arrows to show this movement of electrons. Also suggest the nature and structure of the likely intermediates that will be formed along this reaction pathway.

The electron rich centre of the π electrons cloud attacks the electropositive end of the dibromine molecule. In so doing the carbon/carbon double bond breaks, leaving one carbon with a positive charge. The bromine/bromine bond also breaks and forms a bromide anion, thus conserving charge. Both of these bond fissions are heterolytic in nature.

Indicate, using an appropriate three dimensional representation, the structure of the bromocarbonium species, and on the diagram show the bonds about which there is free rotation, and indicate the direction of the empty p orbital which exists.

Notice that the lone pairs which are on the bromine atom are quite close in space to the empty orbital which is on the carbon carrying the positive charge.

Suggest, a possible movement of electrons which may occur, bearing this juxtaposition in mind, and clearly indicate the position of the charges.

If one of the lone pairs of electrons on the bromine atom is donated into the empty orbital and so forms a dative bond. As a result the bromine would become positively charged as it was the donor of the electrons, and the carbon atom would then become neutral.

Indicate the stereochemical structure of this three membered species, and also indicate the direction of any polarization which may be in existence along the bromine/carbon bonds.

If the second bromine species, which is the bromide anion, reacts with a carbon atom, suggest the likely line of attack, and also show the electron movement which will occur as a result.

The bromide ion attacks the carbon from the direction opposite the bridged bromine atom, *i.e.* from the backside, because the carbon/bromine antibonding orbital projects in this direction. This mode of addition is called *anti*-addition; this is clearly seen from the conformation of the final product before any rotation of the carbon/carbon bond has occurred. As the bromide ion attacks the $\delta+$ carbon, that carbon/bromine bond breaks in order that the octet of electrons is not exceeded on the carbon. This mechanism explains why the final product is the 1,2-dibromo adduct and not the 1,1-dibromo adduct.

The pathway as outlined above is a two step process with a positively charged intermediate. Suggest what may be the result of performing the addition reaction of bromine with ethene, in the presence of chloride ions.

The bridge bromonium ion may be attacked by a chloride ion instead of a bromide ion, and so the mixed bromochloro product may be formed. This type of side reaction actually occurs, and is indicative that the route as suggested above is actually the pathway that is followed in practice.

Suggest what the product will be when one molecule of bromine is added to one molecule of ethyne, write down the equation for such a reaction.

$$H\text{-}C \equiv C\text{-}H + Br_2 = HBrC = CBrH$$

The 1,2-dibromo product is formed. At first sight one would expect bromine to react faster with ethyne than with ethene in forming the single addition product. However, this is not the case. Bearing in mind what you already know about the electron distribution in alkynes and also the nature of the intermediates in the reaction pathway, suggest two reasons to explain the fact that alkenes react faster with bromine than do alkynes.

The first possible reason is that the electrons in the alkynes are held more closely to the nucleus than is the case in alkenes, as evidenced by the difference in acidity of hydrogens which are attached to each type of carbon. The second possible reason is that in the reaction pathway of the alkyne, the bridged bromonium species is anti-aromatic and hence unstable, which in turn would lead to a higher activation energy of this reaction pathway.

This analysis of the simple addition of an electrophilic bromine molecule to a symmetrical alkene or alkyne has highlighted many points. Firstly, there is the induction of a temporary dipole of the soft nucleophile by the π electrons of the carbon/carbon double bond. Secondly, there is the heterolytic fission of the bromine molecule and the subsequent formation of the cyclic bromonium ion. Thirdly, this cyclic intermediate places certain restrictions on the line of attack of the second reagent and controls the stereochemical and structural consequences for the product.

Lastly in this section we will look at the addition of symmetrical molecules to asymmetrical substrates and study some of the further stereochemical consequences of *anti*-addition.

Maleic acid (*cis*-buta-2-endioic acid) reacts with bromine to give 2,3-dibro-mosuccinic acid (2,3-dibromobutandioic acid). Write out the possible permu-tations for this reaction and so deduce the resultant stereochemistry of the products.

This reaction proceeds via a bridge bromonium ion. The *anti*-addition of the bromide may take place at either end of the ring and so two products are formed in a 1:1 ratio. These are the *threo d,l* pair of stereoisomers. The resultant mixture is not optically active because each enantiomer is produced in equal amounts.

Now predict the outcome of the addition of bromine to fumaric acid which is the *trans* isomer of the above acid.

In this case the *meso* isomer is formed. For the sake of completeness we will continue this exercise by adding an asymmetrical molecule, AB, to fumaric acid and deduce the stereochemistry of the product.

The result this time is the *erthreo d,l* pair. Thus this reaction is stereoselective, because only one set of products is produced from each reaction; but,

rotation
⟶

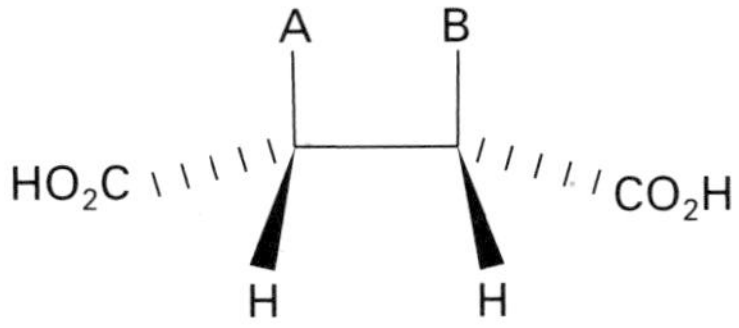

rotation
⟶

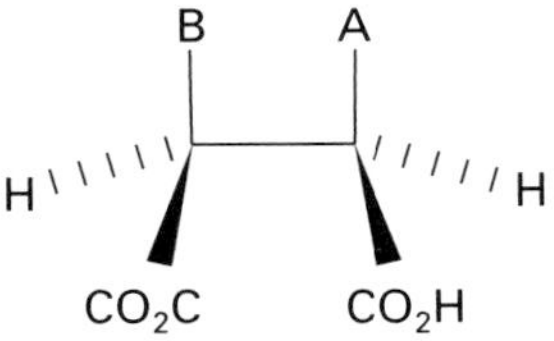

further it is stereospecific, because a given isomer leads to one product pair while the other isomer leads to the opposite product pair.

We will now move on and look more closely at the rate of the addition and the stereochemical consequences of the addition of asymmetrical molecules to asymmetrical substrates.

12.3.2 The Rate of Addition

In these reactions the attacking species has been an electrophile, with an intermediate which is positively charged. Bearing these two points in mind, suggest what the effect would be of an electron donating substituent on the double bond.

The more electrons that are pushed into the π orbital, the greater the attraction will be towards the incoming electrophile, and also the greater the stabilization of the resultant cation.

Suggest an order of reactivity for ethene, propene and phenylethene.

$$\text{phenylethene} > \text{propene} > \text{ethene}$$

Suggest why the phenylethene enhances the rate so much.

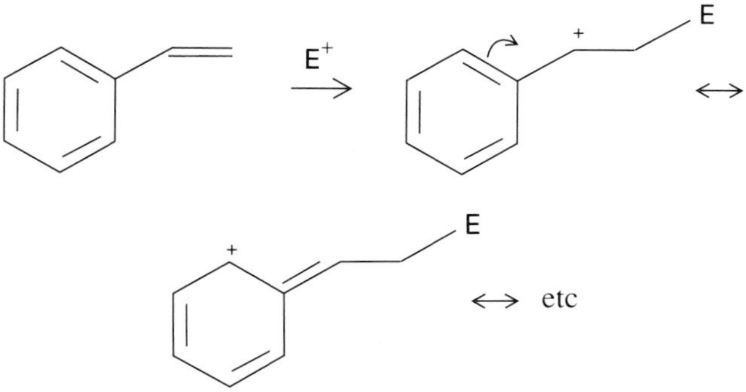

The phenyl group is capable of delocalizing the positive charge much more effectively by mesomeric mechanisms than the alkyl groups may do by inductive mechanisms alone. It will be apparent that the more inductive groups there are feeding electrons into the double bond the more likely there is to be an attack by an electrophile and the less likely there is to be an attack by a nucleophile.

Draw out the intermediate carbonium ion that results from the addition of a proton to ethene and phenylethene. Then suggest which of these two intermediates will have the longer half life.

The carbonium ion of phenylethene will have the longer half life because it is the more stable of the two.

Suggest what may be a direct stereochemical consequence of this longer half life.

As the carbonium ion has a longer half life, there will be more time for rotation to occur about the carbon/carbon bond and this in turn would lead to a more even mix of the *anti* and *syn* products.

12.3.3 The Orientation of Addition

We will use the addition of hydrogen bromide to ethene as an example of the addition of a small asymmetrical molecule to a symmetrical substrate.

Write down the overall equation for this addition reaction.

$$H_2C = CH_2 + HBr = H_3C\text{-}CH_2Br$$

Indicate the direction and nature of the dipole in the hydrogen bromide molecule and so suggest which end of this molecule is attacked by the electron rich π orbitals of the ethene. Then draw a reaction mechanism which illustrates the movement of electrons that results in the formation of the charged intermediate.

The bromine is more electronegative than the hydrogen and so the hydrogen carries a δ+ charge. This means that it is the hydrogen that is attacked by the π electrons to form the carbonium ion, and a free bromide ion results. Notice that this time there cannot be a bridged carbonium ion formed because there is no atom which has either a lone pair of electrons or is large enough to bridge the gap easily to form a three membered ring.

Suggest how this reaction pathway will continue and draw the structure of the product.

The addition of the bromide ion results in the final product, which is bromoethane.

The hydration of an alkane may be catalysed using an acid which has a weakly nucleophilic conjugated base, such as dilute aqueous sulphuric acid. Write down the overall equation for the hydration of ethene, and name the product.

$$H_2C = CH_2 + H_2O = H_3C\text{-}CH_2OH$$

Ethanol is the product. Now suggest a mechanism for this reaction.

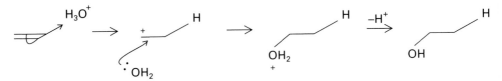

In this case the acid donates a proton to the ethene to form the carbonium ion, which is then attacked by a water molecule to give a protonated form of the ethanol molecule. This then in turn loses its extra proton to the solvent. The acid is merely acting as a catalyst.

Now let us look at the addition of hydrogen bromide to an asymmetrical substrate such as propene. Suggest the two likely products of this addition reaction.

Hypothetically the bromine may join to the terminal carbon or to the central carbon. In reality only one product is formed in this reaction, and to understand why, we need to look more closely at the possible intermediates. Suggest the structure of the two possible carbonium ions that may result from the addition of a proton to this alkene.

One of these is a primary carbonium ion while the other is secondary. Which of these two carbonium ions is the more stable, and why?

The secondary carbonium ion is the more stable, because it has two electron donating groups inductively feeding electrons towards it.

Given this difference in stability of the charged intermediate, suggest what will be the overall product of the addition of hydrogen bromide to propene.

$$H_3C\text{-}CH = CH_2 + HBr = H_3C\text{-}CH_2\text{-}CH_2Br$$

2-bromopropane is the result, and not the 1-bromopropane. We may invoke the Hammond postulate to explain this result. Thus due to the greater stability of the secondary carbonium ion there was a lower activation energy barrier preceding it, in contrast to the higher activation barrier preceding the less stable primary carbonium ion. Hence, the secondary carbonium ion is formed much more quickly, and once formed it reacts irreversible to give the product.

This explanation for the formation of the more stable carbonium ion intermediate underlies one of the basic guidelines of electrophilic addition reactions, which is called the Markovnikov Rule. This rule states that the positive portion of the reagent goes to the side of the double or triple bond which has more hydrogens. This is an example of regioselectively, *i.e.* the orientation of the addition product is determined by the structure of the substrate.

A further illustration of this may be seen in the addition of the interhalogen compound, bromochloride, to propene. Suggest what is the product and draw out the mechanism for this reaction.

The bromine is more positive than the chlorine, and so becomes attached to the terminal carbon, forming the more stable secondary carbonium ion. This is then attacked by the chloride ion to form the 1-bromo-2-chloropropane.

Hydrogen halide may also add to alkynes. Write down the mechanism for the addition of one molecule of hydrogen bromide to one molecule of ethyne.

This molecule is still unsaturated and so it may add another molecule of hydrogen bromide. Write down the mechanism for this step, clearly indicating the orientation of the addition of this second molecule of hydrogen bromide, and also suggest whether this second step is faster or slower than the corresponding addition of hydrogen bromide to ethene.

The second addition follows the Markovnikov principle. This second addition is about 30 times slower than the corresponding addition of hydrogen bromide to ethene because of the electron withdrawing capabilities of the halogen. Notice that the *geminal* product is formed and not the *vicinal* product.

12.3.4 Some Other Electrophilic Additions

Another interesting and useful reaction is hydroxylation. This is the addition of a molecule of hydrogen peroxide across a double bond to give a

1,2-diol. Write down the overall equation for the hydroxylation of propene.

$$H_3C\text{-}CH=CH_2 + H_2O_2 = H_3C\text{-}CHOH\text{-}CH_2OH$$

This reaction may be performed in a number of different ways, and a degree of control over the resultant stereochemistry may be introduced. One method involves the initial formation of the epoxide, followed by hydration to give the 1,2-diol.

Write a balanced equation for the reaction of propene with a general peroxyacid, RCO_2OH which results in the formation of the epoxide.

The mechanism for the formation of this epoxide intermediate follows the same general route of electrophilic attack as the addition of a bromine molecule. Suggest a mechanism for the formation of the epoxide and clearly indicate the movement of electrons.

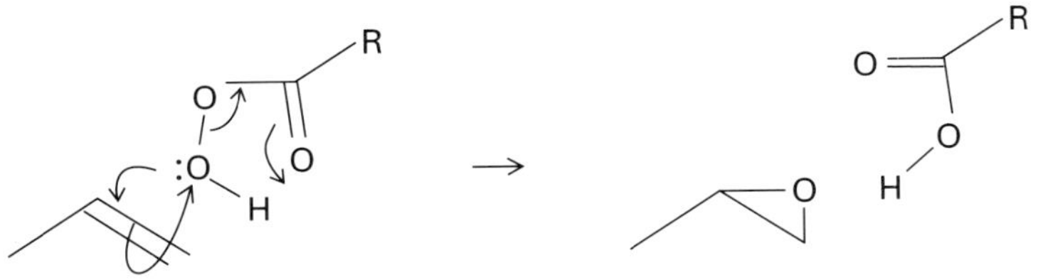

Note that the leaving group in this case is the carboxylate anion, which then further reacts with the protonated epoxide to form the corresponding conjugate acid, *i.e.* the carboxylic acid, and the neutral epoxide.

This epoxide may undergo nucleophilic attack in either acid or base catalysed conditions to yield the 1,2-diol.

Write down the reaction mechanism for the formation of the 1,2-diol under these different conditions.

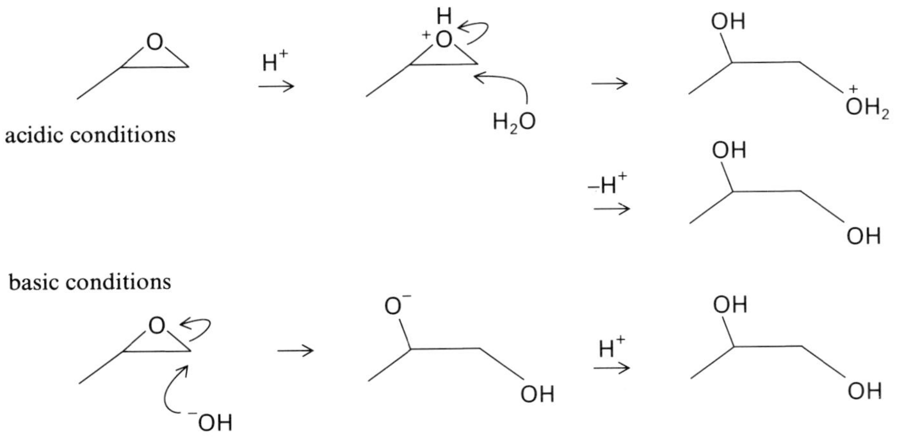

acidic conditions

basic conditions

In each case note that the attack of the nucleophile is from the opposite side to the oxygen bridge, and so the resultant addition of the two hydroxyl units is *anti*. Under basic conditions the nucleophile is an hydroxyl ion, and the resultant anion picks up a proton from the solvent; while under acidic conditions the epoxide is activated by protonation of the epoxide oxygen, which in turn induces electrons away from the carbons and so is prone to attack by a water molecule, which after nucleophilic attack loses a proton to the solvent to give the product. Overall there has been an *anti*-addition of the symmetrical molecule H_2O_2.

Earlier we looked at the formation of 2-bromopropane from the addition of hydrogen bromide and propene, and also at the formation of ethanol from ethene reacting with dilute sulphuric acid. Suggest what would be the product of the reaction between propene and dilute sulphuric acid.

$$H_3C\text{-}CH = CH_2 + H_2O = H_3C\text{-}CHOH\text{-}CH_3$$

Synthetically it would be useful to be able to control the regiochemistry of this addition so that it would be possible to form propan-1-ol. In the next section we will see that there is a way to control the regiochemistry of the addition of hydrogen bromide to give 1-bromopropane. It would be possible to convert this compound into the corresponding alcohol by substituting the bromine functional group for the hydroxyl functional group. Such a substitution is called a functional group interchange, FGI. There is another way in which this overall conversion from propene to propan-1-ol may be accomplished. This is by the indirect hydroboration using diborane, followed by oxidation of the resultant trialkylboron with alkaline hydrogen peroxide. For the present purposes we are only concerned with the reaction between diborane and the alkane. Diborane is generated *in situ* but reacts as if it were the monomer BH_3, which is a Lewis acid. Write down the mechanism for the addition of BH_3 with propane.

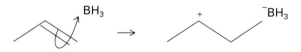

The electrons from the π orbital attack the empty orbital on the boron trihydride to form a secondary carbonium ion, *i.e.* the normal Markovnikov addition; but note this time the centre to which the electrons are being attracted is the boron atom and not a hydrogen. This is because the boron has an empty orbital into which the electrons from double bond move.

To complete this reaction there is first a transfer of a hydride ion from the boron to the positive carbon; secondly, this monoalkylated boron compound then undergoes two further alkylations to give the trialkylated product; and thirdly, on the addition of alkaline hydrogen peroxide the carbon/boron bond is broken and a hydroxyl group is substituted. Write down a structure for each of these intermediates.

Hence the overall regiochemistry is the addition of water in an anti-Markovnikov manner, but it was brought about by a Markovnikov addition of boron to the double bond, followed by the replacement of that boron with the hydroxyl group.

This polar mechanism is in contrast to the non-polar radical mechanism for the addition of hydrogen bromide that we will study in the next section. We will discover that the radical mechanism gives rise to addition with the anti-Markovnikov regiochemistry. There is a further difference between the polar and the non-polar reactions, in that the radical addition is usually *syn* in nature, assuming that the radical has a short half life, while the polar reaction is *anti* in nature.

12.4 Radical Addition

We have looked at the addition of electrophiles to carbon/carbon double bonds, and these reactions follow the Markovnikov regioselectivity. From a synthetic point of view it would be useful if it were possible to select the other regioselectivity, *i.e.* anti-Markovnikov addition.

If we take the addition of hydrogen bromide to propene as an example, this change in regiochemistry may be achieved by the use of a peroxide, *e.g.* $(RO)_2$, which promotes the formation of radicals. Given that $(RO)_2$ readily undergoes homolytic fission to give the RO• radical, suggest how this may react with HBr to give another reactive radical. Once this new radical is formed, suggest how this may react with the alkane, paying particular attention to the relative stability of any possible carbon radical intermediates, and so predict the regiochemistry of the final product.

$(RO)_2 \xrightarrow{\Delta} 2RO•$

$RO• + HBr \longrightarrow ROH + Br•$

In this case the formation and subsequent attack by the bromine radical on the alkane leads to the most stable carbon radical species, which is the secondary radical. This in turn picks up a hydrogen atom from another hydrogen bromide molecule and forms another bromine radical to continue the reaction. This route produces 1-bromopropane and not 2-bromopropane, *i.e.* the anti-Markovnikov orientation. This change in regiochemistry caused by the introduction of some radical initiator is called the peroxide effect, because peroxides are often used as the radical initiator.

12.5 Nucleophilic Addition

So far we have looked at reactions where the attacking species on the alkene was either an electrophile which sought out the electron rich π orbitals, or was a radical which is also an electron deficient species. However, it is possible for a nucleophile to attack an alkene.

Suggest what sort of substituents on an carbon/carbon double bond would favor the attack of a double bond by a nucleophile.

The approaching nucleophile is electron rich and is seeking a positive centre, thus anything which reduces the π electron density of the double bond would favor attack by a nucleophile, such as cyano, nitro, or other -M groups, or even strongly -I groups such as fluorine.

Write down the general equation for the formation of the intermediate after the attack by a nucleophile on an alkene.

$$R_2C = CR_2 + X^- = [R_2C\text{-}CR_2X]^-$$

The resultant anion needs to be stabilized, and this may be done, as usual, more effectively by -M groups rather than -I groups.

Suggest the intermediate which is formed by the attack of the cyanide ion, CN^-, on $HPhC = CPhCN$, taking particular care to determine which end of the double bond the cyanide anion attacks.

The orientation of the addition of the cyanide ion is determined by preferential formation of the more stabilized carbanion.

One reaction of particular synthetic utility is when a nucleophile is made to attack the cyanoethene molecule. Write down the mechanism for the attack of phenol on this molecule when the reaction is performed under basic conditions.

$$PhOH + B: \longrightarrow PhO^- + BH^+$$

The phenoxide anion is formed from the phenol by the action of the base, and the anion then attacks the cyanoethene, which after reprotonation from the solvent, forms an adduct in which a three carbon unit has in effect been added to the phenol. This three carbon unit has a terminal functionality that may be used in further synthetic steps. This reaction is called cyanoethylation and may be performed with many other nucleophiles such as amino anions, alkoxide anions, and even carbon anions. When it is performed with carbon anions it is called the Michael reaction, which in general is the attack by a carbanion on a substituted alkene, and is particularly useful in forming carbon/carbon bonds.

12.6 Conjugated Addition

We saw in the introduction that where the original substrate contains more than one double bond along a chain, then if there is only partial hydrogenation, there will be a number of different products, depending upon where the hydrogen has added. If the double bonds are not conjugated to each other, then the mechanism of addition across anyone of them will follow a similar mechanism to one outlined above. However, if two double bonds are conjugated to each other, that is separated by only one single carbon/carbon bond, then a new possibility arises.

Write down the two possible products which may be formed from the addition of one molecule of bromine to one molecule of 1,3-butadiene.

$$H_2C = CH\text{-}CHBr\text{-}CH_2Br \text{ and } H_2CBr\text{-}CH = CH\text{-}CH_2Br$$

The bromine molecule may add across one of the double bonds as originally depicted in the starting molecule, so as to yield the 1,2-dibromobut-3-ene. Alternatively it may add across the terminal carbons, and with a concomitant rearrangement of the double bond to yield the 1,4-dibromobut-2-ene. The former addition is called 1,2-addition; while the latter is called 1,4-addition. It is additions of the latter type with which we are concerned in this section.

Write down the initial intermediate that is formed by the addition of a bromine cation to 1,3-butadiene.

Initial attack always occurs on the carbon at one end of the conjugated system, and so in this case occurs at a terminal carbon. Consider the structure of the carbonium ion that would result if the bromine cation had attacked at one of the central carbons, and so suggest a reason why the terminally substituted carbonium ion is favored.

In this latter case, the carbonium ion formed cannot be stabilized by delocalization with the other double bond. Using the principle that the more stable intermediate has a lower the activation energy barrier preceding it, it may be deduced that the terminally substituted carbonium ion will be formed more quickly.

Write down the delocalized structure for the terminally substituted carbonium ion.

The C2 and C4 carbons both bear a partial positive charge, which together add up to a single unit of positive charge. Suggest what will be the next step in the addition reaction.

The anionic bromine may attach either the C2 or the C4 carbon and so form either the 1,2-adduct or the 1,4-adduct respectively. It may be discovered experimentally that the 1,2-adduct is the kinetic product while the 1,4-adduct is the thermodynamic product.

In a simple addition reaction of a bromine to a carbon/carbon double bond it is suggested that a cyclic bromonium ion intermediate exists. In the present case, a different cyclic bromonium ion could possible exist. Suggest a structure for it.

If this five membered cyclic bromonium ion did exist, what would be the resultant geometry of the double bond in the final adduct?

Such a five membered cyclic intermediate would give rise to a *cis* double bond in the final adduct. However, it is observed experimentally that the adduct is formed only with the *trans* geometry. Thus, even though this five membered cyclic intermediate may appear attractive on paper, it does not exist in practice. It is a salutary lesson that nature does not always proceed via a structure which looks possible, and further that one must always base a proposed mechanism on experimental observations.

However, what is the nature of the intermediate in this type of conjugated addition? It is observed that if a sample of pure 1,4-dibromobut-2-ene or pure 1,2-dibromobut-3-ene is taken and then heated the same equilibrium mixture is obtained in each case. A possible explanation for this is that the interconversion occurs via an ion pair. Suggest the structure of such an ion pair.

Such an ion pair would be capable of forming either the 1,2-adduct or the 1,4-adduct as well as providing an intermediate structure through which either of the two final adducts could pass when forming an equilibrium mixture on heating.

If an asymmetrical molecule adds to an asymmetrical conjugated olefin, then the issue of the orientation of addition arises.

Draw out the possible intermediate structures, and their canonical structures, which would result from the addition of a proton to 1,3-pentadiene.

I

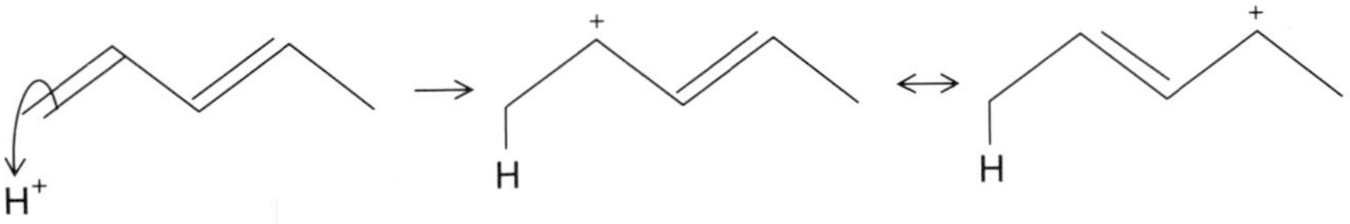

II

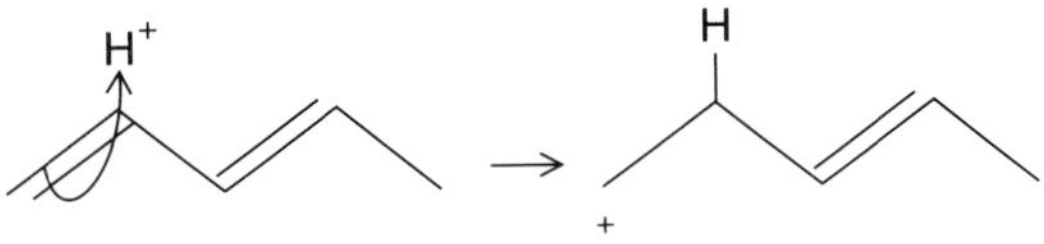

III

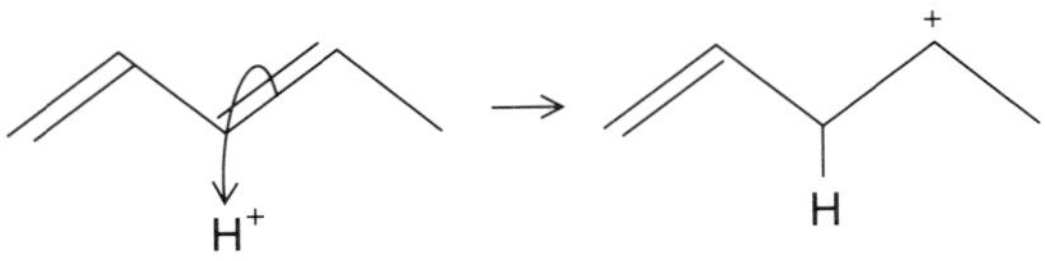

IV

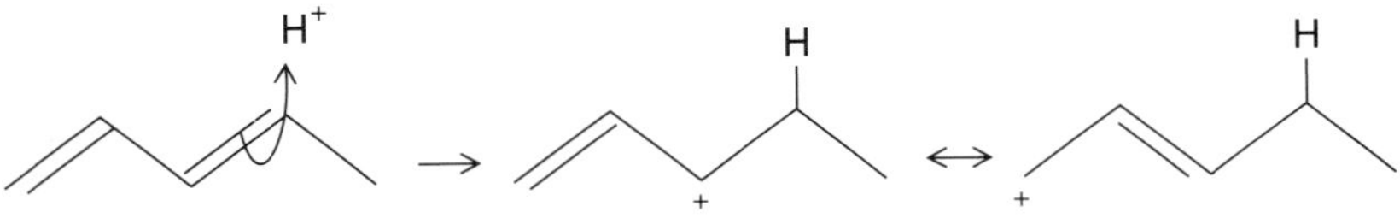

Firstly, it is again apparent that terminal addition, *e.g.* I and IV, gives rise to intermediates which have more favorable delocalization possibilities. Secondly, one may distinguish between the two terminal addition adducts, by the fact that I is more stable because there is greater alkyl substitution on the carbons bearing the positive charge. It is observed that such an intermediate is preferred, and hence the orientation of the addition may therefore be deduced.

We will now look at addition reactions to carbon/oxygen double bonds.

13

Addition to Carbon/Oxygen Double Bonds

13.1 Introduction

In the previous chapter we looked at addition reactions to carbon/carbon multiple bonds. In this chapter we will look at addition reactions to another very common multiple bond system, the carbon/oxygen double bond. This system differs quite significantly from the carbon/carbon system, primarily because there is a permanent dipole across the bond. In the carbon/carbon double bond system such a dipole only occurred if it was induced by the substituents that were attached to it.

13.2 Structure and Reactivity

The general formula for a molecule which contains a carbon/oxygen double bond is $R^1R^2C=O$. If either or both of the alkyl groups is a hydrogen then the molecule is an aldehyde; but if both are alkyl groups then the molecule is a ketone. The $C=O$ group itself is called a carbonyl group. In this chapter we will be looking at the reactions of a carbonyl group which is joined to hydrogens or alkyl groups, *i.e.* aldehydes or ketones. Such carbonyl compounds undergo addition reactions. If one of the groups which is attached to the carbonyl carbon is strongly electronegative, such as a chlorine or hydroxyl group then the carbonyl group will perform different types of reactions that will be studied later.

In the previous chapter we looked at addition to the carbon/carbon double and triple bonds in a variety of species. This homoatomic multiple bond usually has little or no dipole across it. On occasions when the carbon/carbon double bond is not symmetrical Markovnikov's Rule is invoked in order to determine the orientation of addition of an asymmetrical reagent.

The situation is rather different when we come to consider the carbon/oxygen double bond. This heteroatomic multiple bond has a permanent, and clearly defined, polarity. Write down the structure for this system and indicate the polarity, and so deduce what are the two expected resonance structures.

Note that the oxygen is both -I and -M. Thus there is a dipole along the σ bond. Furthermore, the electrons in the π bond may rearrange to give a resonance structure which has a full charge separation. The carbon is thus significantly δ+, with the oxygen correspondingly δ-.

Given this charge distribution, suggest which end of the carbon/oxygen double bond will be attacked by an incoming nucleophile and which by an incoming electrophile.

The oxygen will always be attacked by the electrophile, while the carbon will always be attacked by the nucleophile. A consequence of this is that there is never any doubt about the orientation of attack of an asymmetrical molecule, and so there is no need for a rule similar to Markovnikov's Rule in order to predict the regiochemistry of addition.

In principle the attack could be initiated by either an nucleophile on the carbonyl carbon or an electrophile on the carbonyl oxygen. In practice, the only electrophile of any significance is an acid, either a proton or a Lewis acid. Write down the intermediate that results from the addition of, first, a proton and second, $AlCl_3$.

Proton addition

Lewis acid addition

This addition is rapid and reversible and in each case results in a full positive charge residing on the carbonyl carbon, and so this atom has been activated towards a subsequent nucleophilic attack.

Not only may an addition reaction to a carbonyl group be catalysed by an acid, but it may also be catalysed by a base. In this case, the base reacts with the initial reagent, HX, to form the anion, X⁻, which increases the concentration of the stronger nucleophile. Correspondingly, an acid has the opposite effect and increases the concentration of the weaker nucleophile, HX, by protonation of X⁻, but also may make the carbonyl carbon more prone to nucleophilic attack as we saw above. The consequence of this dependance on the acid/base concentration means that most carbonyl addition reactions have an optimum pH at which they should be performed.

Taking the common case of the carbonyl group, $R_2C=O$, being attacked by a nucleophile, X^-, write down the result of the first step in the addition reaction.

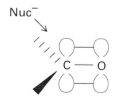

During the addition, the electron density on the carbonyl carbon increases due to the influx of electrons from the nucleophile. Hence suggest whether the rate of attack by a nucleophile would be favored by electron withdrawing groups or electron donating groups, and so suggest whether methanal would react more or less quickly than propanone.

Nucleophilic attack of a carbonyl carbon is increased by electron withdrawing groups being attached to that carbon, because they enhance its electron deficiency and so it exerts a greater attraction towards the incoming nucleophile. Thus methanal reacts more quickly than propanone, because the methyl groups of the later exert a $+I$ effect and so reduce the electron deficiency on the carbonyl carbon.

Apart from the electronic consequences of the substituents, there may also be steric ramifications. Taking the same example of a general nucleophile, X^-, attacking a general carbonyl compound, $R_2C=O$, consider the changes which occur in the steric relationships between the groups as the nucleophile approaches and forms the anionic intermediate.

In the original carbonyl compound, there is a standard sp^2 hybridization with a bond angle of 120° between the groups; while in the anionic intermediate there is a sp^3 hybridization with a bond angle of approximately 109°. Hence in the intermediate the groups are closer together, and so there is an increase in the steric crowding of the molecule. A consequence is that if there are bulky substituents on the carbonyl group, then the intermediate will suffer from steric crowding and so will be less favored than otherwise would be the case, and in turn will be formed more slowly.

Now consider which line of approach the nucleophile will follow when it approaches the carbonyl carbon.

The incoming nucleophile will approach along the line which offers least resistance. Hence it will enter from either above or below the plane which contains the carbonyl group, and approach along a line which passes between the alkyl substituents towards the carbonyl carbon.

Now that we have discussed the reaction in general terms we will move on and look at some individual reactions.

13.3 Hydration

Many addition reactions that a carbonyl group undergoes are reversible, with the equilibrium constant being related approximately to the rate of addition. One of the simplest addition reactions is when one molecule of water adds to a ketone, $R^1R^2C=O$, resulting in the formation of a hydrate. Write down the equation for this reaction.

$$R^1R^2C=O + H_2O = R^1R^2C(OH)_2$$

The rate of this reaction may be increased by either raising or lowering the pH of the solution. If the reaction is performed at high pH, suggest what the first step would be.

$$\begin{array}{c}R^1\\ \\ R^2\end{array}\Big\rangle\!=\!O \quad \overset{\bar{O}H}{\longrightarrow} \quad \begin{array}{c}R^1\\ \\ R^2\end{array}\!\!\!\!\bigg\rangle\!\!\!\begin{array}{c}O^-\\ \\ OH\end{array}$$

At high pH there is a high concentration of hydroxyl ions which are very nucleophilic, and so they readily attack the positive carbonyl carbon. Suggest what is the second step of this pathway.

$$\begin{array}{c}R^1\\ \\ R^2\end{array}\!\!\!\bigg\rangle\!\!\!\begin{array}{c}O^-\\ \\ OH\end{array} \quad \overset{H^+}{\longrightarrow} \quad \begin{array}{c}R^1\\ \\ R^2\end{array}\!\!\!\bigg\rangle\!\!\!\begin{array}{c}OH\\ \\ OH\end{array}$$

The ionic intermediate is then protonated rapidly from the solvent to form the neutral hydrate.

Under acidic conditions, suggest what the first step would be.

$$\Big\rangle\!=\!O \quad \overset{H^+}{\longrightarrow} \quad \Big\rangle\!=\!\overset{+}{O}H$$

In this case the carbonyl oxygen is attacked first by the electrophilic proton donor. Draw the resonance structures for this cationic species.

$$\Big\rangle\!=\!\overset{+}{O}H \quad \leftrightarrow \quad \Big\rangle\!\!\!\!\overset{+}{}\!\!-OH$$

It is clear from these resonances structures that by protonating the carbonyl oxygen the carbonyl carbon has been made more electropositive, and so more susceptible to attack by a nucleophile. Suggest what will be the second step of this mechanism.

In this case one of the solvent molecules, *i.e.* water, is nucleophilic enough to attack the carbonyl carbon, now that it has been made more susceptible to attack as a result of the protonation of the carbonyl oxygen. Finally, a proton is lost from the water molecule that attacked the carbonyl which results in the overall addition of a water molecule.

This illustrates that the rate of attack on the carbonyl carbon is dependant on both the nucleophilicity of the incoming reagent and the positive nature of the carbonyl carbon, the more positive it is then the greater the rate of attack. Suggest whether, generally, a ketone will react faster or slower than an aldehyde.

Generally, an aldehyde would react faster because it has only one electron donating group reducing the positive nature of the carbonyl carbon. Correspondingly, methanal reacts even faster than ethanal, so much so that methanal hydrates quite rapidly even at pH 7, while propanone only hydrates very slowly at neutral pH, and requires either acid or base catalysis for this reaction to proceed at a significant rate.

Suggest what would be the effect upon the rate of attack if an aryl group such as phenyl was substituted instead of an alkyl group, and give your reasons for any change in the observed rate.

In this case the rate would decrease. In the initial carbonyl compound it is possible for the positive charge to be delocalized around the aromatic ring. However, in any adduct, there is no positive charge to delocalize, and so this stability is no longer possible. Thus such an adduct will be disfavored, which reflects itself in a reduction in the rate of formation. The decrease in the rate is also observed for any other system that can conjugate with the carbonyl group.

Bearing this in mind, suggest what would be the hydration product of trichloroethanal, and suggest whether it would be formed quickly or slowly.

Electron releasing groups, such as alkyl groups, reduce the rate of attack by reducing the positive nature of the carbonyl carbon. Aryl groups decrease the rate because they stabilize the positive charge in the carbonyl but not in the adduct. In contrast, electron withdrawing groups increase the rate of attack, because in the carbonyl compound they destabilize the system by enhancing its electron deficiency, but in the hydrate they help to stabilize it by reducing the surfeit of electrons on the carbonyl carbon, that originate form the lone pairs of the oxygens. So in the case of trichlorethanal, the rate is so great and the final product so stable that the crystalline hydrate may be isolated.

Another interesting hydrate is that formed from cyclopropanone. Suggest why this hydrate may be stable.

In this case even though there has been an increase in the number of groups around the former carbonyl carbon, there has been a release in the bond strain which was associated with having a sp^2 hybridized carbon in a three membered ring. There is less strain in having sp^3 hybridized carbons in a three membered ring. Note that this is in direct contrast to the situation for a normal carbonyl compound where the steric crowding would increase in going from a sp^2 arrangement to a sp^3 arrangement.

One further example provides another reason as to why the hydrate may be favored in certain circumstances. The hydrate of diphenylpropantrione may also be isolated as a crystalline adduct. Suggest a reason for the stability of this hydrate.

In this case the middle carbonyl carbon has two electron withdrawing groups next to it, and furthermore, once it has been hydrated it is possible for both hydroxyl groups to hydrogen bond with the remaining carbonyl groups, which further stabilizes the hydrate.

So, in summary, for any general carbonyl compound the rate of addition may be affected by general acid or base catalysis. Either, because of the concomitant effect on either the nucleophilicity of the nucleophile; or, on the activation of the carbonyl carbon by protonation of the carbonyl oxygen. Further, electron donating groups reduce the rate, while electron withdrawing groups increase the rate of addition. There may be other effects such as the amount of increase in the steric crowding in going from sp^2 to sp^3 hybridization or the presence of hydrogen bonding in the product.

13.4 Other Addition Reactions

The addition of alcohols follows a similar pattern to the addition of water. Draw the mechanism for the addition of ethanol to ethanal, under general acid catalysis conditions.

The product is called a hemiacetal, and they are more stable if there are electron withdrawing groups attached to the former carbonyl carbon. Interesting, even though the formation of the hemiacetal is general acid catalysed, the subsequent formation of the acetal, in which both hydroxyl groups of the hydrate have been substituted for alkoxide groups, requires specific acid catalyst. Thus instead of the rate determining step being the protonation of the carbonyl oxygen, it is now the step following the rapid protonation which is the rate determining one, *i.e* the loss of a water molecule. Write the mechanism for the formation of the acetal from the hemiacetal formed above, indicating the fast and slow steps.

Note that the carbonium intermediate has a resonance structure which is the alkylated equivalent of a protonated carbonyl. The subsequent attack by the alcohol molecule and the loss of the extra proton are both fast steps.

This reaction is in fact quite rare when conducted with normal monofunctional alcohols. However, with 1,2-diols the reaction is facile. Write the complete reaction sequence for the formation of the cyclic acetal starting from ethanal and 1,2-ethandiol, under acidic conditions. Also suggest why the product between one molecule of each reagent is favored over the potential product that results form two molecules of the diol and one of the aldehyde.

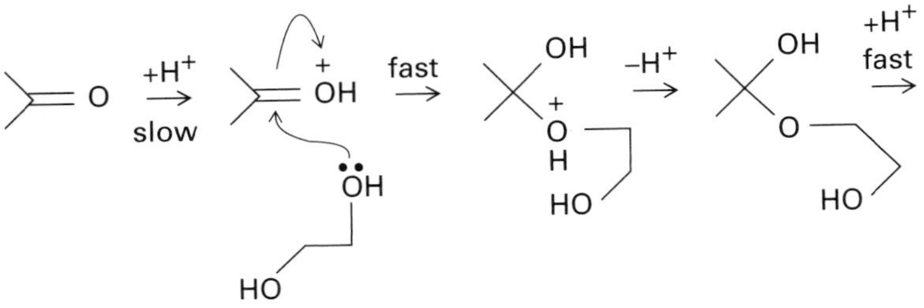

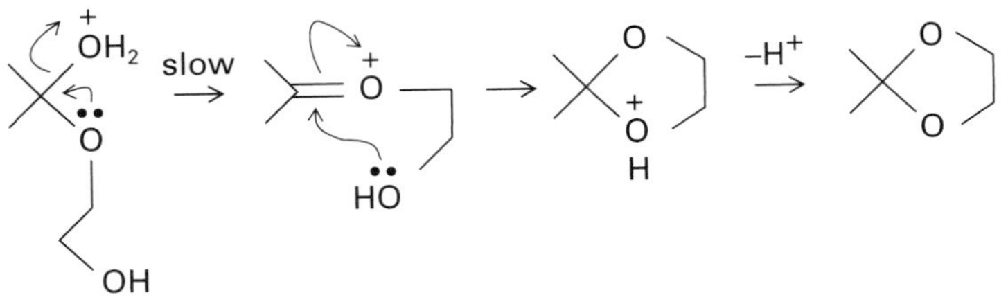

212

Michael Edenborough

The formation of the hemiacetal is general acid catalysed while the formation of the acetal is specific acid catalysed as indicated above. The formation
of the cyclic acetal is favored over the formation of acetal derived from one
aldehyde and two molecules of the alcohol, because, the formation of a five
membered ring is sterically facile; secondly, the approach of a third molecule
would be hindered by the presence of the crowded intermediate; and thirdly,
ring closure is entropically favored compared to the addition of a further
alcohol molecule. The formation of a cyclic acetal in this way is a very useful
reaction, because once formed the acetal is resistant to attack by a base and
so it acts a protecting group for the carbonyl group; and yet this protecting
group may be easily removed later by treatment with dilute acid to reveal the
carbonyl group once again.

Another very common addition reaction to the carbonyl group is that of
the bisulfite anion. In the days before the ready availability of spectroscopic
analysis this was an important reaction, because the crystalline product is
easily formed and purified. The melting point could then be determined, and
so the original carbonyl compound could be identified. Suggest the mechanism for this addition and hence the nature of the product.

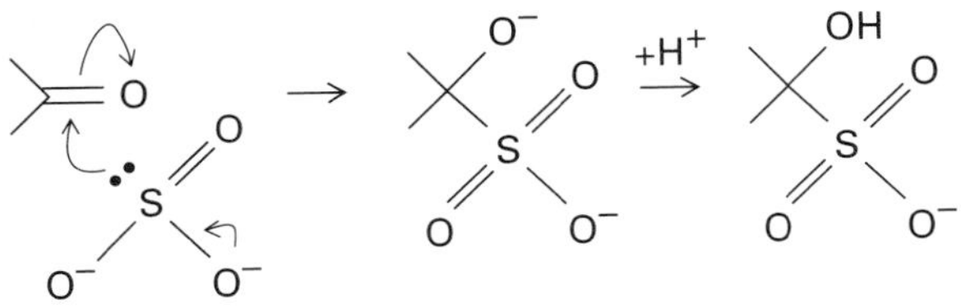

The attacking species is the sulfur atom. The oxygen atom is harder and
thus more basic in nature, while the sulfur is softer and so more nucleophilic
in nature. Then applying the maxim that "soft attacks soft and hard attacks
hard", one may deduce that as the carbon atom of the carbonyl group is to

be attacked, it will be done by the sulfur. Further it appears that the attacking species is the sulfite dianion, rather than the bisulfite anion, since even though the latter is present in much larger concentrations, the former is far more nucleophilic. The final product is the salt of the sulfonic acid and not the alkyl ester of sulfite which would have resulted from an attack by the oxygen.

We will now turn our attention to the addition of carbon nucleophiles to the carbonyl group. This represents a very important group of reaction that have a wide synthetic utility.

13.5 Addition of Carbon Nucleophiles

13.5.1 Cyanide, Alkynyl and Alkyl Anions

The multitude of reactions that occur in organic chemistry may be divided into many different categories along many different lines; however, of all the divisions along synthetic lines, there are two principal groups. The first concerns the conversion of one functional group into another, the so-called functional group interchange, FGI. The second division concerns the formation of new carbon/carbon bonds, and is of particular interest, because by this means small simple molecules may be built into larger and more complex molecules of greater value.

Much of synthetic organic chemistry is concerned with the methods of executing and controlling carbon/carbon bond formation. One of the simplest reactions involves the addition of a cyanide ion to a carbonyl group. Write the first step in the addition reaction between a cyanide ion and a general ketone, $R_2C=O$.

The addition of the cyanide anion is slow, but it is followed by the fast addition of a proton, from either HCN or a protic solvent, to form the cyanohydrin.

It is observed that this reaction requires base catalysis, suggest a reason why this may be so.

The base is needed to form the cyanide anion, because its conjugate acid, HCN, is not a strong enough nucleophile to attack the carbonyl carbon.
This simple reaction has introduced an extra carbon atom into the original molecule. Furthermore, this extra carbon is contained within a functional group which may be subsequently developed, for instance, by hydrolysis to

reveal a carboxylic acid functionality, or by reduction to an amine. This synthetic route is the basis of the Kiliani method for extending the carbon chain of a sugar.

Another simple, yet important, reaction is the addition of an acetylide anion, C_2H^-, to a carbonyl group. The anion may be formed by the reaction between a strong base, such as $NaNH_2$, and an alkyne. Write down the complete reaction sequence from the starting materials to the final product.

$$RC_2H + \bar{N}H_2 \longrightarrow RC_2^- + NH_3 \uparrow$$

The alkyne moiety of the adduct may subsequently undergo further reactions, for example reduction by Lindlar catalyst to a double bond, which results in the overall formation of an allylic alcohol from a carbonyl group.

In both of the previous two examples the carbanion has not required stabilization. Furthermore, the group that has been introduced has been capable of further synthetic elaboration. The cyanide and alkyne moieties are said to be masked functionalities, because they are capable of revealing other functional groups by further simple reactions.

The last group of reactions that we will considered in this subsection involve alkyl anions, *e.g.* $C_2H_5^-$. Unlike the previous two anions which have been considered, it is not very easy to form these anions directly from the conjugate acid, *e.g.* C_2H_6. This is because there is no stabilization of the negative charge. However, a functional equivalent may easily be formed. The reaction between metallic magnesium and an alkyl halide gives rise to an organometallic compound which has a general formula of RMgX. These compounds are called Grignard reagents. Their precise structure is open to debate but they react as a source of negatively polarized carbon moieties, *e.g.* $R(\delta^-)Mg(\delta^+)X$.

It appears that two molecules of the Grignard reagent react with the carbonyl molecule when adding an alkyl group. Suggest a possible mechanism for this addition.

If you are having difficulty writing this mechanism, try arranging the three molecules in a cyclic six membered transition state.

In this cyclic transition state one Grignard molecule donates its alkyl group to the carbonyl carbon. The other Grignard molecule activates that carbon by acting as a Lewis acid in complexing to the carbonyl oxygen. The resultant magnesium alkoxide compound is hydrolysed in the aqueous acidic work-up to yield the alcohol.

The cyclic transition state invoked receives support from the observation that Grignard reagents that have β-hydrogens tend to reduce the carbonyl compound to corresponding alcohol without adding the alkyl group. Suggest a mechanism for this alternative reaction.

Note that in this case the Grignard reagent is converted to the corresponding alkene.

A different side reaction may occur when a Grignard reagent react with a sterically hindered carbonyl compound that possesses an α-hydrogen. Suggest what may be the transition state for this reaction and so suggest what would be the products.

In this case the Grignard reagent complexes with the carbonyl oxygen as usual, but because of the steric crowding due to the bulky side group on the carbonyl carbon there is not sufficient room for a second Grignard reagent to approach. Instead the α-hydrogen from the carbonyl compound is transferred to the Grignard reagent, resulting in the loss of an alkane and the formation of an enol magnesium derivative.

The addition of the alkyl anion is usually irreversible and proceeds in high yield. The use of organometallic reagents in synthetic organic chemistry is common, and other examples would include those derived from zinc or lithium. The latter in particular are useful, because they tend to undergo fewer side reactions, preferring instead to perform the simple addition across the carbonyl double bond.

13.5.2 Carbonyl stabilized Carbon Anions

In the previous sub-section the carbon anion was either inherently stable, *e.g.* the cyanide or ethynyl ions, or arose from the heterolytic cleavage of an organometallic bond, *e.g.* the Grignard reagent. In this sub-section we will look at reactions that involve carbanions in which the stabilization is provided by a carbonyl group. This is one of the commonest methods of providing such stabilization, and as such there are many examples, often identified by the discover's name or the trivial name of the compound which is formed. This tendency to give particular reactions the names of individuals is common in organic chemistry. Once the names are learnt, then it becomes a very quick way of communicating information, even though initially it can be somewhat confusing and mystifying.

The first example we will consider is the aldol reaction, in which two molecules of ethanal, on treatment with a base such as NaOH, react together to form aldol, 3-hydroxybutanal. Write down the mechanism for the removal of a proton from ethanal to give rise to the carbanion, and also indicate the stabilization that is available for this ion.

First, note that this is a simple Brönsted-Lowry acid/base reaction that gives rise to the enol anion. Secondly, the proton that is removed is the one that is α to the carbonyl group. This is an absolute requirement: no α-hydrogen, no enolization. Thirdly, the initial carbanion that is formed is stabilized by the -M effect of the carbonyl group, which gives rise to the negative charge residing on the carbonyl oxygen. It is possible to protonate this resonance structure, and the resultant compound is called an enol, which is a tautomer of the carbonyl compound in its keto form. Write down both tautomers of ethanal, and label the enol and keto form respectively.

keto ⇌ enol

The anion is potentially capable of reacting either through the oxygen or the carbon, *i.e.* it is an ambident nucleophile. When this anion is used to attack another carbonyl compound, then is invariable reacts via the carbon, *i.e.* soft nucleophile attacking a soft electrophile. Write the mechanism for the attack of the anion of ethanal on another molecule of ethanal, and the subsequent protonation to form the neutral adduct.

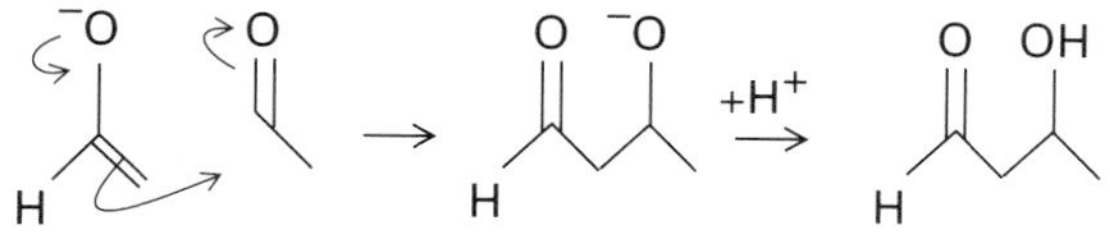

In this case a 1,3-hydroxy carbonyl compound has been formed. This product has another α-hydrogen which may be removed by the base, and then one of two things may then happen. Either, this new anion may react with another molecule of ethanal and so form a trimer, or else a hydroxyl group may be eliminated and so form an α,β-unsaturated carbonyl compound. The detailed mechanism for the elimination will be covered in the next chapter, for the present purposes just indicate the product that would result from the elimination of the hydroxyl group.

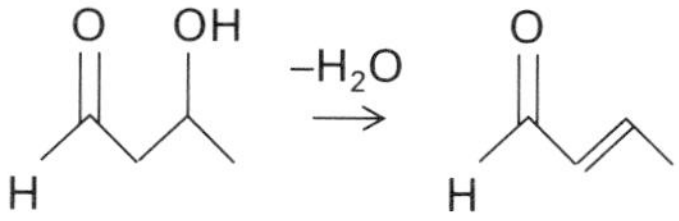

This α,β-unsaturated carbonyl compound is called crotonaldehyde and hence the aldol reaction followed by elimination is called the crotonaldehyde reaction. In summary the formation of crotonaldehyde and aldol from ethanal under basic conditions is as follows:

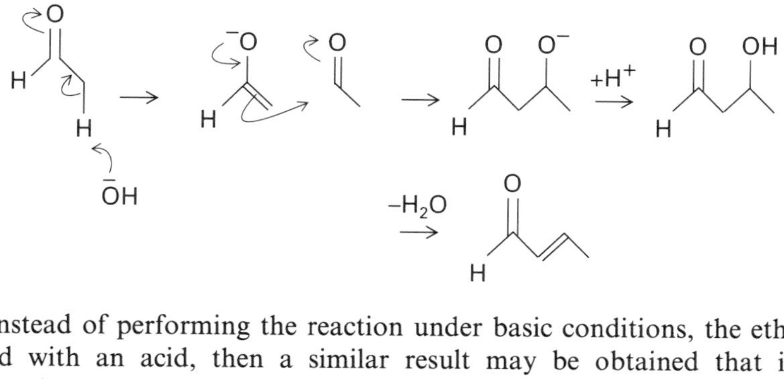

If instead of performing the reaction under basic conditions, the ethanal is treated with an acid, then a similar result may be obtained that in turn attacks the protonated, and hence activated, carbonyl compound to form aldol and then crotonaldehyde. It is more usual for the dehydration step to occur under acidic conditions.

It is not unusual for a compound to undergo a reaction in either basic or acidic conditions, however, particular care must be taken to ensure the intermediates that are invoked are appropriate to the conditions used. In this case the end result is the same, but this is not necessarily so and it is possible for different products to be formed depending upon the pH of the reaction mixture.

If there are two carbonyl compounds present then a cross condensation may occur, but it is of little synthetic value as there are four possible products. However, if one of the carbonyl compounds lacks an a-hydrogen then the reaction may prove useful. An example is the Claisen-Schmidt condensation of an aromatic aldehyde, *e.g.* benzaldehye, with a simple aldehyde or ketone, *e.g.* ethanal, in the presence of 10% aqueous KOH. Write the mechanism for this reaction, and also suggest whether dehydration would occur and why.

Dehydration always occurs in this situation because of the possibility of forming a conjugated system.

As was mentioned earlier in this sub-section this area of synthetic chemistry is replete with reactions that are named after their discover. A few more examples will now be considered.

In the Claisen ester condensation, the anion is formed from an ester directly by the action of a base. This anion then reacts with another ester. Write down the reaction sequence and suggest what will be the product. Also suggest what would be a suitable base.

The product this time is a 1,3-ketoester compound. A suitable base would be the alkoxide that formed the ester, because it does not matter then if the ester undergoes transesterification, that is where one alkoxide moiety is exchanged for another. Note, that instead of the anionic dimer just picking up a proton, that the alkoxide anion is eliminated to give the carbonyl group. If both ester groups are in the same molecule then an internal condensation reaction is possible. This is called the Dieckmann cyclization and works best when the ring formed contains five, six or seven members.

If the ester has a halogen at the α position then there are two common variations that are synthetically useful. The first is the Reformatsky reaction in which an a-halo ester is treated with zinc to form a masked carbanion, and this then reacts with an aldehyde or ketone. Write down the general reaction sequence for this reaction and suggest what will be the product.

The product is a 1,3-hydroxy ester.

In the second reaction sequence the anion is formed directly from the a-halo ester and not by way of an organozinc derivative. Write down the reaction sequence for the addition of this anion to a general ketone, $R_2C=O$.

In this case, there is an oxygen anion that is *vicinal* to a halogen atom, which is usually bromine. Instead of this intermediate dehydrating as occurs in the crotonaldehyde reaction, suggest a different line along which it may proceed, bearing in mind that halogens are good leaving groups.

Here, instead of the oxygen anion being eliminated to give rise to an alkene derivative, it attacks the carbon bearing the halogen and displaces that to form the epoxide ring. This reaction is called the Darzen's condensation.

In the Knoevenagel reaction the anion is derived from a compound of the general formula CH_2X_2, where X is a -M group, such as an ester. The resultant anion then reacts with a different carbonyl compound. The base employed is usually the alkoxide moiety of the ester. Normally dehydration occurs. Write the general reaction sequence for this condensation.

A variation on this reaction is when the anion is formed from the diester derivative of 1,4-butandioic acid (succinic acid). In this case the reaction proceeds via a cyclic ester intermediate, a lactone. Write down the reaction sequence for the reaction between a general ketone, $R_2C=O$, and the anion of succinic acid, suggesting how the lactone intermediate may be formed.

The remaining α-hydrogen to the side chain ester may then be removed and the ring opened up to reveal an α,β-unsaturated ester with a carboxylate group on the methylene side chain. Write the mechanism for these remaining steps.

This reaction is called the Stobbe synthesis.

The anion may be formed from an acid anhydride by using the carboxylate anion of the corresponding acid as the base. This anion may then be reacted with an aldehyde such as benzaldehyde to yield as an initial product a mixed anhydride. This in turn may be hydrolysed to give an α,β-unsaturated acid. Write the steps of this reaction sequence.

This synthetic route is called the Perkin reaction.

The last reaction we will look at, like the first is named after the product that is formed, rather than its discover. Benzaldehyde may undergo attack by a cyanide ion to yield an anion. Write down the mechanistic steps for the formation of this anionic intermediate.

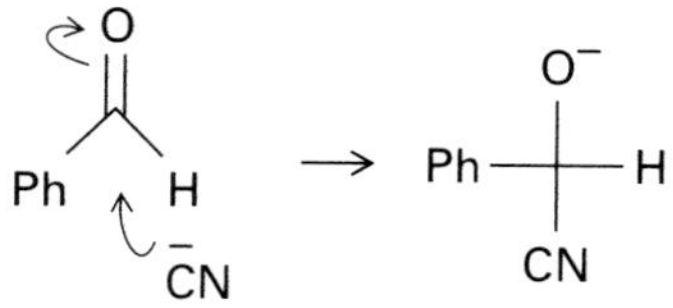

So far this reaction is proceeding along the same lines as the cyanohydrin formation that we looked at in the previous sub-section. Now, instead of simple protonation of the anionic oxygen, the hydrogen on what was the carbonyl carbon may be removed, because it is fairly acidic, and may then protonate the oxygen. The resultant carbanion may then attack another benzaldehyde molecule. Write down these reaction steps.

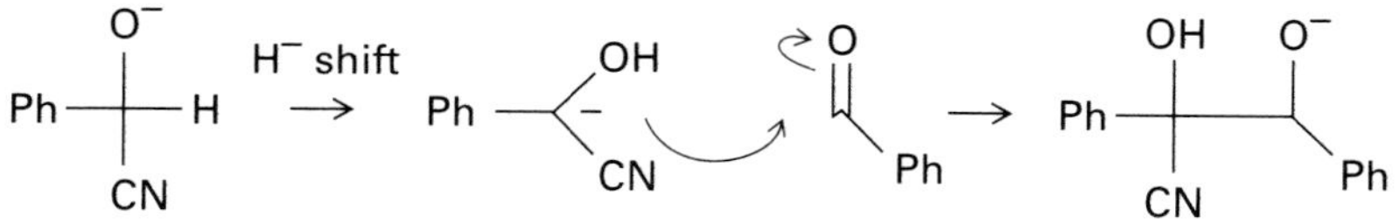

There is now another rearrangement of the acidic hydrogens so as to place the anionic charge on the oxygen *geminal* to the cyanide group. This then reforms the initial carbonyl group and in so doing eliminates the cyanide anion. Write these final steps and identify the product.

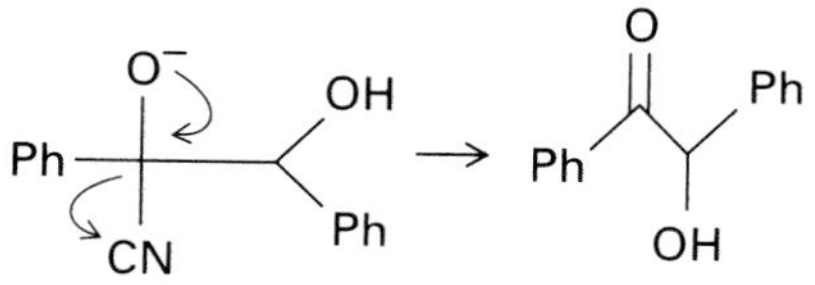

The product is a 1,2-hydroxyketone, and when starting with benzaldehyde is called benzoin, which gives this reaction its name.

In this sub-section we have seen how a carbonyl group has facilitated the formation of a carbonanion. This function may also be performed by other groups, such as a nitro group. Write the tautomeric forms of nitromethane.

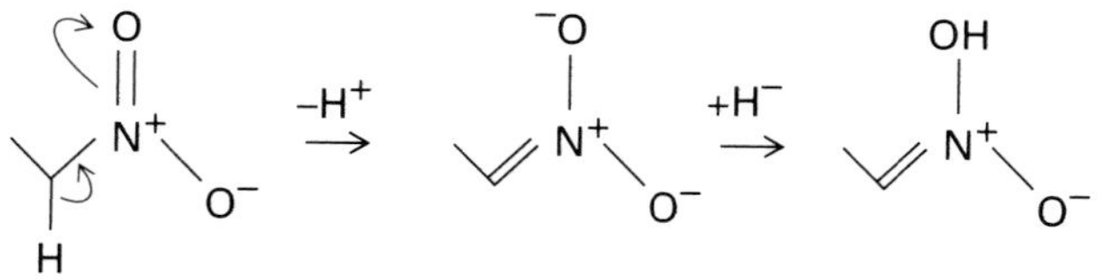

nitro   aci

Anions stabilized in this way may undergo similar reactions to the ones that have been outlined above for carbonyl stabilized anions.

Finally it should be noted that nearly all of the above reactions, with the notable exception of the Grignard additions, are reversible and so under the appropriate conditions α,β-unsaturated carbonyl compounds may be cleaved.

13.6 Stereochemical Consequences

At the beginning of this chapter we saw that the orientation of the addition of a nucleophile is readily determined, because of the clearly defined polarity of the carbonyl group. Another simplification is that the stereochemistry of the addition is of less importance, because it is impossible to tell if the addition has been *syn* or *anti*. Show that this is true by the addition of HY to a general ketone $R^1R^2C=O$ in both the *syn* and *anti* manners.

The product is the racemic mixture.

However, if either of the groups that are attached to the carbonyl carbon, are themselves chiral then there is a possibility of differentiating between the attack from one side or the other of the carbonyl group. It is observed that the carbonyl oxygen orientates itself so to be diametrically opposed to the largest substituent. Draw the Newmann projection along the bond between the carbonyl carbon and the chiral α-carbon. The assign to the groups that are attached to that carbon the letters S, M and L to designate the small, medium and large groups respectively. Suggest from which face the incoming nucleophile will preferably attack.

The nucleophile will prefer to approach on the face that offers least steric hindrance, and this is between the large and small group. This empirical observation is called Cram's rule, and has been widely investigated for Grignard reactions. The greater the degree of steric interference, the greater the degree of selection is found.

Other empirical guidelines have been formulated, *e.g.* Prelog's rule, that attempt to predict and rationalize the addition of HY to such a system. However, for all of these it must be remembered that only the relative direction of addition of the anionic component with respect to the rest of the substrate is considered. The orientation of the addition of the proton is ignored in all cases.

13.7 Conjugated Addition

In α,β-unsaturated carbonyl compounds, the question arises, as it did with conjugated dienes, whether the addition will be 1,2 or 1,4. With Grignard reagents the nucleophilic addition is usually 1,4. However, due to the tautomerism of the resultant enol it gives the same result as would 1,2 addition across the carbon/carbon double bond. Write down the reaction sequence for the 1,4-addition of RH to $H_2C=CH\text{-}CH\text{-}CH=O$.

Similarly the electrophilic addition of HBr is 1,4, but appears to be 1,2 across the carbon/carbon double bond. Write down this reaction sequence.

If the reaction is reversible then the thermodynamic product will pre-dominate, which is usually the 1,4 addition, because the carbonyl group is more stable than the carbon/carbon double bond. If, however, the reaction is irreversible then the kinetic product will dominate and this may be influenced by the steric demands at the respective sites.

When an α,β-unsaturated carbonyl compound is treated with a carbanion, particularly one that is stabilized by a carbonyl group itself, the resultant 1,4 addition with the formation of a new carbon/carbon bond is called the Michael reaction. Like the Claisen ester condensation, it is possible for this reaction to occur internally with a suitably substituted compound. In which case a ring will be formed, and it is the Robinson annulation reaction. Write down the complete reaction sequence for this ring closure reaction between cyclohexanone and but-1-en-3-one.

This reaction concludes the chapter on addition reactions, we will now examine those reactions which perform the opposite synthetic function, *i.e.* elimination of a part of the molecule.

14

Elimination Reactions

14.1 Introduction

So far we have looked at substitution reactions and addition reactions. We will now turn our attention to elimination reactions. The overall result of an elimination reaction is, not surprisingly, the opposite of an addition reaction. However, when we come to study the mechanisms that are involved in elimination reactions, we find that there they are not closely related to the reverse of addition reactions. In contrast elimination reactions are closely related mechanistically to substitution reactions. In fact, elimination and substitution reactions often compete with each other in the reaction mixture.

In a general elimination reaction, a small part of the molecule is lost to give two new molecules. Usually, one part has a greater degree of unsaturation than was present initially. Often this is achieved by losing two groups, A and B, that were adjacent to each other in the starting material. In this case a double bond is formed in the remaining part of the molecule, and another molecule, A-B is formed form those parts that were eliminated. Write the general equation for the loss of the A-B molecule, from the starting material, $R_2CA\text{-}CBR_2$.

$$R_2CA\text{-}CBR_2 = R_2C\!=\!CR_2 + A\text{-}B$$

This type of elimination is called β-elimination, or 1,2-elimination. If there was already a double bond between the α and β atoms, then a triple bond will be formed in the elimination product.

It is possible for both groups, A and B, to be lost from the same atom. Write the general equation for the lose of the molecule A-B from the starting material, $R_3C\text{-}CABR$, and so identify the initial product.

$$R_3C\text{-}CABR = R_3C\text{-}CR + A\text{-}B$$

In this case the initial product is a carbene. If the groups, A and B had been lost from a nitrogen atom, then the initial product would have been a nitrene, $R_3C\text{-}N$. This type of elimination is called an α-elimination, or 1,1-elimination. The carbene or nitrene which is formed may undergo a number

227

of further reactions. If there is hydrogen on the β-carbon then a new double bond may be formed by a hydride shift. Write this step of the reaction.

There is a third type of elimination which is called γ-elimination, or 1,3-elimination, in which a three membered ring is formed. Write down the general equation for this reaction.

There is a final type of elimination mechanism that is called an extrusion reaction, in which a segment of a chain is extruded to form a smaller chain, *e.g.* A-X-B = A-B + X. In this reaction type, unlike the ones which we have seen so far, there is no increase in the overall level of unsaturation of either part.

The commonest type of elimination reaction that occurs in solution is the 1,2-elimination. However, it is also possible for 1,1-elimination to occur under certain conditions. In the gas phase, when the reaction is normally initiated by heat, and so is called pyrolysis, there are two common pathways. The first is via a cyclic intermediate and the second involves a free radical pathway. These two pyrolytic elimination are rather different in nature, from those that occur in solution, and so we will discuss them separately. We will start by looking in some detail at the 1,2-elimination reaction that occurs in solution.

14.2 Bimolecular 1,2-Elimination Reactions

14.2.1 Anti-Elimination

There is a tendency in some elementary organic textbooks to indicate elimination reactions by encircling the two groups, A and B, that are closest to each other, and then showing the final eliminated product. This type of "lasso" chemistry predicts that both groups are eliminated from the same side of the starting material, *i.e.* that there has been *syn*-elimination. In practice, this is found to be rather rare. Instead, what is observed is that the two groups that are eliminated come from opposite sides of the molecule, *i.e.* *anti*-elimination. We will now examine why this is the case.

The fundamental bimolecular elimination reaction is labelled the E2 reaction. This reaction is very similar to the SN2 reaction that we studied in an earlier chapter, and like that pathway is a bimolecular, one step reaction.

However, instead of the incoming group arriving at the same time that the leaving group is departing, which results in the overall substitution of the SN2 reaction, in the E2 reaction, both groups leave at the same time, to give an overall elimination.

Write the equation for the E2 reaction between a general base, B, and the substrate, $R_2CH\text{-}CXR_2$.

$$R_2CH\text{-}CXR_2 + B = R_2C = CR_2\ BH^+ + X^-$$

This reaction has resulted in HX being eliminated. Compare this elimination reaction with the SN2 reaction by writing the equation for the substitution of an X group by a nucleophile, Nuc⁻, on the substrate, $R_2CH\text{-}CXR_2$.

$$R_2CH\text{-}CXR_2 + Nuc^- = R_2CH\text{-}CR_2\text{-}Nuc + X^-$$

Nucleophiles and Brönsted-Lowry bases are very similar. The species X⁻ is called the nucleofuge. Suggest what are the similarities of these two reactions, and then try to suggest what are the differences that distinguish one from the other. Consider the nature of the attacking species and hence consider the target of the attack.

In the case of the substitution reaction, the attacking species is acting like a nucleophile, which means that it is acting as a soft base, and so it attacks the soft centre, which in this case is the δ+ charged carbon. In the case of the elimination reaction, however, the attacking species is acting as a hard base, and thus attacks the hard centre, which is the α-hydrogen, whose σ bond electrons then act almost like an internal substitution reaction to eliminate the leaving group. Thus the fundamental difference between the substitution reaction and the elimination reaction, is that in the former the attacking species is soft in nature, while in the latter it is hard. Hence elimination via the E2 mechanism will be favored over substitution when a hard basic reagent is employed, and *vice versa* when a soft nucleophilic reagent is employed.

Now write the mechanism for an E2 elimination reaction, that combines all the points that have been discussed so far.

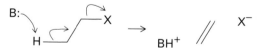

Of course, no attacking species acts purely in one way or another, but each has a character that is somewhere along a continuum between one extreme and the other. This is the reason why in any given reaction mixture substitution and elimination reactions compete with each other.

The stereochemistry of this reaction may be investigated by studying the elimination of HBr from 1,2-dibromo-1,2-diphenylethane. It is observed that the *meso* form, gives only the *cis* isomer of the alkene. Suggest what conformation of the transition state would explain this result. Use the sawhorse notation to illustrate your answer.

meso   cis

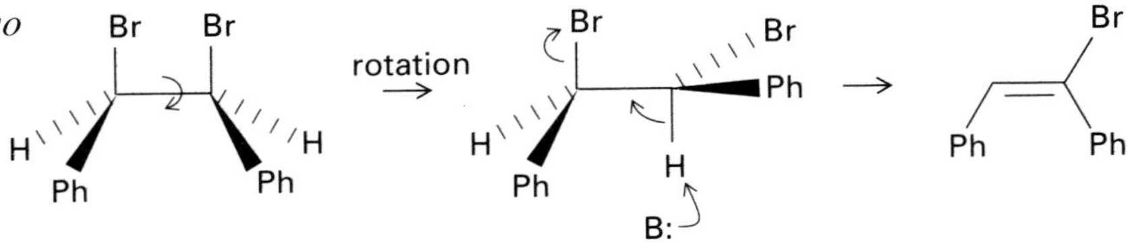

R,S

When the hydrogen and the bromine are *anti-periplanar* to each other, and the attacking species is also in that same plane, then the *cis* isomer of the alkene will be formed. This type of elimination is called *anti*-elimination. If the *d,l* isomeric pair of the same compound reacts in the same way, what will be the stereochemistry of the resultant alkene from each optical isomer?

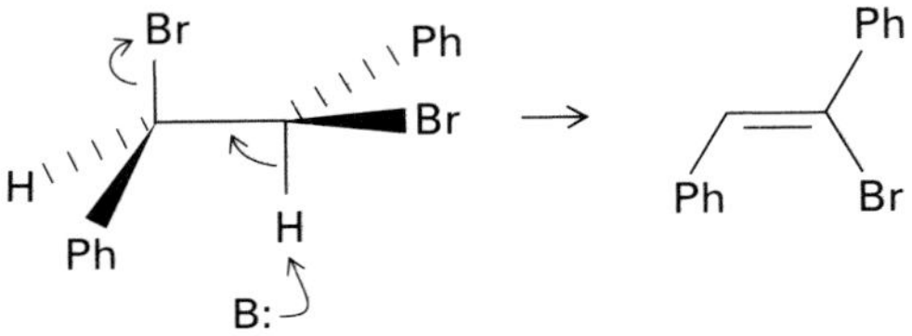

R,R   *trans*

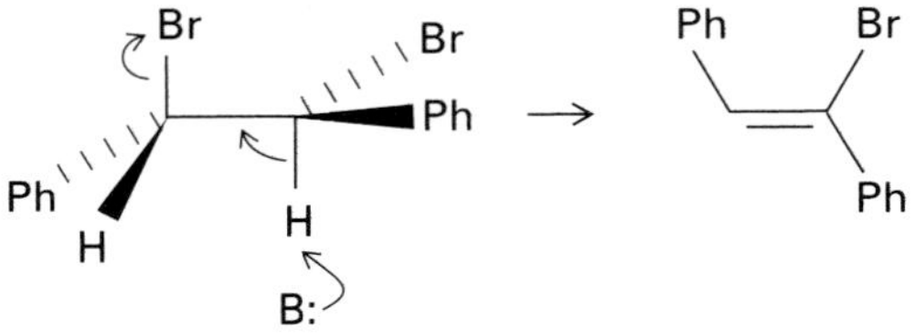

S,S   *trans*

In each case the *trans* isomer will result.

If instead of eliminating HBr from this compound, a bromine molecule is eliminated by the action of an iodide ion, what would be the resultant geometry of the alkene from the *meso* isomer and from the *d,l* isomeric pair?

meso   *trans*

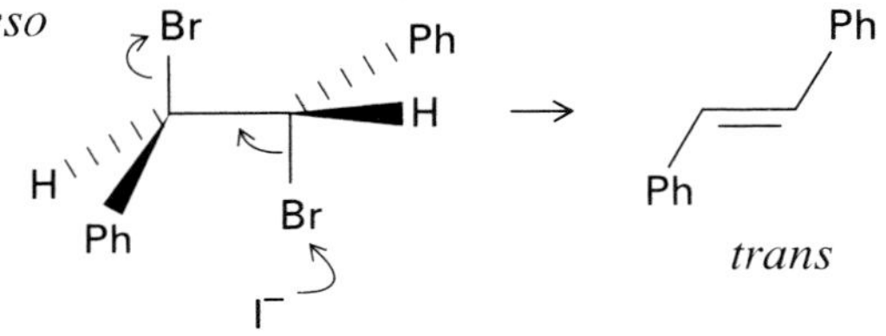

R,S

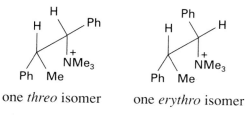

In this case the *meso* compound gives the *trans* isomer as a result of *anti*-elimination, while the *d,l* pair gives the *cis* isomer.

In general, *anti*-elimination will result in an *erythro d,l* pair giving the *cis* isomer, and the *threo d,l* pair results in the *trans* isomer. The *threo* isomer, however, tends to react significantly faster than the *erythro* isomer. Taking the elimination of $HNMe_3^+$ from (1,2-diphenylpropyl)trimethylammmonium ion, $PhMeCH\text{-}CHPhNMe_3^+$ by the ethoxide ion, as an example, suggest a reason why this is the case. Illustrate your answer using the sawhorse projection.

one *threo* isomer   one *erythro* isomer

In this case, the *threo* isomer may achieve the required *anti-periplanar* geometry with the phenyl groups staggered with respect to each other. In the case of the *erythro* isomer, the phenyl groups are *gauche*. Thus for elimination to occur in the erythro isomer, it must initially adopt a disfavorable conformer, while in the case of the *threo* isomer, elimination may occur from the most favorable conformer. This in effect, increases the concentration of the conformer which is capable of undergoing the desired reaction and so results in an increased rate of reaction. This is called the eclipsing effect, and has wide applicability.

So far we have only considered the sterochemistry of the transition state and how that relates to the stereochemistry of the product. This is most readily investigated when there is only one β-hydrogen, because this means that the resultant carbon/carbon double bond can only be formed in one place. If there is more than one β-hydrogen, then the double bond may be formed in different places depending on which hydrogen has been removed.

The first guideline in deciding where the double bond will end up in the product is Bredt's rule. This states that a double bond will not be formed at a bridgehead, unless the ring is large enough. In order to determine whether the ring is large enough we use the S number, that is the sum of the number of the atoms in the bridges that form the bicyclic system. What is the S number for the norbornyl system?

The norbornyl system is a [2.2.1] bicyclo system and so the S number is five. Bridgehead double bonds only form if the S number is seven or greater. There is an additional requirement, in that the double bond may only exist in a ring which contains eight atoms or more, *i.e.* a derivative of *trans*-cyclooctrene or higher homologue will be formed.

The next principle may be illustrated by investigating which isomer is formed on the elimination of 4-bromopent-2-one. Suggest what are the two possible alkene products, and then suggest which one will be formed in practice.

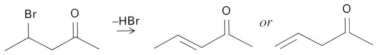

Elimination could occur to yield the unconjugated alkene. However, that is not observed in practice, instead elimination nearly always occurs so as to give the conjugated system. The side chain in which the elimination did not occur is distinguished by using a prime symbol, so in this case the C5-methyl group has the β'-hydrogens, while the C3-methylene group has the β-hydrogens, one of which is actually eliminated to give the conjugated product.

In *anti*-E2 reactions that occur on an aliphatic chain, where the nucleofuge is neutral, or if elimination occurs from a six membered ring, it is observed that the most highly substituted alkene is preferentially formed. This is called Zaitsev's rule and is generally true when the nucleofuge is a good leaving group, *i.e.* the C-X bond is weak and the resultant anion, X⁻, is very stable. When the leaving group is positively charged, then the opposite orientation is found, *i.e.* the double bond that is formed goes towards the least highly substituted carbons. This is called the Hofmann product, and is generally favored when the leaving group is poor and the C-X bond is strong.

An example of a reaction that shows this latter regiochemistry is the Hofmann exhaustive methylation, which is sometimes called the Hofmann degradation. In this reaction an amine is methylated with methyl iodide until the quaternary ammonium iodide is formed. This is then treated with silver oxide to convert the iodide to the hydroxide. Write out the final elimination step which occurs on heating.

The hydroxide ion removes the β-hydrogen in a normal *anti*-E2 mechanism. When this reaction is performed with a cyclohexyl derivative then Zaitsev's rule is followed instead.

There are many explanations that attempt to rationalize the difference between these two rules for the regiochemistry of the elimination; however, none of them are completely satisfactory. Some suggest that for positive nucleofuges, the E2 reactions, becomes paenecarbanionin in nature, *i.e.* more like the E1cB process (which is discussed in more detail below), and as such the cleavage of the C-H bond is more important, and so groups which stabilize the incipient carbanion, such as primary carbons, will favor this route. When the nucleofuge is neutral, the reaction becomes more paenecarbonium like, and so the cleavage of the C-X bond becomes more important which means that groups that result in a more stable alkene will favor this route, *e.g.* a highly substituted carbon.

Another explanation centres on the steric interactions, rather than the electrostatic ones, and suggests that an reason similar to the eclipsing effect which was mentioned earlier. It also has been pointed out earlier, that very sterically demanding bases will not always remove the most acidic proton, but are sometimes limited to those which are more accessible, *i.e.* on the terminal methyl groups. The removal of such a proton would normally result in Hofmann regiochemistry. Hofmann regiochemistry is favored as the size of the leaving group increases. However, Zaitsev elimination is favored as the number of β-alkyl substituents increases, but if these β-substituents are capable of stabilizing a negative charge, then Hofmann elimination once again becomes the favored option.

For a substrate that is undergoing a bimolecular reaction, as we have noted earlier, there is a competition between the substitution and elimination pathways. As the amount of branching increases in the carbon chain, would this favor or disfavor the elimination pathway over the corresponding substitution pathway?

For bimolecular reactions, increased carbon branching favors the elimination pathway, to the extent that in highly branched substrates there may be very little substitution reaction that occurs.

This may be illustrated by re-examining a reaction that we have already studied. Earlier we saw that in the elimination of a bromine molecule on the treatment of iodine on a 1,2-bibromide, the reaction followed a normal *anti*-E2 pathway which means that the *threo* isomer yields the *trans* isomer. If this elimination reaction is performed when either, or both, of the bromine atoms is attached to a primary carbon, instead of a secondary or tertiary carbon as was the case when we first looked at this reaction, then a different stereo-chemistry results. When this experiment is performed with deuterated deriva-tives, the *threo* isomer gives the *cis* isomer. Suggest an explanation for this result.

Br ⤴ Br → Br ⤴ H D → D D

H ⟍ ⟍ 'H H ⟍

D D D

meso I⁻ I⁻ *cis*

Contrary to first appearances the elimination mechanism has not changed; instead there is now an initial substitution reaction by the iodide anion of a bromide anion, with a concomitant inversion of configuration at that centre, that is then followed by an elimination reaction with the normal *anti* stereo-chemistry. This illustrates the point that substitution reactions occur more easily at a primary carbon than at a secondary or tertiary carbon, at which sites elimination reactions are preferred. Suggest why this may be the case.

There are two reasons. The first is that for a tertiary carbon centre there is a large release of crowding strain on elimination, while for a primary carbon centre there is little such effect. Secondly, the resultant alkene is increasingly more stable as the number of substituents increases on the carbon/carbon double bond.

This example also illustrates one of the pitfalls of attempting to elucidate a mechanistic pathway. The use of alkyl groups, such as methyl and ethyl, to label certain carbon atoms changes their character from primary to second-ary, and this may have an effect on the mechanism that is being studied. The use of deuterium labels may be preferred, but requires the initial stereo-selective synthesis of such compounds, and this may be difficult. Further-more, it must be possible to distinguish between the various isotopic products, otherwise no information will be gained.

We have already briefly mentioned that elimination is favored by a reagent which is hard in nature, while substitution is favored by one which is soft. The concentration of the base also has an effect on the relative amounts of substitution to elimination reaction that occurs. For the bimolecular pathway, suggest what would be the effect of increasing the concentration of the external base.

Increasing the concentration of the base favors the elimination over the substitution pathway. The size of the attacking reagent will also affect this distribution of reaction types: the larger the reagent, the harder it will for it to approach the carbon centre to perform a substitution reaction. However, it will still normally be able to remove the β-hydrogen in order to perform the E2 elimination.

Furthermore, it is found that decreasing the polarity of the solvent increases the proportion of elimination to substitution. Thus alcoholic KOH is used for elimination reactions, while aqueous KOH is used to effect the substitution of a functional group.

An example of this concerns the elimination reaction in which a *geminal* dihalide, R_2CH_2-$CHCl_2$, is treated with alcoholic KOH. Suggest what may be the product from this reaction.

$$R_2CH_2\text{-}CHCl_2 + KOH = R_2CH=CHCl + H_2O + KCl$$
$$R_2CH=CHCl + KOH = R_2C\equiv CH + H_2O + KCl$$

After the first elimination of HCl, the reaction may be repeated on the alkene to give the alkyne. If instead of a *geminal* dihalide, a *vicinal* dihalide is used, the same alkyne may be produced. There is also a different elimination reaction which may compete with the normal second elimination step. Suggest what the alternative side product may be which results from the double elimination of HCl from the general dihalide, R_2CH_2CHCl-CH_2Cl.

$$R_2CH_2CHCl\text{-}CH_2Cl + KOH = R_2CH_2CHCl=CH_2 + H_2O + KCl$$
$$R_2CH_2CHCl=CH_2 + KOH = R_2CH=C=CH_2 + H_2O + KCl$$

The second elimination may occur in the opposite direction and so give rise to an allene.

A change of solvent may have a great effect on the strength of the base. In particular, if the solvent is changed from a polar, hydroxylic to one which is aprotic, bipolar such as DMF or DMSO, then the base strength increases. Suggest why this is the case.

In the latter type of solvent there will be no hydrogen bonds, and so there is no envelope of solvent molecules hindering the attack of the base on the proton. This also has the effect of increasing the charge density on the base, as it is not distributed over a solvent shell. This in turn increases the hardness of the base, and so favors elimination over substitution. Furthermore, in a solvent that has a low polarity, the stability of any charged intermediate would be reduced, and so bimolecular over unimolecular reactions would be favored, and this in turn favors elimination.

What would be the effect upon the relative ratios of substitution to elimination, of raising the temperature at which the reaction is performed?

In a substitution reaction, the number and nature of the products is the same as the starting materials, and so there is little change in entropy, and thus a change in temperature has little effect. For elimination reactions, the entropy is increased as there are more products than there are starting materials, and so raising the temperature favors these products, and so favors the elimination pathway.

We will now look at the special factors that are involved in the bimolecular *syn*-elimination reaction.

14.2.2 Syn-Elimination

In the previous sub-section, we saw that *anti*-elimination requires that the two groups that are to be eliminated are *anti-periplanar* to each other. If there is a deviation from this angle then a different mechanism is observed.

If *exo*-2-bromonorbornane, which is stereospecifically deuterated in the *exo*-α-position to the halide, is treated with a base, then an elimination reaction occurs. However, in the carbon product there is no deuterium, *i.e.* deuterium bromide has been eliminated. Account for this observation.

In this case *syn*-elimination must have occurred in order for the deuterium to be lost from the norbornyl skeleton. The hydrogen which was in the *endo*-α-position has a dihedral angle of 120° to the halide, and as this atom was not eliminated with the halide in the normal *anti*-elimination mechanism, it appears that for the E2 *anti*-elimination mechanism to operate the dihedral angle between the groups must be very close to 180°. It is generally found that *syn*-elimination is slower than *anti*-elimination and so it is usually only observed in significant amounts when the *anti*-elimination is so disfavored, *e.g.* by the impossibility of achieving a dihedral angle of 180°, that the *syn*-elimination reaction may compete successfully.

There are some exceptions to this general observation. For example, *syn*-eliminations reactions have been found to be promoted by weakly ionizing solvents, but only when the leaving group is neutral. For the elimination reaction of HX, from the general substrate $R_2CH\text{-}CXR_2$, where X is neutral, *e.g.* a bromide group, using potassium *t*-butoxide as the base, suggest a reason why *syn*-elimination would be favored when the polarity of the solvent is low, *e.g.* benzene.

When the solvent polarity is low, then an increased amount of ion pairing occurs, and as such the cyclic transition state may easily be formed, and this results in *syn*-elimination.

If the leaving group is positive, though, the opposite is found to be true, *i.e.* solvents of low polarity now favor *anti*-elimination. Suggest why this is the case.

The explanation is the same, in that a solvent of low polarity still favors ion pairing. However, in this case the base forms a cyclic transition state with the substrate directly, without the involvement of its corresponding cation, *i.e.* the transition state is a five and not a six membered ring, and so the base is now in an optimal position to remove the *syn*-hydrogen.

If it is possible for the regiochemistry of the double bond to be in doubt, in the *syn*-E2 reaction, the elimination nearly always follows the Hofmann regiochemistry. This is in contrast to the situation of the *anti*-E2 reaction in which either the Zaitsev or the Hofmann elimination could occur, depending on the circumstances.

Now that we have studied the bimolecular elimination reaction in some depth we will turn our attention to the unimolecular version.

14.3 Unimolecular 1,2-Elimination Reactions

In the previous section we looked at the 1,2-bimolecular elimination reaction. We will now look at the unimolecular version of that reaction, that is given the label E1. This reaction bears many similarities to the S1 reaction that we studied in an earlier chapter.

Remind yourself of the mechanism for the S_N1 substitution reaction.

$$CR_3X = CR_3^+ + X^-$$
$$CR_3^+ + Y^- = CR_3Y$$

This reaction occurs in two steps, the first of which is the rate limiting loss of the leaving group to give the carbonium ion. The second step involves the attack by the incoming nucleophile to result in overall substitution. The first step is unimolecular, and so the concentration of the other reagent is not involved in the rate equation.

If, instead of the addition of an incoming nucleophile in the second step, the β-hydrogen is lost, then the overall result would be a unimolecular 1,2-elimination reaction. Write the first ionisation step, and then use the elimination step in the second stage.

This is the mechanism for the E1 elimination reaction. The similarities with the SN1 reaction are plain. Like the E2 and SN2 reactions that often compete with each other, the E1 and SN1 reactions also often compete with each other.

The first step is exactly the same in the SN1 and E1 mechanisms, and thus those considerations that apply to the first step of the SN1 mechanism apply equally to the first step of the E1 mechanism. These factors include the stability of the carbonium ion intermediate, the loss of stereochemistry at that centre if no other factors intervene, and the possible rearrangement of the carbonium ion intermediate.

In the normal E1 mechanism, the heteroatom leaves first to form the carbonium ion intermediate. However, it is possible for the hydrogen to leave first. This would result in a carbanion being formed. Write this reaction step, and then deduce what the second step must be in order to result in an overall elimination.

This reaction type is called the E1cB mechanism, which stands for unimolecular elimination conjugate base reaction, because the conjugate base of the starting material is being formed as the reactive intermediate. It is sometimes called the carbanion mechanism. As this mechanism results from the removal of a hydrogen cation, it is not surprising that it is favored by those substrates which possess an acidic hydrogen. Thus would you expect the E1cB mechanism to be more or less prevalent in reactions that result in a carbon/carbon double or a carbon/carbon triple bond?

To form a carbon/carbon triple bond, the proton must be removed from a sp^2 hybridized carbon, rather than a sp^3 hybridized carbon, and as the former is more acidic than the latter, one would expect the E1cB mechanism to be preferred in the former case.

The intermediate of the E1cB mechanism is a carbanion, and thus any factors that stabilize such an ion should favor this mechanism. We have already noted above that formally elimination reactions are the reverse of addition reactions. However, we also noted that the mechanisms involved in elimination reactions were similar to substitution reactions than addition reactions. This is because, normally elimination reactions proceed via a carbonium ion or a single step similar to the manner in which substitution reactions occur. However, there are addition reactions that proceed via a carbanion intermediate, for example the Michael-type reaction, in which a carbanion adds to an α,β-unsaturated carbonyl compound. Write the Michael-type addition between the anion formed from the diester of propandioic acid and a general α,β-unsaturated carbonyl compound.

Now write the reverse of this mechanism.

This pathway follows the E1cB route. Furthermore, it is an example of the microreversibility of a reaction, *i.e.* that a reaction may proceed in both directions, and along exactly the same pathway at every step of the way.

Another factor that favors the E1cB mechanism is a poor nucleofuge. If the nucleofuge left easily then, either a normal E1 mechanism would occur, or it would leave at the same time as the β-hydrogen was being removed by the base.

The E1cB mechanism is rare in practice when the elimination reaction would result in a carbon/carbon double bond. When a carbon/oxygen double bond is to be formed then it is far more common. For example, the E1cB mechanism is found in the reverse of the cyanohydrin formation reaction. You will recall that the forward reaction involves the addition of a cyanide ion to a carbonyl group. Write the pathway for the reverse reaction, *i.e.* the elimination reaction.

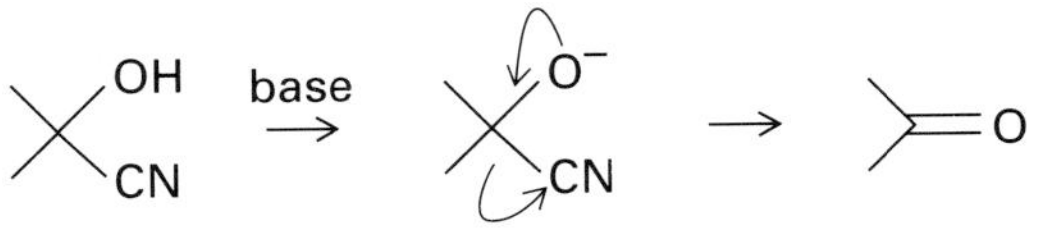

The starting molecule has an acidic proton; an α group capable of stabilizing a carbanion; and, a nucleofuge that is strongly bonded to its carbon atom.

It is clear from this discussion that the E1, E2 and E1cB mechanisms are all closely related. In all of these mechanisms, one groups leaves with its electron pair (the nucleofuge) and the other group without its electron pair, and that is often a proton. The difference arises in the exact timing of the departure of these two groups. In the first mechanism, *i.e.* E1, the nucleofuge leaves first; while in the last, *i.e.* E1cB, the proton leaves first, and thereafter the nucleofuge. In the E2 mechanism, the two groups depart simultaneously, *i.e.* it falls half way between the two other extremes. As we have already seen in the case of nucleophilic substitution reactions, it is very rare that a reaction has only the characteristics of the idealized extreme, instead it is far commoner for any given reaction to exhibit intermediate characteristics. This is also the case for the E2 reaction, which rarely falls equally between the two extremes cases, instead is may resemble the E1 reaction more closely, or the E1cB reaction more closely. In the former case, it is said to be paenecarbonium, while in the later paenecarbanion. It is possible to distinguish between these various possibilities by studying the effect on the rate of the elimination reaction by isotopic substitution. Either the hydrogen may be replaced by a deuterium atom, or else the atom within the nucleofuge that is bonded to the carbon may be substituted for a different isotope from its normal one.

An example of an E2 mechanism which is paenecarbanionic is the base elimination of water from 1,3-hydroxylcarbonyl. Normally water is eliminated under acidic conditions in which the hydroxyl group is protonated to from the more favorable leaving group, $\leftarrow OH_2^+$. When the first stage of the aldol reaction is performed between two molecules of ethanal, the initial product is aldol itself, which contains a 1,3-hydroxyl carbonyl arrangement. Under the normal reaction conditions, elimination of water occurs to yield the conjugated unsaturated carbonyl compound, crotonaldehyde. Suggest what is the mechanism for this reaction.

In this case the -M aldehyde group makes the hydrogen more acidic, and also helps stabilize the incipient carbanion in the transition state. -M groups generally increase the rate of elimination, by the E1cB or E2 mechanism, when adjacent to the hydrogen atom, because even though they may conjugate with the double bond that is being formed when either adjacent to the nucleofuge or the hydrogen atom, they may only increase the acidity of the hydrogen and help stabilize that carbanion when adjacent to that hydrogen atom. All -M groups, like phenyl or carbonyl, when adjacent to the hydrogen atom, or even -I groups, like the halogens, favor the E1cB route; while +M group, or even +I groups like alkyl groups, disfavor this pathway.

In E1cB reactions, there is usually little ambiguity as to the regiochemistry of the resultant compound, because usually there is only one hydrogen that is sufficiently acidic to be removed.

In the case of the E1 mechanism where there is doubt over the regiochemistry of the resultant double bond, the orientation of the double bond in the final product may be deduced from the relative stabilities of the final unsaturated product. From heat of combustion data, it is found that the more highly substituted the double bond is, then the greater the stability of that alkene. In the case of 2-bromobutane, suggest what will be the resultant alkene after the elimination of HBr.

But-2-ene will result to the exclusion of but-1-ene, as the secondary carbonium ion is significantly more stable than the primary carbonium ion, *i.e.* the thermodynamic product. Thus in E1 reactions, Zaitsev's elimination is preferred. However, it does not apply if the predicted product would suffer steric strain.

What would be the effect of an alkyl or an aryl group at the α-carbon on the E1 pathway?

In the case of the E1 pathway a carbonium ion intermediate is being formed, and this may be stabilized by both an aryl or an alkyl group, and so both of these substituents will favor the E1 pathway when attached to the α-carbon.

What would be the effect on the relative rates of elimination to substitution of increasing the amount of branching in the carbon chain of the substrate that is reacting in a unimolecular manner?

Generally, the more branching in the carbon chain, the more elimination occurs.

In unimolecular reactions, the base that removes the proton in the second stage of the reaction is usually the solvent, and not a separate component of the reaction mixture. Thus suggest what would be the effect of adding an external base to the reaction mixture.

The effect would be to move the reaction towards a bimolecular pathway as the concentration and strength of the base increased. At low concentration of a weak base, the unimolecular pathway is favored, so long as the solvent is ionizing. Under these conditions substitution is favored over elimination as well.

The nature of the leaving group has quite a large effect on the particular pathway that is followed. Suggest whether a very good leaving group would favor the E1 pathway or the E2 or the E1cB pathways.

The better the leaving group, the more the E1 pathway is favored. Synthetically, the most important leaving group is the $\leftarrow OH_2^+$ group. If the nucleofuge is $\leftarrow Cl$, $\leftarrow Br$, $\leftarrow I$ or $\leftarrow NR_3^+$, then reaction tends to proceed via the E2 route. The E1cB pathway is favored by poor leaving groups.

The E1, E1cB and E2 elimination pathways are all 1,2-elimination reactions. These account for the vast majority of elimination reactions that occur in solution. We will now briefly examine some other pathways that occur under particular conditions.

14.4 1,1-Elimination Reactions

If 1-chloro-1,1-dideuteriobutane is treated with a very strong base such as phenyl sodium, then even though some of the resultant alkene contains two deuteriums on the terminal carbon atom, a higher percentage of the alk-1-ene contains only one deuterium atom. This deuterium atom is located on the terminal carbon atom. The expected product, with two deuterium atoms, which is formed in smaller amounts, results from the normal E2 mechanism. Suggest a mechanism that accounts for the formation of the major product.

Under strongly basic conditions, we have seen that the E1cB mechanism may operate. In that case the base removes the β-hydrogen. In this case, the phenyl sodium removes the deuterium α to the chlorine. This deuterium is the most acidic. The resultant anion may then loss the chlorine anion to give a carbene that then inserts its lone pair into the β-carbon C-H bond. This, in effect, results in an 1,2 hydride shift to give the mono-deuterated alkene. This type of reaction is called the 1,1 or α-elimination, and is less common than the 1,2-elimination.

So far we have mainly studied elimination mechanisms that occur in solution. They are characterized in that they tend to involve ionic intermediates, like the E1 and E1cB mechanisms, or at least require an ionic reagent like the E2 mechanism. There is another class of elimination reactions that occur when compounds are heated. These are called pyrolytic elimination reactions, and usually proceed via a cyclic transition state or by a pathway that involves radicals.

14.5 Pyrolytic Elimination Reactions

This type of elimination reaction often occurs in the gas phase, without the addition of another reagent. This point distinguishes this type of reaction from those reactions that we have already studied, because all of the other types required a base, be it an additional external base, or the by action of the solvent.

If carboxylate esters are heated, they readily form an alkene and a carboxylic acid. Suggest a possible mechanism for this reaction, illustrating your answer with the general ester, $RCH_2\text{-}CH_2\text{-}OCOR$.

In this case there is a six membered cyclic transition state, in which the acid and alkene are formed. There is a requirement for a *cis*-β-hydrogen in order for this reaction pathway to proceed, if the reaction involves the loss of a molecule of the type HX.

We saw earlier that for the E2 *anti*-elimination, an *erythro* isomer would result in the *cis* alkene isomer. For the *erythro*-1-acetoxy-2-deutero-1,2-diphenylethane, suggest which isomer of the alkene will be formed on pyrolysis.

The *trans* isomer of the alkene stilbene is formed, as indicated by the presence of the deuterium in the product. In this case elimination is via a cyclic transition state and so the opposite stereochemical is obtained other than that achieved in the E2 mechanism. This cyclic pyrolytic elimination is labelled the Ei mechanism, which stands for intramolecular, or internal, elimination.

In those versions of the Ei mechanism that involve a four or five membered cyclic transition state, then there is a requirement that all the atoms are co-planar. However, this restriction does not apply for the six membered cyclic transition state.

Methyl xanthates, $R_2C-O-CSSMe$, may be easily synthesized from the corresponding alcohol by treatment with NaOH and CS_2, followed by MeI. These compounds readily undergo pyrolytic elimination to give an alkene, and initially $MeSC=OSH$ which fragments to MeSH and COS. This is called the Chugaev reaction. The xanthate group is quite large, and so when it is present in a six membered ring, it adopts an equatorial, rather than an axial, position. Bearing this in mind, suggest what will be the products from the *cis*-β-methyl derivative of the methyl xanthate cyclohexane.

On the unsubstituted side of the cyclohexane ring, there is both a *cis* and *trans* hydrogen with which the xanthate may form the cyclic transition state for the Ei mechanism. However, on the methyl substituted side, there is only a trans hydrogen, and yet the xanthate is still able to form a cyclic transition state. This latter transition state resembles the *trans* decalin molecule. So six membered cyclic transition states may be formed with both *cis* and *trans* β-hydrogens, because of the capability of a six membered ring to form an unstrained puckered ring of either configuration.

When we looked at *anti*-E2 bimolecular eliminations, we saw that the Hofmann exhaustive methylation followed that pathway. There are occasions when it proceeds via different route. Normally, the hydroxide attacks the β-hydrogen in order to remove it and initiate the E2 mechanism. If, however, the quaternary ammonium derivative is highly hindered so that this is not possible, then the hydroxide anion removes a proton on the ammonium salt. Suggest how this reaction proceeds and then write the whole reaction pathway.

This is an example of a five membered ring version of the Ei mechanism. Also it is an example of a synthetic reaction that proceeds via two different pathways depending upon the exact conditions. These pyrolytic elimination reactions are valuable synthetically, because there is normally no opportunity for the starting material to rearrange, which is always a possibility when the elimination occurs via the E1 route, because the carbonium ion is an intermediate.

The other type of pyrolytic elimination reaction involves free radicals. The steps are similar to those that we studied for free radical substitution reactions, *i.e.* there is an initiation step, that is followed by several propagation steps, and then there are some termination steps. Free radical elimination is found in polyhalides and primary monohalides. For the general primary monohalide, R_2CHCH_2X, suggest what will be the first step.

$$R_2CHCH_2X = R_2CHCH_2\bullet + X\bullet$$

The initiation step is the thermally induced homolytic cleavage of the C-X bond to give two free radicals.

Suggest a couple of possible propagation steps.

The two that have been chosen here, represent a typical hydrogen abstraction that gives rise to the alkyl radical, and a fragmentation reaction of that radical to regenerate more halide radicals and the desired final product.

Suggest a termination reaction that also produces the desired product.

This disproportionation reaction produces more of the starting material and also some of the final product.

Free radical elimination reactions in solution are rather rare, instead they tend to be confined to the pyrolysis of polyhalides as mentioned originally.

We will now briefly examine extrusion reactions. An example is the formation of a cyclopropane from 1-pyrazoline initiated either by heat or light. Suggest a mechanism for this reaction.

The cleavage of the weak C-N bond and the formation of the nitrogen/ nitrogen triple bond coupled with the great increase in entropy help drive this reaction. The initial diradical that is formed by the loss of the nitrogen atom, ring closes to give the cyclic product.

The great driving force, both enthalpically and entropically, that is provided by the loss of a nitrogen atom, may be used to produce some strained compounds that would be difficult to make in any other way. For example 3H-pyrazole on photolysis yields cyclopropene. This reaction is thought to proceed via a diazo intermediate, that may then form a carbene. Suggest what is the mechanistic pathway for this extrusion reaction.

In this case the 3H-pyrazole is stable to heat, and the reaction only proceeded on irradiation with light. Light may be used to initiate another extrusion reaction in which the carbonyl group is lost from a chain. Suggest the route for the conversion of R-CO-R to R-R and C=O.

In this case the C-C bonds on each side of the carbonyl group cleave, in what is called a Norrish type I reaction, and the resultant radicals then rejoin and so effect the extrusion of the CO molecule.

That ends this chapter on elimination reactions. We will now look at sequential reactions, in which an addition followed by an elimination, or *vice versa*, leads to an overall substitution reaction.

15

Sequential Addition/
Elimination Reactions

15.1 Introduction

We have now dealt with the three principal reaction mechanisms that occur in organic chemistry, *i.e.* substitution, addition and elimination reactions. The remaining chapters in this part of the book will deal with some of the other types of reactions that may be encountered. In this chapter we will look at the sequential addition/elimination reactions that result in an overall substitution occurring on the original substrate. The division between those simple substitution reactions that we have already studied and the sequential variety is sometimes rather fine. For example in the S_N1 unimolecular substitution reaction, there is an initial loss of the leaving group to form the carbonium ion intermediate. This intermediate is then attacked by the incoming group, resulting in the final substituted product. This is normally considered under simple substitution reactions, because the intermediate is charged and only exists for a short time. Many of the reactions in this chapter have intermediates that are neutral and also have a full compliment of electrons, *i.e.* they are not radical or carbene intermediates. Some of these intermediates may be isolated, however, many are very reactive and so may only be produced *in situ*.

It is apparent that a addition followed by an elimination reaction which will result in an overall substitution occurring. However, the order of the reactions may be reversed, *i.e.* an elimination reaction followed by an addition reaction, will have the same result. Examples of both types of sequences exist.

15.2 Addition/Elimination Reactions

In an earlier chapter we dealt with S_N2 reactions in aliphatic compounds. In that chapter you will recall that the normal nucleophilic pathway could not operate at carbon centres that were unsaturated, because of the impossibility of the inversion of the configuration of the carbon involved. This does

249

not mean that nucleophilic substitution at a vinyl carbon is impossible, just that if it is to occur there must be a different pathway by which it may be achieved.

The first combination mechanism that we will study is the addition-elimination mechanism that allows for the substitution at a vinyl carbon. When 1,1-dichlorethene is treated with ArS⁻, in the presence of a catalytic amount of ethoxide anion, the product is the 1,2-dithiophenoxy derivative. Suggest a route for this reaction.

After the first addition, which yields the saturated compound, two elimination reactions occur to give the alkyne derivative. This then undergoes a second addition to result in the rearranged substitution product.

We will now turn our attention to SN2 reactions in aromatic compounds. It is immediately apparent that the mechanistic pathway must be different from that observed in aliphatic compounds, because the incoming nucleophile cannot approach the leaving group from the back side. Furthermore, the approach of the nucleophile is inhibited by the delocalized electrons that circulate above and below the aromatic ring. It is for this latter reason that electrophilic attack at an aromatic ring is usually the favored pathway. However, nucleophilic attack on an aromatic ring does occur under some conditions. Suggest what may be the intermediate when the methoxy nucleophile MeO⁻, attacks the ethyl ether derivative of 2,4,6-trinitrophenol.

The incoming nucleophile can only approach from the side. The extra negative charge is delocalized into the aromatic ring with the help of the strongly -M nitro groups. This reaction pathway is labelled the SN2(aromatic) mechanism. The intermediate that is formed is called a Meisenheimer complex, which may be isolated as a red crystalline solid with the counterion of the alkoxide. On the addition of some protic acid, the complex may fragment with the elimination of either the ethoxy or methoxy anions. If a methoxy anion is eliminated, then the starting material is formed; if the ethoxy anion is eliminated this would result in an overall substitution of an ethoxy group for methoxy group. So overall a substitution at the *ipso* position in the aromatic ring may been achieved.

A further example is the attack of an amine on a 2,4-dinitrohalobenzene. For this reaction suggest whether the fluorine derivative would react more or less quickly than the bromine derivative and why.

The fluorine derivative reacts about a 1,000 times faster than the bromine derivative, which is the opposite to that which would occur in a normal SN2 reaction. This is because the highly electronegative fluorine significantly increases the positive character of the carbon to which it is attached, and this increases the Coulombic attraction for the incoming nucleophile. Furthermore, the fluorine helps in stabilizing the anionic intermediate. 2,4-Dinitrofluorobenzene is used to label the amino terminal of polypeptides, because this reaction is so facile. Furthermore, once it has reacted, the product is very resistant to subsequent hydrolysis.

We have already seen in a previous chapter that simple carbon/carbon double bonds do not readily undergo the simple addition of a nucleophile so as to form a tetrahedral intermediate. This is because in such a case the negative charge would have to be carried by a carbon atom. If there are sufficient electron withdrawing groups then it is possible to form this intermediate, which could then eliminate another group so as to reform the carbon/carbon double bond and result in an overall substitution reaction. However, this is not observed in practice.

Carbon/oxygen double bond compounds, *i.e.* carbonyl compounds, often undergo sequential addition elimination reactions. They proceed via the tetrahedral mechanism and in this case the negative charge may be carried by the oxygen atom. The tetrahedral intermediates may be isolated when a suitable substrate is used. For example, suggest what is the initial stable product that results from the addition of hydroxylamine, NH_2OH, to a general carbonyl compound.

This intermediate is similar in nature to the cyanohydrin product that is formed by the addition of a cyanide anion to a carbonyl compound. However, in the latter case, no further reaction may occur, yet in this case there is the potential for further reaction. Suggest what are the next steps, and what is the final product.

$$
\underset{\substack{\\ \text{NHOH}}}{\overset{\text{OH}}{\diagup}} \xrightarrow{+H^+} \underset{\substack{\\ \text{NHOH}}}{\overset{\overset{+}{O}H_2}{\diagup}} \longrightarrow \overset{+}{=}NHOH \xrightarrow{-H^+} =N\diagdown_{OH}
$$

In the final product the carbonyl oxygen has been substituted for the =NOH group. This is called an oxime. This type of tetrahedral substitution reaction may be performed with many nitrogen derivatives, such as hydrazine, NH_2NH_2. Write down the product that is formed from the reaction of a carbonyl compound with hydrazine.

$$
=O \;+\; HN_2-NH_2 \;\longrightarrow\; =N\diagdown_{NH_2}
$$

This type of compound is called a hydrazone, and is the first intermediate in a very important reaction called the Wolff-Kishner reduction. This reaction, along with the Clemmensen reduction, is one of the principal methods for reducing a carbonyl compound to the dihydro-derivative, *i.e.* $R_2C=O$ to R_2CH_2. In the Wolff-Kishner method, the carbonyl is heated with hydrazine hydrate and a base, which is usually sodium or potassium hydroxide. After the initial formation of the hydrazone suggest what are the subsequent steps in this reduction.

The loss of a nitrogen molecule helps drive this reaction. This reaction is usually performed in refluxing diethylene glycol, and is then called the Huang-Minlon modification.

Another important class of derivatives are formed from ammonia, NH_3, and primary amines, NH_2R. Write down the product that results from the reaction with a primary amine.

$$
=O \;+\; RNH_2 \;\longrightarrow\; =N\diagdown_R
$$

This is called an imine, and usually they are not very stable, However, if one of the substituents on the carbon is an aryl group then they become quite stable and are known as Schiff bases. For all of these derivatives, the original nitrogen compound possessed at least two hydrogen atoms, so that the nitrogen could end up with a double bond to what was the carbonyl carbon. If a secondary amine, NHR_2, is used instead, suggest what will be the product.

$$\text{(CH}_3)_2\text{C=O} \quad R_2NH \quad \longrightarrow \quad (CH_3)_2C(OH)(NR_2)$$

In this case, the intermediate is incapable of losing a water molecule as occurred in the previous cases which would allow the formation of the carbon/nitrogen double bond. However, it is capable of losing water in a different manner. Suggest how this may occur and hence suggest what will be the final product for this combination of reagents.

$$(CH_3)_2C(OH)(NR_2) \quad \xrightarrow{-H_2O} \quad \text{enamine}-NR_2$$

This compound has some similarities to the enol form of a carbonyl compound, and it is called an enamine. As this reaction is performed under reversible conditions the thermodynamic isomer is always formed, *i.e.* the double bond is formed on the most highly substituted side. These compounds are of great synthetic importance. Furthermore, their use compliments the use of the anions formed from carbonyl compounds. This is because enols are produced under strongly basic, irreversible, conditions, and so the proton that is normally removed is the kinetically favored one. This hydrogen atom may be different from the one that is eliminated under thermodynamic conditions in the formation of the carbon/carbon double bond in the enamine.

So far we have looked at the substitution of the carbonyl oxygen in these addition/elimination sequential reactions at a carbonyl centre. There are many reactions in which after the first addition, it is not the carbonyl oxygen that is eliminated as a water molecule, but instead one of the other groups that was joined to the carbonyl carbon that leaves. We have already looked at some of these reactions in the nucleophilic substitution chapter under ester hydrolysis. In that case the alkoxide group, $\leftarrow OR$, was replaced by the hydroxyl group, $\leftarrow OH$, so as to form the carboxylic acid.

Some of the mechanisms for ester hydrolysis were simple substitution reactions. However, others proceed via a different route which is also displayed in the hydrolysis of an acid chloride, $RC=OCl$, when it is treated with water. Suggest what the product will be, and the mechanism whereby this is reached.

In the first step the water adds to the carbonyl group. Instead of this intermediate losing the carbonyl oxygen, the chloride anion is a better leaving group and so departs. This results in the re-formation of the carbon/oxygen double bond. The result is the overall substitution of one of the groups which was originally attached to the carbonyl carbon. This type of substitution reaction is common so long as the bond to the carbonyl carbon is not from another carbon or hydrogen, *i.e.* not a ketone or aldehyde. This is the general mechanism whereby the various carboxylic acid derivatives such as acid chlorides, acid anhydrides and amides, are interconverted.

Attack does not always occur at the carbon centre. It is possible for the attack to occur at the conjugated position. You will recall that this was also possible for nucleophilic substitution. For example suggest what is the intermediate when tetrachloroquinone is treated with hydroxide anions.

Carbonyl compounds are capable of undergoing another completely different sort of reaction, in which the carbonyl oxygen is replaced, not by a heteroatom as we saw above, but by a doubly bonded carbon group. When triphenylphosphorus is treated with a secondary aliphatic halide, a normal substitution reaction occurs, resulting in the replacement of the halide by phosphorus. This phosphonium salt adduct may then be treated with a very strong base, such as phenyllithium, to produce an ylid. An ylid is a compound in which there is a positive charge adjacent to a negative charge. Write down this reaction sequence.

The ylid may then be used to attack a carbonyl compound. Suggest how this occurs and the nature of the intermediate.

The zwitterionic intermediate, which is called a betaine, then forms a four membered cyclic intermediate. This is possible because the phosphorus is a third row element and thus has available d sub-orbitals, and so is not so limited by the octet rule as is the carbon atom. This intermediate, which is called an oxaphosphetane, then fragments to form the desired compound. Write down the mechanism for this step.

The complete synthetic pathway is called the Wittig reaction, and it is a very useful synthetic tool, because the position of the carbon/carbon double bond is known. This method may also be used to produce a carbon/carbon double bond in a position which would be difficult by any other means, *e.g.* *exo*-cyclic bonds, or 1,4 dienes.

We will now look at a few sequential reactions which occur in the opposite order, *i.e.* elimination and then addition.

15.3 Elimination/Addition Reactions

The first reaction we will study in this section is similar to one which we looked at in the previous section. We saw earlier a reaction which involved the substitution of 1,1-dichloroethene by the ArS⁻ anion, with a catalytic amount of ethoxide anion. If the concentration of the ethoxide anion is increased, and the substrate altered to the *cis*-1,2-dichloroethene, then the first reaction is an elimination step, instead of an addition step. Suggest what the subsequent steps may be and the geometry of the final product.

Each elimination is followed by an addition reaction to result in the overall production of the *cis*-1,2-dithiophenyloxy derivative. There is overall retention of the configuration, because both the elimination and the addition reactions are *anti* in nature. This reaction pathway is called the elimination-addition mechanism.

In aromatic systems, a rather similar sequence is found under certain extreme conditions. Thus chlorobenzene may be converted to aminobenzene when treated with sodium amide in liquid ammonia. If this reaction is performed upon *p*-chloromethylbenzene, then two compounds are produced. Both the *meta* and the expected *para* amino derivatives may be isolated. The substitution in the unexpected position is called *cine* substitution. Suggest a mechanism whereby this result may be explained.

The NH_2^- anion is a very strong base and removes a proton from the aromatic ring. This anion then eliminates a chloride anion, to leave an aromatic species which also contains a triple bond. This is called an aryne intermediate. This may be attacked by another NH_2^- anion in either the *meta* or *para* positions to give, after protonation, the two products. This pathway is called the benzyne mechanism.

Benzyne intermediates may be formed in a large number of different ways. If *o*-aminobenzoic acid is treated with nitrous acid, the resultant diazonium salt may decompose to form benzyne itself. Write out this reaction sequence.

Here two gaseous products are formed, which help drive the reaction.

Another method of producing benzyne is from the decomposition of 1-aminobenzotriazole, after it has been oxidized with lead tetraacetate. The oxidation converts the amino group to the nitrene intermediate, and this fragments. Suggest how this occurs.

In this case two molecules of nitrogen are formed. In all cases the benzyne goes on to perform an addition reaction to complete the remainder of the elimination/addition sequence, which thus results in an overall substitution occurring.

That concludes this section, as well as this chapter, on sequential reactions that result in overall substitution. We will now look at rearrangement and fragmentation reactions in the next chapter and then finally some redox reactions that cannot be more conveniently categorized under a different heading.

16

Rearrangement and Fragmentation Reactions

6.1 Introduction

Sometimes a reaction occurs in which the final product is totally different in structure from the starting materials. This may have resulted from a rearrangement of the carbon skeleton of the original molecule. Often such reactions appear at first sight to be very complicated. This, however, is a false impression. Rearrangement reactions follow the same basic principles that were outlined in Part I, and have already been illustrated in the preceding chapters of Part II. Rearrangement reactions require a source of excess electrons and a sink to receive them; there is still the overall push-me-pull-you picture that characterizes curly arrow mechanistic chemistry.

We will also consider in this chapter fragmentation reactions, *i.e.* those reactions in which the molecule falls apart to yield smaller units. Again, these reactions often appear difficult at first sight because the skeleton of the molecule has changed so greatly. Again, this type of reaction follows the same basic principles that have already been covered.

For the purpose of this chapter, the rearrangement reactions will be divided by the nature of the principal bond that is being broken in the first part of the reaction, *e.g.* carbon/carbon, carbon/nitrogen, carbon/oxygen and lastly other bonds. This chapter will give a large number of examples, because this is the easiest way to become familiar with rearrangements, and so to become confident in drawing the mechanisms.

16.2 Carbon/Carbon Rearrangements

Rearrangement reactions usually involve a 1,2 shift, *i.e.* the migrating group, M, goes from the migration origin, A, to the adjacent atom, which is called the migration terminus, B. The vast majority of rearrangements are nucleophilic in character. This means that the migrating group moves with its electron pair. Such rearrangements may also be called anionotropic. If the migrating group moves without its electron pair, then it is described as an electrophilic or cationotropic rearrangement, or if the migrating group is hydrogen then prototropic. Radical rearrangements are also possible, but like electrophilic rearrangements, these are rare. We will only consider nucleophilic rearrangements because they are so much more common.

Nucleophilic rearrangements formally occur in three steps. These steps may be illustrated by the rearrangement of a carbonium ion. A typical example is provided by the neopentyl system. It will be recalled that the SN2 mechanism is slow, because of the steric hindrance that the *t*-butyl group in the β-position provides. Also normal elimination is impossible, because there is no β-hydrogen.

Write down the first formed carbonium ion that results after protonation of neopentyl alcohol.

This is the first step, *i.e.* formation of the initial carbonium ion. These ions may be formed in many different ways: protonation of an alcohol, loss of nitrogen from a diazonium ion, addition of a proton to an alkene, or the interaction of a Lewis acid with a halide derivative.

The resultant carbonium has only six electrons, *i.e.* an open sextet. This provides the electron deficient sink that will accommodate the excess electrons that are held by a nucleophilic migratory group. Suggest what will be the next step.

The second step is the actual migration. The migratory origin becomes electron deficient, while the migratory terminus regains its octet. In simple molecules it is usually clear which bond is migrating, but in more complicated molecules it may not be immediately obvious. In order to avoid ambiguity a double headed arrow with a small ring at its tail is used. The ring encloses the bond that is migrating.

Suggest what will occur on the final step. There are two common endings.

Either the migratory origin will complete its octet by undergoing a nucleophilic substitution, or else there will be a loss of a proton to result in elimination.

This rearrangement of a carbonium ion is called the Wagner-Meerwein rearrangement, and proceeds via a Whitmore 1,2 shift. In this case the electrons in the bond that migrated, provided the push; while, the open sextet of the migratory terminus provided the pull.

In the actual migration step, it is possible for the migratory group to become detached from the original molecule and join another molecule before it moves to the migratory terminus. Such a reaction is called an inter-molecular rearrangement. If the migratory group does not become detached, then it is called an intramolecular rearrangement. These two types may be distinguished by cross over experiments, where two similar sets of reagents are allowed to react in the same vessel; the presence of intermolecular rear-rangements would be inferred from the presence of a product that contain a group from the other starting material.

Another important carbon skeleton rearrangement involves the movement of double bonds in an alkene derivative. Write down the carbonium which initially results from the loss of a bromide ion from 1-bromobut-2-ene.

This carbonium ion is stabilized by the conjugated double bond, and so forms a delocalized carbonium ion. The delocalized cation may now be attacked in two positions by an incoming nucleophile. Suggest what are these two positions, and so identify the two alternative substitution products.

Either the incoming nucleophile will attack the terminal carbon to give the un-rearranged substitution product; or else, it will attack the C-3 of the original compound, to give the rearranged substitution product. This is called the allylic rearrangement, because it occurs when the functional group is in the allylic position. In the case of substitution reactions, when the attack by the incoming nucleophile has occurred at the conjugated position this is indicated by using a prime sign, ('), *i.e.* SN1' or SN2'. In the case of the SN1' reaction, the longer the half life of the intermediate carbonium ion, the greater the amount of conjugated attack that will occur. This will result in a smaller product spread, *i.e.* the un-rearranged to rearranged products will be produced in more equal amounts.

When $CH_3CH=CHCD(OH)CH_3$ is treated with thionyl chloride, $SOCl_2$, the monochloro product has all of the deuterium in the vinylic position. Suggest a mechanistic pathway which accounts for this observation.

You will recall that thionyl chloride usually reacts via a SNi mechanism. In this case the intermediate is attacked by the displaced chloride ion at the conjugated position, *i.e.* SN2' manner, before the sulfur complex fragments to allow the internal attack by the second chloride anion.

If this chloro product is refluxed with ethanoic acid, two products are formed in approximately equal proportions. In one, the deuterium is still in the vinylic position; while in the other, the deuterium label is on the carbon attached to the oxygen. Suggest a pathway which accounts for this.

In this case the ethanoic acid is a solvent with a high dielectric constant, but is a weak nucleophile, and so it provides ideal conditions for a long lived carbonium ion intermediate. This is then statistically attacked at the terminal and central carbons to give the 1:1 ratio of substituted products, *i.e.* equal amounts of SN1 and SN1' reactions.

Wagner-Meerwein and allylic rearrangements are very common whenever a carbonium ion is formed. For example, in the simple Friedel-Crafts alkylation, the alkyl group always rearranges to give the most stable carbonium ion that then adds to the aromatic ring. In order to prevent this rearrangement, the Friedel-Crafts acylation process is used instead. Here the acyl cation does not undergo the Wagner-Meerwein rearrangement and so the alkyl substitute on the aromatic ring has the same carbon skeleton as the starting material. Another advantage is that as the acyl derivative de-activates the aromatic ring, there is greater control, and usually only one group adds to the ring. In the case of normal alkylation, the ring becomes more activated with the addition of each alkyl group, and so poly-alkylation isfavored.

In the Wagner-Meerwein rearrangement reaction the excess of electrons is provided by the bond that existed between the migratory origin and the migrating group itself. In the next example, the electrons that provide the electronic push come from a lone pair located on an oxygen atom. *Vicinal-diols* rearrange in the presence of acid to give the related carbonyl compound. Thus write down the mechanistic pathway for the rearrangement of pinacol, $Me_2COHCOHMe_2$, under acidic conditions.

The product is called pinacolone, and so this reaction is called the pinacol-pinacolone rearrangement. There are many variations of this reaction. They all depend upon placing a positive charge on the carbon adjacent to the carbon bearing the hydroxyl group. So, for example, β-amino alcohols on treatment with nitrous acid will perform a similar rearrangement. This version is called the semipinacol rearrangement. Write down the mechanistic pathway for this reaction when $Ph_2C(OH)CHMeNH_2$ is treated with nitrous acid. What are the stereochemical consequences at the migratory terminus?

There is inversion of the configuration at the migratory terminus in this reaction. What does this indicate about the first two steps of the migratory process, *i.e.* the formation of the carbonium ion and the migration?

In this case, as there is no racemization at the migratory terminus, the carbonium ion cannot have a separate existence, but instead the departure of

the leaving group is assisted by the migratory group moving across simultaneously, in the same way that a neighbouring group would interact.
However, unlike a neighbouring group interaction, the migrating group
cleaves from the migratory origin and bonds to the migratory terminus. This
results in an inversion of the configuration at the migratory terminus similar
to an SN2 type mechanism. This anchimeric assistance often results in an
increased rate of reaction.

So far in the examples given, only one type of group could migrate. When
there is a choice of which group may migrate, there are many factors that may
determine the answer. Apart from the stability of the various possible carbonium ions, and the configuration limitations that may be present in the molecule that may override other factors, it has been observed that aryl groups
tend to migrate more easily than alkyl groups.

The semipinacol reaction may be modified so that a ring expansion takes
place. For example, draw the mechanism for the ring expansion of 1-amino-
methyl-cyclopentanol.

This version is called the Tiffeneu-Demyanov ring expansion reaction.

Ring expansion may also be achieved in a different manner. Take the
geminal dibromide compound, 1,1-dibromobicyclohexane [3.1.0], and predict
what will occur when it is treated with aqueous silver nitrate.

Even though the five and three membered rings have expanded to give a
six membered ring, this reaction involves a formal ring contraction for the
three membered ring has resulted in the allylic cation, which is then hydro-
xylated by the nucleophilic attack of the solvent water.

A cyclohexadienone system that has two alkyl groups in the 4 position
undergoes a rearrangement on treatment with acid. Suggest what will be the
product, and also what is the driving force behind this reaction.

This reaction is called the dienone-phenol rearrangement. The driving force is the formation of the aromatic ring.

Earlier we saw that *vicinal* diols perform a rearrangement when treated with acid. If an α-diketone is treated with a base, it will undergo a rearrangement. Suggest what will be the reaction pathway when PhCOCOPh, benzil, is treated with sodium hydroxide solution.

The product is the sodium salt of the α-hydroxy carboxylic acid, which in this case is benzilic acid. This reaction is called the benzil-benzilic acid rearrangement.

The closely related α-haloketone when treated with an alkoxide anion undergoes a rearrangement. Suggest what is the pathway for this mechanism and so deduce what will be the product.

This reaction is called the Favorski rearrangement. Again there are many variations, for example, instead of an alkoxide anion, a hydroxide anion or even an amine may be used, in which case the salt of the carboxylic acid or the amide will be formed respectively. This reaction may also be used to effect a ring contract. Write down the mechanism for the reaction of 2-chlorocyclohexanone with an alkoxide anion.

The product is the expected ester.

In an earlier chapter we saw how to reduce the carbon chain by one methylene unit using the Hunsdeiker reaction. The carbon chain may be extended by one unit by using the Arndt-Eistert synthesis. In the first step an acyl halide is treated with diazomethane to form the α-diazo ketone. This is then treated with water and silver oxide. The resultant product is the free acid. If an alcohol is used instead of water, then the related ester is formed. Suggest what is the pathway for this reaction.

Note that in this reaction, the migration and the completion of the octet of the migratory origin atom occur at the same time. The electron deficient carbon species is a carbene. This is the best way of extending a chain by one unit if the carboxylic acid is available. The process of extension in this manner is called homologization. Again, like the Favorski rearrangement, it may be used to effect a ring contraction if one starts with the appropriate α-diazo ketone.

The actual rearrangement part of the Arndt-Eistert synthesis is called the Wolff rearrangement. The direct product of this rearrangement step is the ketene that then undergoes addition to yield the final desired product.

We will now look at a few reactions in which there is migration from a carbon atom to a nitrogen atom.

16.3 Carbon/Nitrogen Rearrangements

In the Hofmann rearrangement an unsubstituted amide upon treatment with sodium hydroxide and bromine yields the corresponding amine with one fewer carbon atoms. The first intermediate is the N-bromo amide, which then undergoes the three steps of the rearrangement reaction simultaneously. Suggest the pathway for this reaction, and how the final product is achieved.

The product of the rearrangement reaction is the isocyanate, which is similar to the ketene intermediate that was obtained in the Wolff rearrangement. In this case the isocyanate is hydrolysed under the reaction conditions to yield the amine. If NaOMe is used instead of NaOH then the carbamate, RNHCOOMe, is formed instead. This reaction is sometimes called the Hofmann degradation, but this causes confusion with the Hofmann exhaustive methylation, whereby an quaternary ammonium hydroxide is thermally cleaved to yield an alkene, as this latter reaction is also sometimes called the Hofmann degradation. It is probably better to avoid the term Hofmann degradation altogether.

The Hofmann rearrangement proceeds quite smoothly, even when the amide carbon is joined to a bridge head carbon, as in 1-norbornyl carbox-yamide, which may be easily converted to 1-aminonorbornane. The ease of this reaction indicates that the migrating group moves with retention of con-figuration, because in this case there is no possibility of inversion or racemi-zation, and if such was required, then the reaction should be very difficult to perform on such a substrate.

The basic principle behind the Hofmann rearrangement has a large number of variations, many of which are named reactions in their own right. For example, the pyrolysis of acyl azides to isocyanates. Write down the pathway for this reaction.

There is no evidence for the existence of the free nitrene in this reaction, unlike in the Wolff reaction where there was evidence for the existence of the free carbene, so probably the steps are concerted in this case. The great power of the driving force which is provided by the nitrogen leaving may be shown in that a variation of this reaction may be used to expand an aromatic ring. Indicate the pathway that is followed when an aryl azide is heated with phenylamine.

In the Lossen rearrangement, O-acyl derivatives of hydroxamic acids, RCONHOCOR, give isocyanates on treatment with hydroxide ions, that in turn may be hydrolysed to the amine.

In the Schmidt reaction the treatment of a carboxylic acid with hydrogen azide (hydrazoic acid) also gives the amine, via the isocyanate, when catalysed by an acid, such as sulfuric acid. The first step is the same as the $A_{AC}1$ mechanism to form the carbonyl cation, and so isfavored by hindered substrates. The protonated azide undergoes the rearrangement reaction.

There are a number of further variations to the Schmidt reaction. For example, the reaction of hydrogen azide with a ketone. Suggest the mechanism for this reaction.

In this case there is an elimination reaction to form the azide, which then rearranges, and after tautomerization yields the desired product. In essence this reaction has resulted in the insertion of a NH unit in the carbon chain. If the ketone was cyclic, then this reaction may be used to expand the ring and so form a lactam.

The last carbon/nitrogen reaction that we will study is slightly different in its mechanism, but is still yields an amide as the product. In this case an oxime is treated with PCl_5, in concentrated sulfuric acid. Suggest what is the pathway in this case.

This is called the Beckmann rearrangement. Suggest whether the *syn* or the *anti* alkyl group to the hydroxyl group mitrates. The *anti* group is the one that migrates in this reaction. If the reaction is performed on an oxime of a cyclic ketone, then ring enlargement occurs and gives the corresponding lactam.

We will now look at some reactions in which the migratory terminus is an oxygen atom instead of a nitrogen atom.

16.4 Carbon/Oxygen Rearrangements

When a ketone is treated with a peracid, in the presence of an acid catalyst, the related ester of the ketone is produced by the equilivalent of an insertion of an oxygen atom. This is the Baeyer-Villiger rearrangement. If a cyclic ketone is treated in this manner a lactone with a ring size one larger than the original ketone results. The mechanism is similar to that with diazomethane (Arndt-Eistert synthesis) and with hydrogen azide (Schmidt reaction). Suggest a pathway for the formation of the ester.

This reaction proceeds in high yield and so is a good synthetic reaction.

Another example of carbon/oxygen rearrangement is the Claisen rearrangement in which an allyl aryl ether on heating gives an *o*-allylphenol. Suggest a mechanism for this reaction.

The mechanism here is not a Whitmore 1,2 shift, but instead is a concerted pericyclic [3,3] sigmatropic rearrangement, that results in a cyclic ketone, that tautomerizes to give the phenol.

Now that we have covered the major rearrangement reactions, we will now look at a few fragmentation reactions.

16.5 Fragmentation Reactions

Fragmentation of a molecule may occur in many different ways. The following examples will give you a feel for some of the different mechanisms that may be invoked.

The fragmentation reaction that is most often first encountered is the iodoform reaction. This is because it forms the basis of a very easy qualitative test for the presence of a methyl ketone, or alternatively the $CH_3CHOH\rightarrow$ grouping. If a methyl ketone is treated with iodine and sodium hydroxide, then a yellow precipitate of iodoform, CHI_3, is rapidly produced. Write the mechanistic steps for this reaction.

Each time the terminal methyl group has a hydrogen substituted for another iodine atom, the remaining hydrogens become more acidic due to the -I effect of the halogen. Eventually the trisubstituted methyl group is capable of existing as an anion and so may act as a leaving group. This method may be used to cleave methyl ketones to yield the carboxylic acid. This type of reaction is called an anionic cleavage.

Another example of this type of reaction is the decarboxylation of alipha-tic acids. For the decarboxylation to proceed easily there must be an electron withdrawing group capable of stabilizing the negative charge. An example would be an α-carbonyl compound, RCOCOOH. Suggest a mechanism for this reaction.

The mechanism is essentially either a SE1 or a SE2 process. There are many variations on this theme. For example the Darzen's condensation produces an α,β-epoxy carboxylic acid, which when heated often undergoes decarboxylation. Suggest a mechanism for this reaction.

The initial product of the decarboxylation tautomerizes to give the carbonyl compound. This is called the glycidic acid rearrangement.

If the electron withdrawing group is a β-carbonyl group, then a different mechanism may operate. Suggest what this may be.

In this case it is possible for there to be a cyclic transition state, that results in the loss of carbon dioxide directly. The compound that is produced directly from the decarboxylation then tautomerizes to yield the final product.

1,3-diketones and β-keto esters may be cleaved under basic conditions, in a reaction pathway that is essentially the reverse of the Claisen condensation. Write the pathway for the ring opening of 3-ketohexanone by hydroxide anions.

This cleavage reaction occurs for any compound which has a 1,3 arrangement of carbonyl groups.

Primary and secondary aliphatic nitro compounds may tautomerize under basic conditions to the salt of the aci isomer. These isomers may in turn be hydrolysed by sulfuric acid, and so result in the fragmentation of the original molecule. Suggest how this reaction may proceed, and what is the final product.

Firstly, the base produces the salt of the nitro compound. Then on addition of the sulphuric acid the aci form of the original nitro compound is formed, which is hydrolysed to the carbonyl compound. This reaction is called the Nef reaction.

The last example is a rather unusual fragmentation reaction that involves the rearrangement of a hydroperoxide. In this case the hydroperoxide R_3C-O-O-H under acidic conditions gives a ketone and an alcohol. Suggest a mechanism for this reaction.

The migratory terminus is an oxygen atom. The unstable hemiacetal intermediate readily hydrolyses to give the final products.

That concludes this chapter on rearrangement and fragmentation reactions. The final chapter is on redox reactions.

17

Redox Reactions

17.1 Introduction

This is the final chapter in this part of this book. We have already come across some reactions that formally involve an oxidation or a reduction process of the carbon atoms in the molecule, *e.g.* the formation of a *vicinal* diol from a double bond is an oxidation reaction, while the addition of a hydrogen molecule is a reduction reaction. However, such reactions proceeded by clearly defined mechanisms which fit conveniently into other general mechanistic divisions, *e.g.* addition or elimination reactions, and so they were considered under those headings. There remains, however, a number of reactions which do not comfortably fit into any of the previous reaction types that we have studied so far, and so they will now be considered together in this chapter.

Redox reactions in organic chemistry do not have the premier importance which they have in inorganic reactions. In the latter, all reactions may be grouped into four major divisions, namely redox, acid/base, complexation and precipitation; even then, the last two types may be considered to be merely variations on the more general division of ligand interactions with a central ion.

In inorganic compounds, every ion of a particular element usually has the same oxidation state, because every atom of a given element has a similar spatial position relative to its neighbours, and so reacts in the same manner. This results from the non-directional nature of the Coulombic interactions which characterize ionic bonding.

As the compound becomes more complicated this idea breaks down. In particular when directional covalent bonding is involved within a complex ion; so, for example, in the thiosulfite anion, $S_2O_3^{2-}$, even though it is numerically possible to assign the same integral oxidation state to each sulfur, *i.e.* $+2$, in fact the two sulfur atoms occupy different positions in space relative to the oxygen atoms within the complex ion, and so have different redox properties. In this case it is better to assign an oxidation state of $+4$ to the central sulfur atom and zero to the sulfur atom to which it is doubly bonded, as this more closely reflects their individual properties, as revealed by the different paths that they follow when this anion undergoes redox reactions.

This highlights the approach that should be adopted when assigning oxidation numbers in covalent molecules. Each and every atom should be assigned an individual oxidation state that recognizes the relative position of any given atom to all of its neighbours, is likely to be different from the position occupied by any other atom, whether that other atom is of the same element or not.

This chapter is sub-divided along the lines of whether the overall process is an oxidation or reduction of the organic component in which we are interested. Of course there must always be a corresponding reduction or oxidation of another reagent in order to ensure that there is an overall balancing of reduction and oxidation. In some organic reactions the organic component of interest undergoes both oxidation and reduction, *i.e.* there is a disproportionation reaction. These will be considered at the end to the chapter.

In common with the rest of this book, it is not intended to give an exhaustive coverage of all possible reduction and oxidation reactions which are of synthetic utility in organic chemistry. Instead the aim is to give a selection of reactions that will illustrate the major mechanistic pathways. In the case of redox reactions for organic molecules, there are a large number for which the mechanism has not been studied in any detail, or if it has, no consensus has arisen as to the mechanism. This is even true for such an important synthetic reaction as the Clemmensen reduction of a carbonyl group to a methylene unit by a zinc-mercury amalgam with concentrated HCl. The detailed mechanism of many metal/acid couples is not know; however, there are still reactions that may be studied from a mechanistic view point.

17.2 Reduction Reactions

17.2.1 Electron Donors

Many redox reactions in organic chemistry involve single electron transfers, *i.e.* are radical reactions. We have already looked at a number of radical reactions that resulted in overall substitution. We will now study a few more radical reactions that do not conveniently fit under that heading. The addition of a single electron to an atom amounts to the reduction of that atom's oxidation state by one unit.

In the previous chapter, one of the first reactions that we studied was the pinacol-pinacolone rearrangement, in which a symmetrical diol gave a carbonyl compound. There are many ways to make a vicinal diol, *e.g.* the addition of a H_2O_2 to a carbon/carbon double bond. Such reactions were considered in the chapter on addition reactions, even though they are clearly oxidation reactions. Here we will consider a fundamental different approach.

Indicate the products that would result from a homolytic cleavage of the central C-C bond in a symmetrical 1,2-diol, $R_2COHCOHR_2$.

Now remove a proton from the oxygen, and also a single electron from the resultant radical anion, and then suggest what may be the starting material.

After removing a proton and a single electron from the oxygen anion, a biradical is left, which may combine with itself to form a carbonyl compound, $R_2C=O$. To perform the reverse of this analysis, *i.e.* the synthesis of the desired *vicinal* diol, a ketone reacts with magnesium in dry ether, under anaerobic conditions, *i.e.* without air present. Write down the complete reaction pathway for the synthesis of pinacol itself, $Me_2COHCOHMe_2$, starting with Mg and propanone. Note that as magnesium is capable of forming a divalent cation, it may react with two molecules of the ketone to form a cyclic intermediate, which is then opened up on aqueous acid hydrolysis.

This reaction is called the pinacol synthesis, and is generally applicable. It works best for symmetrical aromatic ketones, because of the high stability of the aromatic ketyl intermediate. This intermediate is blue in color, due to the delocalization of the radical electron around the aromatic ring.

Simple ketones are not the only carbonyl compounds that will accept single electrons, esters may also pick up an electron when sodium metal in dissolved in an inert solvent like xylene. Write down the pathway for this reaction, up until the stage where a dimer has been formed.

So far the reaction has proceeded along similar lines to the pinacol reaction. However, in the present case, the dimer contains two good leaving groups. Write down this next elimination step.

Now a diketone has been formed, but such a compound may react itself with single electrons. Suggest what may be the remaining steps for this compound, and so suggest what will be the final product.

The diketone picks up two more single electrons, and so is further reduced, and after tautomerization yields the 1,2-hydroxy ketone. This coupling reaction between two ester molecules is called the acyloin condensation. Note that in the Claisen ester condensation the resultant product has a 1,3 distribution of functional groups, while in this case the product has a 1,2 distribution of functional groups. This reaction may be used to effect the cyclization of long chain diesters in order to form very large rings. Catenanes were first made utililizing this reaction.

Sodium metal is a good source of free electrons, because it is so electropositive it readily gives up one electron to form the cation. If sodium metal is dissolved in liquid ammonia, the free electron may be solvated by the ammonia molecules. At higher temperatures the sodium reacts to form sodamide, $NaNH_2$, and hydrogen, in a simple inorganic redox reaction. At the lower temperatures required to keep the ammonia liquid, the solvated electron may be put to other uses. One such reaction is to effect the reduction of a benzene ring. Suggest what may be the first step of such a reduction.

The unpaired electron and the lone pair position themselves as far apart on the ring as is possible, *i.e.* *para* to each other. This reaction is usually per-

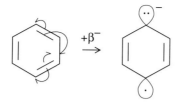

formed in the presence of an alcohol. Suggest what the remaining steps of
this reaction would be and so suggest what will be the product, and so what
is unusual about it.

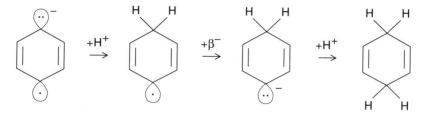

The alcohol acts as a source of protons that react with the lone pairs. This
reaction is called the Birch reduction. The final product is the cyclohexadiene
in which the double bonds are unconjugated. If the alcohol is not present,
then dimers may be obtained as the radical anions react with each other. In
some substituted aromatic systems both electrons are added before the ionic
intermediate picks up any protons so that a diradical is formed initially.

The presence of a substituent on the aromatic ring will influence the final
position of the double bonds, and also the overall rate of reduction. Suggest
what will be the product when anisole is reduced by sodium in liquid
ammonia, and whether the rate will be slower or faster than that experienced
by pure benzene.

The methoxy group is electron donating and so it disfavors the addition of
a further electron to the aromatic ring. Furthermore, it disfavors the posi-
tioning of the lone pair on the carbon to which it is attached. Thus the
methoxy group is found on the non-reduced carbon in the final product.

Similarly, suggest what will be the effect of an amide group on the rate
and regiospecifity of the Birch reaction.

The rate is increased and the amide group ends up at the reduced carbon
site.

The Birch reaction may also be used to reduce carbon/carbon triple bonds that are contained within a carbon chain. Suggest what will be the geometry of the resultant double bond after it has been protonated.

The resultant double bond is *trans*, because in the intermediate the lone pairs repel each other and so are *trans* to each other when they are protonated. This is a way of effecting *trans* hydrogenation. Terminal carbon/carbon triple bonds cannot be reduced by simply using sodium and ammonia because the acetylide anion is formed under those conditions. Instead ammonium sulphate must be added to the reaction mixture.

17.2.2 Hydride Donors

We will now look at some reactions which are performed by hydride ion donors. A very important synthetic reaction involves the reduction of the carbonyl group by a complex metal hydride to give an alcohol moiety. One of the commonest examples is the reduction of a ketone by lithium aluminium hydride, $LiAlH_4$, to give a secondary alcohol. Suggest what the first step of this reaction is.

The aluminium tetrahydride anion donates a hydride anion irreversibly to the carbonyl carbon. The resulting alkoxide anion complexes with the aluminium trihydride. Write this step of the reaction.

It appears that the next step involves the disproportionation of this complex and so yield the tetra-alkoxide aluminium anion while regenerating more of the aluminium tetrahydride anion. The tetrahydride seems to be the only species that donates hydride ions to the carbonyl carbon. In the work-up of the reaction mixture the tetra-alkoxide anion is decomposed by a protic solvent to give the protonated alcohol product as desired.

Another common variation is the reduction of an ester to a primary alcohol. Suggest what is the mechanistic pathway for this reaction.

The reaction is facilitated by the fact that the alkoxide group is a reasonable leaving group under these conditions. The intermediate carbonyl compound is then attacked again so as to effect the addition of the second hydride ion. This now forms an oxygen anion that is capable of complexation with the aluminium, which is destroyed upon the acidic work up, and so liberate the desired product. If lithium aluminium deuteride is used then this method may be used to produced isotopically labelled compounds.

The formation of the intermediate complex may cause problems on occasions. For example, in the synthesis of ethanolamine from the ester of glycine, suggest the structure for the intermediate complex.

This cyclic complex is particularly stable and cannot be destroyed using the normal work up procedures that involve aqueous ammonium sulphate. Instead concentrated hydrochloric acid must be used to give the ethanolamine as its hydrochloride salt.

Lithium aluminium hydride is a very reactive hydride ion donor and may be used to effect a large number of reductions. For example it may be used to reduce nitriles, $R\text{-}C\equiv N$, and primary amides, $RCONH_2$, to the related amine, RCH_2NH_2. It is thought that the reduction of the amide proceeds via the nitrile.

$LiAlH_4$ is very reactive and hence rather unselective. In order to improve its selectivity various derivatives have been made which are less reactive. These derivatives usually have three of the hydroxide ion replaced by alkoxide ions, *i.e.* $AlH(OR)_3^-$.

Instead of a complex metal hydride being the source of the hydride anion, the hydride anion may be donated from a carbon atom. In such a case the reaction is called the Meerwein-Ponndorf-Verley (MVP) reduction. In contrast to the reduction by a complex metal hydride the MVP reduction is

reversible. The MVP reduction is performed by heating aluminium isoproproxide with an excess of propan-2-ol. Two different routes compete, one
involves one molecule of the aluminium isoproproxide per molecule of the
carbonyl to be reduced, while the other route uses two. Suggest the mechanism that involves only one molecule of the aluminium compound.

In this case, propanone is the lowest boiling constituent of the reaction
mixture, and hence by continuously distilling this out of the system the equilibrium may be effectively displaced to completion. The excess of propan-2-ol
exchanges with the mixed aluminium alkoxide to liberate the reduction
product.

As this is reaction involves steps which are reversible, it is not surprising
that the reaction may be forced to proceed in the reverse direction, *i.e.* to
achieve the oxidation of an alcohol by using $Al(OCMe_3)_3$ as a catalyst with
an excess of propanone. In this case the reaction is called the Oppenauer
oxidation.

17.3 Oxidation Reactions

Oxidation reactions include those reactions that involve the removal of
hydrogen from a molecule. In an earlier chapter we saw that hydrogen molecules may be added across a carbon/carbon double bond in the presence of a
metal catalyst such as Pt or Pd. The reverse reaction is also possible, *e.g.*
heating cyclohexane with Pt at 300° gives benzene. It is thought that the
mechanism of dehydrogenation is simply the reverse of the hydrogenation
addition process. The reaction is easier if the ring already contains some
unsaturation. Dehydrogenation may also be performed by quinone, which is
reduced to hydroquinone in the process. Suggest the reaction pathway for
the oxidation of cyclohexadi-1,3-ene by quinone.

Usually the tetrachloro or dichloro-dicyano derivatives of quinone is used, and they are referred to as chloranil and DDQ respectively.

Oxidation reactions also include those reactions that involve the addition of oxygen. We have already seen the addition of H_2O_2 to a carbon/carbon double bond to give the 1,2-diol. Such compounds may undergo further oxidation by treatment with periodic acid, HIO_4, or lead acetate, $Pb(OAc)_4$. These two reagents are complimentary, as the former is best used in an aqueous medium, while the latter is used in organic solvents. The mechanistic pathway for both reagents, however, is similar, and involves a cyclic intermediate. Suggest how this intermediate using $Pb(OAc)_4$ may be formed.

Once formed, this cyclic intermediate is capable of fragmentation to produce two ketone compounds. Suggest how this may occur.

The pathway from a double bond to the *vicinal* diol to the ketone compounds is very efficient and may be used preparatively. The cleavage of *vicinal* diols is also used to elucidate the structure of carbohydrate compounds.

Instead of going via the *vicinal* diol, and then cleaving that molecule with either $Pb(OAc)_4$ or periodic acid, it is possible to cleave the double bond directly to give the carbonyl derivatives. This is achieved by reacting the olefin with ozone, O_3. First, write the structure of ozone.

Ozone is a dipolar molecule, and is capable of performing a 1,3-dipolar addition to a carbon/carbon double bond. Write down this step.

This intermediate is called an initial or primary ozonide (or even a molo-zonide!). One of the oxygen/oxygen bonds now heterolytically breaks, and then the carbon/carbon bond breaks to yield a carbonyl compound and a zwitterion. Write down these two steps.

These intermediates may recombine in a different 1,3-dipolar addition to give the normal ozonide.

This new ozonide may be decomposed by zinc and ethanoic acid to give two carbonyl compounds, either ketones or aldehyde depending on the sub-stituents on the original carbon/carbon double bond. The ozonide may also be reduced by $LiAlH_4$ to give two hydroxyl compounds, or oxidized by H_2O_2 to give ketones or carboxylic acids.

In a totally different reaction, hydrogen peroxide may also be used to effect a rather unusual reaction in which an aromatic aldehyde, *e.g.* a benzal-dehyde derivative, is oxidized to the corresponding phenol. The reaction occurs under alkaline conditions. Suggest what the first step in this pathway may be.

What occurs next is a rearrangement that is similar to the Beayer-Villiger rearrangement to result in an ester. This intermediate is hydrolysed to yield the phenol derivative. Write down these steps of the mechanism.

This reaction is called the Dakin reaction, and requires a hydroxyl or amino group in the *ortho* or *para* position of the aromatic aldehyde.

It is possible to oxidize a methylene group that is adjacent to a carbonyl group to yield another carbonyl group. This transformation is achieved using selenium dioxide. The presence of a base is necessary. Suggest how a selenium intermediate may be formed.

The enolate may attack the selenium to form a selenate ester of the enol. This intermediate may rearrange to reform the carbonyl and place an oxygen on the α-carbon. Suggest how this may occur.

The last proton is removed by the action of the base which initiates a fragmentation reaction to yield the α-keto carbonyl compound. Suggest what is the mechanism for this step, and so suggest what are the products for this reaction.

The α-dicarbonyl and elemental selenium are produced. This reaction is useful because it may be performed under very mild conditions. Another mechanism has been proposed for this reaction that involves the intermediary of a β-ketoseleninic acid, ←CO-CH-Se = OOH, and the selenate ester is not involved. This just highlights the fact that very few proposed mechanisms are actually proved in reality. Often, all that may be said for any particular mechanistic pathway is that it accounts for all the kinetic data, including isotopic substitution experiments, and also for any stereochemical observations. Intermediates can rarely be captured and characterised, but it can sometimes be possible to detect such species spectroscopically.

We will now look at reactions in which both oxidation and reduction occur in the organic components.

17.4 Disproportionation Reactions

In any reaction which involves the oxidation or reduction of an organic component, there also must be a corresponding reduction or oxidation of the other reagent, be it organic or inorganic. Normally we are not interested in the other component of the reaction mixture, so long as the organic substrate is modified in the manner that we desire. Occasionally, however, the organic component undergoes both oxidation and reduction, sometimes even within the same molecule at different functional groups. When there are two identical functional groups, and one is oxidized, while the other is reduced then this redox reaction is called disproportionation. This obviously requires a functional group that is somewhere in the middle of the redox spectrum that exists for the carbon atom, *e.g.* an aldehyde group which may be oxidized to a carboxylic acid or reduced to an alcohol. We have already seen a variation of this theme in the chapter on rearrangement reactions. In the benzil-benzilic acid rearrangement a molecule that initially contained two ketone groups rearranged to give a molecule with an α-hydroxycarboxylic acid arrangement of functional groups. This was an example of an internal disproportionation reaction.

Earlier, we saw how a carbonyl group may be reduced by a hydride ion. In the examples given, the hydride anion came from either a complex metal hydride such as $LiAlH_4$, or from a carbon reagent, such as aluminium isoproproxide with an excess of propan-2-ol. Under certain circumstances the hydride ion that is being used to effect this reduction may come from an aldehyde compound that lacks an α-hydrogen atom, *e.g.* methanal or benzaldehyde. The receiving molecule may be a second molecule of the same aldehyde or a different one. In the former case disproportionation occurs, and the reaction is called the Cannizzaro reaction; while in the latter case it is called a crossed Cannizzaro reaction. The reaction requires a strong base, and the rate law is found to depend on the square of the concentration of the aldehyde plus the concentration of the base used. In the reaction between hydroxide and benzaldehyde suggest what may be the first step.

The anion that results from the addition of the hydroxyl ion then donates a hydride ion to another aldehyde; write down the mechanism of this step.

Overall a carboxylate anion and an alcohol have been formed from two molecules of aldehydes, and so a disproportionation reaction has been effected.

The reaction may occur internally to the molecule, *i.e.* α-ketone aldehydes give the α-hydroxy carboxylic acids on treatment with hydroxide ions.

A variation of this reaction is called the Tollens reaction. In this case a ketone or aldehyde that contains an α-hydrogen is treated with formaldehyde in the presence of $Ca(OH)_2$. If the first formed product of the cross aldol reaction contains another α-hydrogen, then another cross aldol reaction occurs. This continues until there are no more α-hydrogens left. Write down the product that results from the exhaustive condensation of ethanal with methanal.

$$CH_3\,CHO \quad + \quad 4\,HCHO \quad \longrightarrow$$

This intermediate has three hydroxyl groups and one carbonyl group, but now has no α-hydrogens left. Now a cross Cannizzaro reaction occurs that reduces the aldehyde group to give the fourth hydroxyl group. Write down the equation for this step.

This is an example of a disproportionation reaction between two different molecule, but each contained the same functional group initially.

Aldehyde compounds provide a ready example of organic molecules capable of undergoing disproportionation reactions, because it is well known that they occupy the midway position in the oxidation sequence from primary alcohols to carboxylic acids.

They are, however, not the only examples of organic compounds that are capable of undergoing a disproportionation reaction. All that is necessary is to find a compound that occupies a similar midway position in some redox sequence, *e.g.* suggest what is the disproportionation reaction which occurs between two alkyl radicals.

In this case they may react together to form an alkane and an alkene. This reaction occurs in competition with dimerization and other termination reactions in free radical substitution reactions.

That concludes this part of the book in which we have studied the basic mechanisms which are thought to account for the vast array of organic reactions which have been observed. In all of the reaction pathways which we have examined the basic principles which were outlined in Part I have been applied. Now that these fundamental principles have been learnt and put into the context of the basic mechanism, you will be able to write a sensible mechanistic sequence for most of the reactions which you will encounter. There are, of course, more advanced principles which help explain results which would appear anomalous when using just the principles outlined here. Such principles include frontier orbital theory and symmetry controlled reactions which are beyond the aim of this introductory text.

Part III:
Appendices

18

Glossary

Absolute Configuration

Defines the real position in space of all the molecular co-ordinates. Previous refered to the fact that the configuration of the molecule had been related to D or L glyceraldehyde.

Absorption

The process whereby the chemical in question is taken into the body of the compound doing the absorption. In contrast to adsorption.

Acetal

The functional group $RHC(OR)_2$. Sometimes also used for the functional group $R_2C(OR)_2$ which should more correctly be called a ketal.

Acetylides

Containing the RC_2^- unit.

Achiral

A molecule which is not chiral.

Aci

A tautomeric form, $R_2C=N^+(-OH)O^-$, of a nitro compound, R_2CH-NO_2, hence the nitro/aci tautomerism.

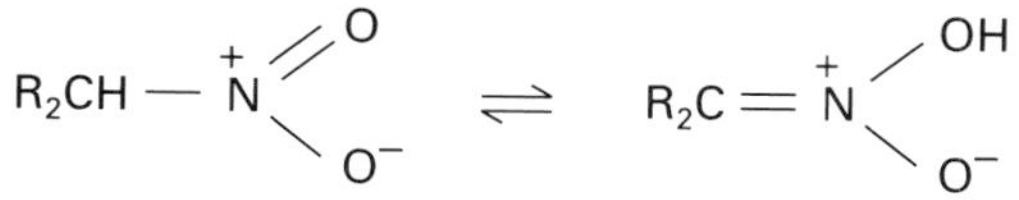

$$R_2CH - \overset{+}{N}\underset{O^-}{\overset{O}{\diagup}} \rightleftharpoons R_2C = \overset{+}{N}\underset{O^-}{\overset{OH}{\diagup}}$$

nitro   aci

Acid

Usually either a proton donor (Brönsted-Lowry definition) or an electron pair acceptor (Lewis definition), but there are other definitions.

Acid anhydride

Compound of the general type RCOOCOR. May be symmetrical or mixed.

Acid halide

Compound of the general type RCOX.

Acrylonitrile

Cyanoethane, $CH_2=CH-C\equiv N$.

Acyl

The functional group $\leftarrow C(=O)R$.

Acylion condensation

The reaction between two molecules of esters in the presence of sodium dissolved in

289

xylene to give a 1,2-hydroxyl ketone. Useful for forming large rings. The acylion group is $R_2C(OH)-C(=O)R$.

Addend The product of an addition reaction.

Adsorption The process whereby the molecule in question only reacts with the surface of the compound which is doing the adsorption. In contrast to absorption.

Alcohol Compound of the general type ROH. Subdivided into primary, secondary and tertiary depending on the number of carbons attached to the carbon which is joined to the OH group.

Alcoholysis The cleavage of large molecule by an alcohol to give two smaller molecules with the overall addition of the alcohol to the original molecule.

Aldehyde Compound of the general type $RCH(=O)$.

Aldimine Compound of the general type $RCH=NR$. Also called a Schiff base or an azomethine.

Aldoketen Compound of the general type $RHC=C=O$.

Aldol condensation The reaction between an aldehyde, ketone or ester with another aldehyde or ketone in the presence of a base to give a 1,3-hydroxyl ketone. Dehydration may follow to give the α-unsaturated ketone, then called the crotonaldedye reaction.

Aldoxime Compound of the general type $RCH=NOH$.

Alicylic A carbon ring compound which may be saturated or unsaturated.

Aliphatic A carbon compound which is open chained and not either cyclic or aromatic in nature. It may be saturated or unsaturated.

Alkali A compound which gives rise to the hydroxyl anion in an aqueous solution.

Alkene Containing the $C=C$ group.

Alkoxide The oxygen anion of an alcohol, *i.e.* RO^-.

Alkyl The generic term for a aliphatic carbon substitutent, often abbreviated to $R\rightarrow$.

Alkylation The addition of an alkyl group, to a carbon (C-), oxygen (O-), or a nitrogen (N-) atom.

Alkyne Containing the $C\equiv C$ group.

Allene Containing the $C=C=C$ group. If there

Allylic position
Alternant

are more than two such double bonds then the compound is called a cumulene.

The methylene group next to a $C=C$ bond.
Used to divide aromatic hydrocarbons into two types, either alternant or non-alternant. In the former the conjugated carbon atoms may be divided into two sets such that no two atoms of the same set are adjacent, *e.g.* naphthalene, while in non-alternant in one set, there is at least one pair of carbons which are adjacent, *e.g.* the shared carbons between the 5 and 7 membered rings of azulene.

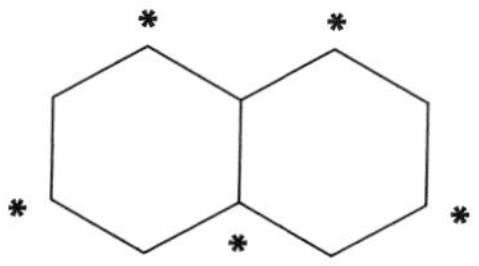

naphthalene

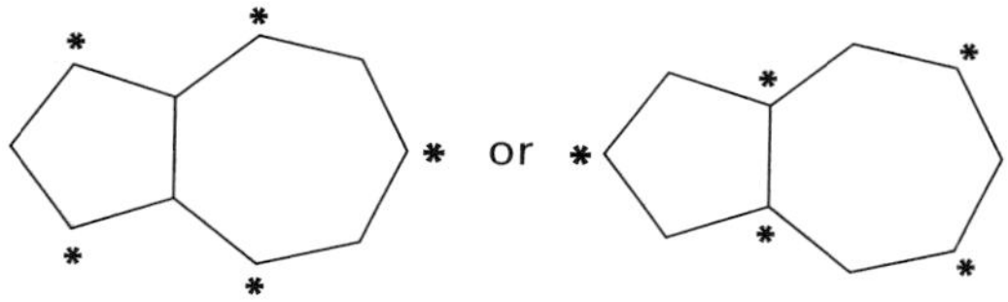

azulene

Ambident nucleophile

A nucleophile which is capable of attacking in two or more different ways to give different products, *e.g.* NCO⁻ may attach via the oxygen to give cyanates, ROCN, or *via* the nitrogen to give isocyanates RNCO.

Amide

The condensation product of a carboxylic acid and an amine. It may be primary, secondary or tertiary depending on whether there are zero, one or two alkyl groups on the nitrogen.

Amidine
Amidoxime
Amine

The NH_2-CR$=$NH grouping.
The RC($=$NOH)-NH_2 grouping.
A alkylated or arylated ammonia compound, maybe primary, secondary, tertiary or quaternary, *e.g.* RNH_2, R_2NH, R_3N, or R_4N^+.

Amino acid

A molecule which has both an amino group and a carboxylic acid group in it. If these two groups are attached to the same carbon

	then it is called an α-amino acid which on polymerization form proteins.
Aminonitrene	The R_2N-N: grouping.
Ammonia	The NH_3 molecule.
Amphoteric	A species which may act as both an acid or a base, *e.g.* water.
Analine	The molecule $C_6H_5NH_2$.
Angle strain	The strain imposed on a bond when a sp^3 hybridized carbon is in either a large or a small ring so that the angle between the bonds in the ring deviates from $109°" >$. Also applies to the strain on sp^2 or sp hybridized carbons in a ring.
Anhydride	A condensation product which has resulted from the lose of a water molecule.
Anion	A negatively charged ion.
Anisole	The molecule C_6H_5OMe.
Anisotropic	A property which is not the same along all directions of space. Opposite of isotropic.
Annellation	In a fused ring system, such as triphenylene, where one ring releases some of its electrons so that adjacent rings may become aromatic.
Anti	A mechanistic description of the addition or elimination of a molecule, AB, to or from a double bond, where A joins or leaves from the diametrically opposite orientation to B. Opposite of *syn*.
Anti-Markovnikov	The addition to a double bond in which the negative part of the adduct joins to the carbon which initially had the most hydrogens, usually the result of a radical mechanism.
Antibonding Orbital	An orbital, which if occupied by electrons leads to the molecule being less stable than if it was not occupied.
Anticlinal	Used to describe the conformer of a 1,2-disubstituted ethane derivative in which the dihedral angle between the groups under consideration is 120°. The other conformerations are called *antiperiplanar*, *gauche* or *synclinal*, and *synperiplanar*, to represent the dihedral angles of 180°, 60° and zero respectively.

antiperiplanar anticlinal

gauche or   synperiplaner
synclinal

Antiperiplanar	*Vide anticlinal.*
Apofacial	*Vide* synfacial.
Aprotic	Used to describe a solvent which does not readily donate protons, *e.g.* THF. Usually such a solvent has no hydrogens attached to heteroatoms.
Arene	A compound based on the benzene ring.
Arenium ion	*Vide* Wheland intermediate.
Arndt-Eistert	The conversion of an acid chloride to the carboxylic acid which is its homologue with an extra carbon in the chain. Performed using diazomethane to give the diazo ketone, followed by heating with silver oxide and then acid hydrolysis. The second step involves the Wolff rearrangement.
Aromaticity	The property of having $2n + 2$ π electrons in a closed circuit of molecular orbitals. The commonest example being benzene, where n equals one, around a six carbon ring. A more precise definition is the ability to sustain an induced diamagnetic ring current, which is called diatropicism.
Aryl	The generic term for an phenyl derivative, which may be substituted or unsubstituted on the aromatic ring.
Ate complex	When a Lewis acid combines with a base to give a negative ion in which the central atom has a higher than normal valence, *e.g.* $Me_4B^-Li^+$. *Vide* also onium salts.
Atomic number	The number of protons in the nucleus of an element.
Atomic orbital	An electronic orbital which is centred on only one nucleus.
Atropisomers	Stereoisomers which may only be separated

because they do not readily interconvert by reason of restricted rotation about a single bond.

Axial   Used to describe a bond which projects vertically from the plane of a cyclohexane ring. Opposite of equatorial.

Azide anion   The N_3^- anion.
Azide   A molecule of the general formula RN_3.
Aziridine   An saturated three membered ring comprising of two carbons and one nitrogen.
Azirine   The 1-azirine has a three membered ring comprising of two carbons and doubly bonded nitrogen. The 2-azirine, in which the double bond is between the carbon atoms, does not exist, maybe because it is predicted to be antiaromatic.

1 – azirine   2 – azirine

Azlactone   A five membered ring compound, which is a useful synthetic intermediate, also called 5-oxazolone.

Azo compound   A compound of the general formula $RN = NR$.

B strain   Stands for back strain, and is the compression strain resulting from the steric crowding of non-bonding groups.
Baeyer-Villiger   The rearrangement of a ketone with *m*CPBA to give the ester.
Banana bond   *Vide* small angle strain.

Base	Usually either a proton acceptor (Brönsted-Lowry definition) or an electron pair donor (Lewis definition), but there are other definitions.
Bathochromic shift	When a given chromophore absorbs at a certain wavelength and the substitution of one group for another causes the absorption to be at a higher wavelength. The opposite to a hypsochromic shift.
Beckmann	The rearrangement of a ketone to give the secondary amide, on treatment with with hydroxylamine, NH_2OH, followed by PCl_5. The reaction proceeds *via* an oxime.
Bent bonds	*Vide* small angle strain.
Benzidine	The rearrangement of hydrazobenzene under acidic conditions to give 4,4'-diaminobiphenyl, benzidine. General for N,N'-diarylhydrazines.

Benzil-Benzilic acid	The rearrangement of a 1,2-diketone aromatic species in the presence of hydroxyl ion to give the α-hydroxyl carboxylic acid.

Benzion	The reaction between two aromatic aldehydes, with no α-hydrogens in the presence of KCN, to give the 1,2-hydroxyl ketone adduct.
Benzoic acid	The molecule $C_6H_5CO_2H$.
Benzoyl	The $C_6H_5CO\rightarrow$ group.
Benzyl	The $C_6H_5CH_2\rightarrow$ group.
Benzophenone	The molecule PhCOPh.
Bisulfite	The product which result from the additon of SO_3^{2-} to a carbonyl group.

Boat conformer	A conformer of cyclohexane.

Bond energy	The average energy needed to break all the bonds of a given type in a certain molecule. So in water it is the average of the two O-H dissociation energies, while in methane it is the average of four C-H dissociation energies.
Bredt's rule	In small bridged bicyclic compounds double bonds are impossible at the bridgehead. Small means that the S number must be less than or equal to six. The S number is the sum of the number of atoms in the bridges of a bicyclic system. If the S number is equal to seven then if the double bond is in a seven membered ring it cannot be formed, while if it is in an eight membered ring it can.
Bromomium ion	The cyclic version of the β-bromo carbonium ion, R_2C^+-$CBrR_2$ ion. The bromine atom dative bonds to the carbonium ion.

Brönsted-Lowry	A theory which defines an acid as a proton donor, and a base as a proton acceptor. Every acid and base is related to its conjugate base and acid respectively.
Brosylates	Compounds of the general type p-Br-$C_6H_4SO_2$-O-R.
Bunte salts	Anions of the general type $RSSO_3^-$.
C-alkyation	*Vide* alkyation.
Cannizario	A disproportionation reaction between two molecules of aldehydes which both have no α-hydrogens in the presence of base to give an alcohol and an acid. Called a crossed Cannizario reaction if the reaction occurs between different aldehydes.
Canonical form	Each of the various electronic structures which are possible for any given molecule is a canonical form, or Lewis structure. Each

form is represented by a wave equation, which is then combined in the valence bond method to give a description of the molecule, called the resonance structure.

$$R-\underset{O^-}{\overset{O}{\diagup}} + R-\underset{O}{\overset{O^-}{\diagup}} \Rightarrow R-\underset{O^{(-)}}{\overset{O^{(-)}}{\diagup}}$$

canonical forms   resonance structure

Carbamate	The molecule of the general formula R_2NCOOR.
Carbamic acid	The molecule H_2NCOOH.
Carbanion	A negatively charged carbon species.
Carbene	The divalent neutral carbon species, which has only six electrons around it, *e.g.* dichlorocarbene, CCl_2.
Carbenium ion	Strictly, the name for the trivalent positive carbon species, CR_3^+; however in practice this species is always referred to as a carbonium ion.
Carbides	Containing either the C^{4-} or C_2^{2-} unit: the former should be called methylides, while the latter should be called acetylides.
Carbocation	Any positively charged carbon species.
Carbodiimide	A molecule with the general formula $RN=C=NR$.
Carbonate	An ester of carbonic acid, $O=C(OR)_2$.
Carbonic acid	The molecule H_2CO_3.
Carbonium ion	Strictly, the name for tetra and penta-coordinated positive ions of carbon; however, in general use as the name for any positive carbon species.
Carbonyl	A carbon double bonded to an oxygen, *i.e.* $R_2C=O$. The R groups are usually hydrogen or carbon chains. If the R groups are otherwise, then the whole grouping is given a new name, *e.g.* carboxylic acid, ester, acid halide, acid anhydride, amide, *etc.*
Carbonylation	The formation of a carbonyl functionality, *e.g.* α to an existing carbonyl group using SeO_2
Carboxylate	The oxygen anion of a carboxylic acid, *i.e.* RCO_2^-.
Carboxylation	The addition of a $\leftarrow CO_2H$ group.

Carboxylic acid   Compound of the general type R-C(=O)OH.

Carboxylic ester   Compound of the general type RC(=O)OR, the condensation product between a carboxylic acid and an alcohol.

Catalysis   A substance which alters only the rate at which the equilibrium of a reaction is reached without altering the position of the resulting equilibrium. The catalysis is not consumed during the reaction. May be either general or specific.

Catenanes   Compounds made up of two or more rings interlaced as links in a chain.

Cation   A positively charged ion

Chain reaction   A reaction which is self perpetuating, *e.g.* the photochlorination of methane. The product quantum yield is greater than one, which means that more molecules of product are formed than quanta of light are absorbed by the reagent. There is typically an initiation reaction, followed by a large number of propagation reactions, ending with a termination reaction.

Chair conformer   The most stable conformer of cyclohexane.

Chelotropic reaction   A reaction in which two σ bonds that terminate at a single atom are made or broken in concert.

Chiral   Used to descirbe a molecule which does not possess a centre of symmetry, nor a plane of symmetry, nor an alternating axis of symmetry. Such a molecule will rotate plane polarized light.

Chloral hydrate   The water addition product of chloral, *i.e.* $CCl_3CH(OH)_2$.

Chloroform   The trival name for trichloromethane.

Chloromethylation   The addition of a ←CH_2Cl group, *e.g.* by the action of methanal and HCl on an arene.

Chromophore   A group which causes the absoption of light.

Chugaev   The thermal pyrolysis of methyl xanthates to give the alkene.

Cimmamic acid	The molecule *trans*-3-phenylpropanoic acid.
Cine	When substitution occurs at a different position on the aromatic ring than at the position of the original group.
cis	A term used to describe the geometric isomer that has the two substitutents in question on the same side of the double bond. In the CIP system the symbol would be Z. The opposite of *trans*.
cisiod	Used to describe a conformer of a diene.

Claisen ester	The reaction of two ester molecules in the presence of the appropriate alkoxide base, followed by mild acid hydrolysis to yield the 1,3 ketone ester adduct. The cyclic version is called the Dieckman reaction.
Claisen rearrangement	The thermal rearrangement of allyl aryl ethers to *ortho*-allylphenols. If both *ortho* positions are occupied then the allyl group migrates to the *para* position.

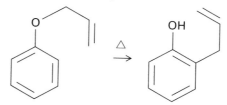

Claisen-Schidt	The cross condensation reaction of an aromatic aldehyde with a simple aliphatic aldehyde or ketone with 10% KOH to yield the dehydrated product.
Clathrate	A host compound that forms a crystal lattice in which a guest molecule fits, the only interaction being by van der Waals forces. The spaces formed by the host compound completely enclose the guest molecule, unlike inclusion compounds.
Clemmensen	The reduction of a carbonyl group using a zinc mercury amalgam and concentrated HCl to give the methylene unit.
Condensation	A reaction in which two approximately equally sized molecules react so as to form one larger molecule with the eliminations of a much smaller molecule, usually water.

Configuration	When two isomers can only be interconverted by the breaking and reforming of covalent bonds.
Conformation	When two isomers may be interconverted merely by free rotation about bonds.
Conjugate acid/base	Two molecules which are related by the addition of a single proton to the Brönsted-Lowry base to give rise to the Brönsted-Lowry acid.
Conjugate system	Alternating single and multiple bonds.
Cope reaction	The thermal cleavage of an amine oxide to produce an alkene and a hydroxylamine. The amine oxide is usually formed *in situ* from the amine and an oxidizing agent. Stereoslectively *syn*, *via* a five-membered Ei mechanism.
Cope rearrangement	The [3,3] sigmatropic rearrangement of 1,5 dienes. If the reagent is symmetrical about the 3,4 bond then the product is identical to the starting material.

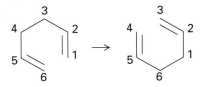

Correlation diagram	*Vide* Woodward-Hoffmann rules.
Covalent bond	A bond which comprises of electrons which are shared more or less equally usually between two nuclei. The electrons occupy molecular orbitals.
Cram's rule	The less hindered side of a carbonyl group will be preferentially attacked by an incoming nucleophile.

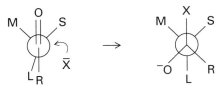

L, M, S, are large, medium & small groups

Cresol	The molecule MeC_6H_4OH. There are three isomers, *ortho*, *meta* and *para*.

Cross-conjugation

A system in which there are at three unsaturated groups, two of which are not conjugated with each other.

O

Crown ether

A large ring ether compound containing several oxygens atoms which possess the property of forming complexes with positive ions, often metallic ions.

Cumulene

Containing more than two carbon/carbon double bonds joined directly to one another, *i.e.* not in conjugation as there is no separating carbon/carbon single bond, *e.g.* $C=C=C=C$. If there are just two such double bonds then the compound is an allene.

Curly arrow mechanism

The movement of electrons, usually in pairs, is indicated by curly arrows to indicate which bonds are being formed and which are being broken within the reacting molecules.

Cyanate

Compound of the general type $R\text{-}O\text{-}C\equiv N$ group.

Cyanic acid

The molecule HNCO.

Cyanide ion

The univalent $\leftarrow C\equiv N^{-}$ anion.

Cyanide

Vide nitrile.

Cyanogen

The molecule $N\equiv C\text{-}C\equiv N$.

Cyanohydrin

Compound of the general type $R_2C(OH)CN$; also called a hydroxynitrile.

Cyanothylation

The addition of a $\leftarrow CH_2CH_2C\equiv N$ group.

Cycloaddition

A type of pericyclic reaction in which there is a concerted addition of two molecules, *i.e.* addition takes place in only one step.

Cycloalkanes

Cyclic version of an open chain alkane.

Cyclopentadienyl ion

The aromatic cyclic anoin $C_5H_5^{-}$. Its salts are called cyclopentadienide salts.

Cyclopropenyl ion

The aromatic cyclic cation $C_3H_3^{+}$. The salts are called cyclopropenium salts.

Darzen's condensation

The coupling of an α-bromocarboxylate ester with a carbonyl under basic conditions to give a glycidic ester, *i.e.* an ester with an α,β epoxide group.

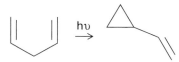

Decarboxylation	The lose of a molecule of carbon dioxide from a molecule.
Dehydration	The loss of a molecule of water from a molecule.
Delocalization	Used to describe the situation in which a molecular orbital is centred on more than two atomic nuclei.
Di-π-methane	The photochemical rearrangement of 1,4 dienes to vinylcyclopropanes.

$$\text{hv}$$

Diastereomer	A stereoisomer which is not an enantiomer.
Diastereotopic	Used to describe the situation when two groups cannot be brought within chemically identical positions by rotation or any other symmetry operation.
Diatropic	*Vide* aromaticity.
Diaziridine	A saturated three membered ring comprising of one carbon atom and two nitrogen atoms.
Diazoalkane	A compound of the general type $CR_2 = N^+ = N^-$.
Diazonium	A molecule containing the $\leftarrow N_2{}^+$ group.
Dieckman	The cyclic version of the Claisen reaction
Diels-Alder	The 1,4 addition of a double bond to a conjugated diene (a $4+2$ cycloaddition) to produce a six membered ring.
Dieneophile	A species which likes to attack dienes.
Dienone-Phenol	

$$H^+$$

	A rearrangment reaction in which a cyclohexadienone that has two substituents at the 4 position, yields on acid treatement a phenol substituted at the 3 and 4 positions.
Diimide	The $HN=NH$ molecule.
Dimerization	The formation of a dimer, *i.e.* a molecule which has the same empirical formula, but double the RMM.
Diol	A compound which contain two hydroxyl groups, usually on different carbons, *e.g.* a *vicinal* diol such as $CH_2OH\text{-}CH_2OH$. If the diol is *geminal*, then it is called a hydrate.
Dismutation	*Vide* metathesis.
Displacement reaction	A substitution reaction in which hydrogen is replaced by another group.
Dissociation energy	The energy necessary to cleave a particular bond to give the constituent radicals, *e.g.* HO-H to give HO. and H. requires 118 kcal/mol; while H-O to give .O. and H. requires 100 kcal/mol. The average of 109 kcal/mol is taken as the bond energy for O-H in water.
E isomer	A sterochemical term used to denote that the two groups with the highest ranking under the CIP system are diametrically positioned across a double bond. If they are on the same side then the system is desonated *Z*. *E* and *Z* are abbreviations of *entgegen* and *zusammen*, meaning opposite and togther respectively.
Eclipsed	The dihedral angle is zero between groups which are on adjacent atoms in an ethane molecule. When the dihedral angle is 60°, then the conformation is called *staggered*.

eclipsed   staggered

Electrocyclic	A type of pericyclic reaction in which there is a concerted cyclization of a polyene, or the reverse which results in ring opening.
Electrofuge	A leaving group which departs without its

Electronegativity	electron pair, *i.e.* the leaving group equivalent of an attacking electrophile. The property of an element which represents the degree of attraction which that element has for electrons. The greater the value the greater the attraction.
Electrophile	A species which likes to attack areas of high electron density, *i.e.* negative sites. The electrophile is usually electron deficient or positively charged or both.
Elimination	A reaction where a small part of the original molecule is eliminated to leave behind the bulk of the molecule, which usually now has degree of unsaturation.
Elimination/addition	Overall a reaction which results in substitution of a group, but which proceeds *via* an initial elimination followed by an addition.
Empirical formula	The ratio of the constituent elements in one molecule of the compound.
Enamine	A compound which contains the $R_2C = CR$-NHR grouping. The nitrogen atom usually is alkyated.

$$\text{NHR}$$

$$\overset{\alpha}{\diagup}\diagdown_{\beta}$$

Enantiomer	A stereoisomer which is superimposable on its mirror image.
Enantiomorph	An alternative name for an enantiomer.
Endo addition	In the Diels-Alder reaction if part of the dieneophile is under the diene such that the hydrogens on the dieneophile adopt the equatorial position in the product, then the addition is *endo*. This allows for maximum overlap of secondary orbitals and is thus

	favored, and so this is the kinetic product. The opposite of *exo* addition.
Endo cyclic	Used to describe a double bond that is between two atoms within a ring, as opposed to one which is external to a ring, which is called *exo*.
Ene synthesis	The addition of a reactive dieneophile to an alkane which has a hydrogen in the 3 position.

Enol	The tautomeric isomer of a ketone or aldehyde, *e.g.* the enol of propanone is $CH_2 = COHCH_3$. Usually involved in the keto/enol tautomerism.
Enol ether	The O-aklyl derivative of an enol, *e.g.* $R_2C = CHOR$. Also called vinyl ether.
Enolate	The oxygen anion of an enol.
Enthalpy	The energy of a system.
Entropy	The disorder of a system.
Envelope conformer	One of the puckered conformations of cyclopentane, the other being the half chair conformation.

Epimer	A diastereomer differing at only one chiral centre.
Episulfide	The sulfur equivalent of an epoxide. Also called a thiirane.
Episulfone	A three membered ring which contains two carbons and a SO_2 group.

Epoxide	A three membered ring compound where two corners are occupied by carbon atoms and the third by an oxygen. Also called an oxirane.
Equatorial bond	In a six membered ring system, those bonds which projected approximately along the

Erythro isomer

plane of the six atoms. The opposite of axial.

The stereoisomer of $C_{abx}C_{aby}$ which is represented by the following diagram using the Fischer projection.

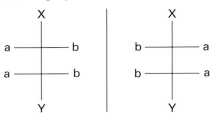

Eschweiler-Clarke

The methlyation reaction of an amine with methanal and methanoic acid to give the N-methyl derivative.

Ester

The condensation product of a carboxylic acid and an alcohol, *i.e.* RCO_2R. Strictly a carboxylic ester, as it is possible for alcohols to form esters with inorganic acids, *e.g.* with carbonic acid to form carbonate esters, $O=C(OR)_2$.

Ether

The molecule of the general formula ROR.

Exhausted aklyation

Used in reference to the aklyation of, for example, an amine derivative, when the nitrogen becomes fully alkylated.

Exo addition

The opposite of *endo* addition, in that the smaller side of the dieneophile is under the cyclic diene and hence there is no opportunity for secondary orbital overlap. This leads to the thermodynamic product being formed.

Exo cyclic — Used to describe a double bond which joins an atom in a ring to an atom which is not within the ring. The opposite of an *endo* cyclic double bond.

F-strain — Face strain is the steric strain present when a covalent bond is formed between two atoms each of which has three large groups.

Favorshii — The rearrangement which occurs when an α-bromoketone, under basic conditions, reacts to form an ester.

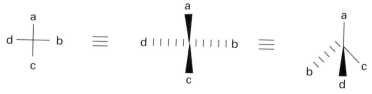

Fehling's solution — Made by dissolving Rochelle salt (sodium potassium 2,3-dihydroxybutanedioate, tetrahydrate) and NaOH in water and adding the mixture to a solution of $CuSO_4$. The Rochelle salt prevents the precipitation of the $Cu(OH)_2$ and the mixture formed is violet due to complexed Cu^{2+} ions. It is reduced by aldehydes to give a red precipitate of Cu_2O. Usually used as a qualitive test for an aldehyde.

Fenton's reagent — A mixture of H_2O_2 and $FeSO_4$. Used as an oxidizing reagent, *e.g.* to hydroxylate aromatic rings but the yields are often poor. *Vide* also Udenfriend's reagent.

Ferrocene — A well known metallocene which comprises of the sandwich compound between an Fe^{2+} ion and two cyclopentadienyl anions.

Field effect — The direct through space polarization of a bond by another charged centre.

Finkelstein — The exchange of one halide for another in aliphatic compounds. It is an equilibrium reaction.

Fischer projection — A notation used to portray the three

dimensional arrangement of the constituent parts of a chiral molecule. The horizontal lines project forwards from the paper, while the vertical lines project backwards.

Formamide
The molecule H_2NCHO.

Formamidine
A compound of the general type $RN=CHNR_2$.

Formylation
The addition of a ←CHO group.

Fragmentation
A reaction in which the carbon moiety is the positive leaving group (*i.e.* the electrofuge) in an elimination reaction.

Free radical
A species that has an unpaired electron.

Friedel-Crafts
The reaction of an alkyl halide, using stiocheiometric amounts of a Lewis acid such as aluminium trihalide, with an aromatic compound to form the alkylated derivative. If the acyl halide is used instead of the alkyl halide then polysubstitution is avoided. This variation is called the Friedel-Crafts acylation reaction.

Friers
The rearrangement of a phenolic ester by heating with a Friedel-Craft catalyst to produce *ortho* and *para* acylphenols.

Frontier orbital

The orbital that is either the highest energy orbital which is occupied (HOMO) or the lowest in energy which is empty (LUMO). *Vide* also Woodward-Hoffmann rules.

Fulvene

A cyclopentadiene with an additional *exo-*cyclic double bond.

Fumaric acid

Trans-butenedioic acid, the *cis* isomer is called maleic acid.

Functional group

That element, or grouping of elements, which will react as a single moiety. Nearly always refers to a heteroatomic element or group such as a halide atom, a hydroxyl or nitro group.

Furan

A five membered heterocyclic ring containing two carbon/carbon double bonds and an oxygen.

Gatterman

The conversion of diazonium salts to the aryl chloride or bromide using copper and HCl or HBr.

Gatterman amide

The synthesis of an amide by the reaction between an aromatic molecule and carbamoyl chloride, R_2NCOCl, in the presence of $AlCl_3$.

Gatterman Koch

The reaction of an aromatic molecule with CO and HCl to give an aromatic aldehyde, in the presence of $AlCl_3$ and CuCl.

Gauche

Vide anticlinal. Also called *synclinal*.

Geminal

The two groups under consideration are attached to the same carbon atom, *e.g.* $\leftarrow CCl_2 \rightarrow >$ is a *geminal* chloride.

General acid catalysis

In the two step reaction where the first step is the protonation of A by the conjugate acid of the solvent to form AH^+, and the second step is the reaction of AH^+ to form the products; then, if the second step is fast and the first step is rate determining, then the reaction is described as being general acid catalysed.

General base catalysis
The same as general acid catalysis, *mutatis mutandis*, but with the first step being the deprotonation of AH.

Glyceraldehyde
The *dextro* or (+) isomer of glyceraldehyde was arbitrarily assigned the configuration shown below by Rosanoff and given the label D. The *levo* or (-) isomer was given the label L. In 1951, by using phase lagged x-ray crystallography, Bijvoet was able to determine that Rosanoff had made the correct choice. When a molecule has been assigned to either the D or L series it is said that its absolute configuration is known. The DL system is little used now having been replaced by the Cahn-Ingold-Prelog system which uses the symbols R and S.

Glycidic ester
The reaction in which a glycidic ester, an α,β-epoxoy carboxylic ester, rearranges under acidic conditions to give, after heating, an aldehyde.

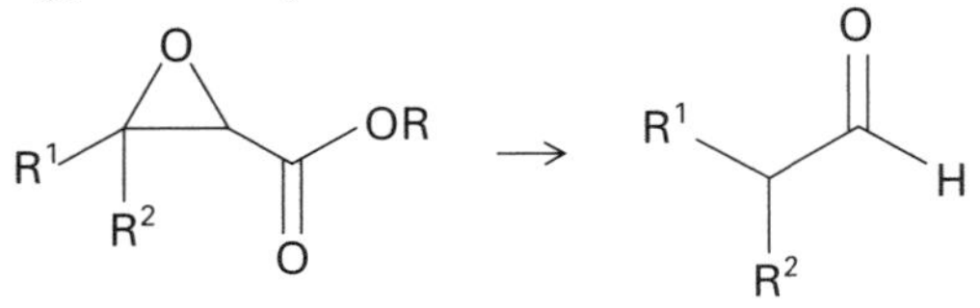

Gomberg-Bachmann
When an acidic solution of a diazonium salt is made alkaline, the aryl portion of the diazonium salt may couple with another aromatic ring.

Grignard reagent
An organobromomagnesium compound with the generic formula of MgBrR, but may more accurately be represented by the Schlenk equilibrium which exists between the following species: $RMgX$, R_2Mg and MgX_2, and "$R_2Mg \cdot MgX_2$".

Guanidine
The compound with the formula $(H_2N)_2C=NH$. A very strong organic base.

Half chair conformer	One of the puckered conformations of cyclopentane, the other being the envelope.

Haloform	A molecule of the general formula CHX_3, where X is a halogen. Used also as the general name for the reaction which is illustrated by the iodoform reaction.
Halogen	The general term for a group VII element, *e.g.* F, Cl, Br or I. The adjective is halo.
Halogen dance	The rearrangement of polyhalobenzenes catalysed by very strong bases, which proceeds *via* an aryl carbanion intermediate (*i.e.* SE1 mechanism).
Halogenation	The addition of a halogen to a molecule.
Halohydrin	A molecule of the general formula $R_2C(OH)X$.
Hammond postulate	This states that for any single reaction step the geometry of the transition state for that step resembles the side to which it is closer in free energy.
Hard acid/base	An acid or a base which is difficult to polarize and hence tends to act as a Brönsted-Lowry acid or base as oppose to a nucleophile or electrophile.
Haworth	A reaction in which an aryl compound is treated with a cyclic anhydride, such as succinic anhydride, and the intermediate Friedel-Crafts product is reduced and then

cyclized *via* an internal Friedel-Crafts reaction to give a 1,2-substituted aromatic compound with a carbonyl group in the side chain. The whole sequence is called the Haworth reaction.

Haworth notation
A type of notation used to describe the 3-dimensional configuration of a compound.

Helicene
A molecule which is chiral due to the whole molecule having a helical shape, *e.g.* hexahelicene, in which one side of the molecule must lie above the other because of steric crowding.

Hell-Volhard-Zelinskii
The reaction in which a carboxylic acid is treated with PBr_3 (formed *in situ* from red phosphorus and bromine) to give the α-bromo carboxylic acid bromide.

Hemiacetal
A molecule of the general type, RHC(OR)OH. Also used to describe a molecule of the general type $R_2C(OR)OH$, which is more correctly called a hemiketal.

Hemihydrate
Formed by the condensation of two hydrates of carbonyl compounds to give the dimeric compounds with the general formula RCH(OH)-O-CHR(OH).

Heteroatom
Either any atom which is different from that under consideration, or any atom which is not a carbon or a hydrogen, but is usually restricted to non-metals, *e.g.* O, S, N or P.

Heterocylic
A cyclic compound that contains a heteroatom within the ring. If the ring is aromatic, then heteroaromatic.

Heterogeneous	An adjective which is usually applied to a catalyst to mean that the catalyst is in a different physical state from that of the reagent whose reaction it is catalysing.
Heterolytic fission	The unequal breaking of a covalent bond so that both electrons are taken by one part, and hence the products are charged.
Hofmann degradation	The reaction of an amide with bromine and sodiuum methoxide in methanol to give the corresponding amine, with the lose of the carbonyl carbon.
Hofmann elimination	The double bond being formed goes mainly toward the least highly substituted carbon.
Hofmann exhaustive methylation	The treatment of an amine with excess methyl iodide to form the quaternary ammonium iodide, which is then converted into the hydroxide by treatment with silver oxide, which on heating in an aqueous or alcoholic solution decomposes to give the alkene.
Homoallylic	Used to describe the situation where there is one carbon unit, usually a methylene group, between the group in question and the carbon/carbon double bond.

v vinylic
a allylic
h homoallylic

Homoaromatic	A compound which contains one or more sp^3 hybridized carbons atoms in an otherwise conjugated cycle, *e.g.* the homotropylium cation, $C_8H_9{}^+$, in which there is one sp^3 carbon, and an aromatic sextet spread over seven carbon atoms.
Homogeneous	An adjective which is usually applied to catalyst to mean that the catalyst is in the same physical state as that of the reagents whose reaction it is catalysing.
Homologization	The addition/removal of a methylene unit to form the homologue.
Homologue	A compound of the same class, but with

	either one more or less methylene units in the carbon chain.
Homolytic fission	The equal breaking of a covalent bond so that both parts take one electron each, so that each product is a radical.
Homotopic	*Vide* stereoheterotopic.
Huang-Minlon	The modification of the Wolff-Kishner reaction in which the reaction is carried out in refluxing diethylene glycol. This version has completely replaced the original method.
Hückel's rule	Electron rings which have $4n + 2$ π electrons will be aromatic, while those which have $4n$ π electrons will be non-aromatic.
Hund's rule	When a number of degenerate orbitals are available and there are not enough electrons to fill them all, then all the orbitals will be half-filled before any of then are fully filled.
Hunsdiecker	The reaction of the silver salt of a carboxylic acid with bromine to give the bromoalkane with one carbon less than the original compound.
Hydrate	A molecule of the general formula $R_2C(OH)_2$, *i.e.* the water addition product of a carbonyl.
Hydration	The addition of water to a molecule.
Hydrazide	A molecule of general formula $RCONHNH_2$.
Hydrazide anion	The R_2NNH^- anion.
Hydrazine	The $H_2N\text{-}NH_2$ molecule, which forms hydroazo compounds when one hydrogen on each nitrogen is substituted, RHN-NHR, and hydrazines, when the substitution is only on one nitrogne, RHN-NH$_2$.
Hydrazoic acid	HN_3.
Hydrazone	A compound with the general formula $R_2C = NH\text{-}NR_2$.
Hydride	The H^- anion.
Hydridization	The hypothetical mixing of atomic orbitals to form hybrid orbitals.Hydroacylation
	The addition of an RCHO group across a double bond, resulting in the addition of hydrogen to one end and an acyl, $RC = O\rightarrow$, group to the other.

Hydroboration	The addition of a HBR_2 group across a double bond.
Hydrocarbon	A compound composed of just carbon and hydrogen atoms, may be alicyclic, cyclic, aromatic, saturated or unsaturated.
Hydrocarboxylation	The addition of HCO_2H across a double bond.
Hydrocyanic acid	The molecule HCN.
Hydroformylation	The addition of HCHO across a double bond, achieved by the reaction of CO and H_2 with a cobalt catalyst.
Hydrogen bonding	The weak bond between a functional group X-H and an atom or group of atoms, Y, in the same or different molecules, *i.e. intra-* or *inter-*molecular respectively. X and Y are usually oxygen, nitrogen or fluorine.
Hydrogenation	The addition of a hydrogen molecule across an unsaturated bond, usually a carbon/carbon bond.
Hydrogenolysis	The cleavage of a functional group and its replacement with a hydrogen atom.
Hydrolysis	The splitting of a large molecule into usually two fragments by the overall addition of water, *e.g.* ester plus water gives alcohol and carboxylic acid.
Hydroperoxide	The ROOH grouping.
Hydroxamic acid	Compound of the general type RCONHOH. The enol form is called hydroximic acid.

keto ⇌ enol

Hydroxide ion	The OH⁻ anion.
Hydroxonium ion	The H_3O^+ cation.
Hydroxyl	The ←OH group.
Hydroxylamine	The molecule NH_2OH.
Hydroxylation	The addition of a molecule of H_2O_2 to give a 1,2-diol.
Hydroxynitrile	*Vide* cyanohydrin.
Hyperconjugation	Delocalization of electrons using the σ molecular bond electrons from a hydrogen/carbon bond, also called a no-bond resonance. Muller and Mulliken call the hyper-

conjugation found in neutral molecule sacrificial hyperconjugation, and this involves no-bond resonance and charge separation not found in the ground state. In the isovalent hyperconjugation found in free radicals and carbonium ions, the canonical forms display no more charge separation than is found in the main form.

Hypohalous acid	An acid of the general formula HOX, where X is a halide.
Hypsochromic shift	The opposite to a bathochromic shift.
I strain	Internal strain which results from changes in ring strain in going from tetrahedral to a trigonal carbon or *vice versa*.
Imidate	*Vide* imidic ester.
Imidazole	An unsaturated heterocycle, also called iminazole or glyoxaline.

Imidazolide	The N-acyl derivative of an imidazole.
Imide	Compound of the general type RC(=O)-NHC(=O)R.
Imidic ester	Compound of the general type RC(OR)=NR, also called imidate, imino ester or imino ether.
Imidine	Compound of the general type RC(=NH)-NH-C(=NH)R.
Imido sulfur	The R-N=S=N-R grouping.
Imidoyl chloride	The R-N=CR-Cl grouping, also called imidoyl chloride.
Imine	The $R_2C=NR$ grouping, also called ketimine.
Iminium ion	An ion of the general formula $R_2C=NR_2{}^+$.
Imino chloride	*Vide* imidoyl chlorides.
Imino ether	*Vide* imino esters.
Iminonitrile	The $RC(C\equiv N)=NH$ grouping.
In situ	Used to describe when reagents are made in the reaction vessel, usually because they are unstable and so cannot be stored.
Inclusion compound	A host compound which forms a crystal

	lattice that has spaces large enough for a guest molecule. The only bonding between the host and the guest compounds are van der Waals forces. The spaces in the lattice are in the form of long tunnels or channels, unlike clathrate compounds in which the spaces are completely enclosed.
Indicator	Usually a water soluble acid/base conjugate pair in which one or both of the species is highly coloured. The pK_a is around the pH which is of interest to the reaction being studied.
Inductive effect	The polarization of one bond caused by the polarization of an adjacent bond acting through the σ bonds.
Initiation	The first step of a free radical reaction in which the number of free radicals increases.
Insertion	A reaction in which another atom is introducted into a chain. Often the species which is introduced is a carbene, nitrene or such like.
Intermediate	A molecular species which actually exists, if only briefly, along a reaction pathway from the starting material to the final product. Such species exist in a well in a free-energy profile, *i.e.* there is an activation barrier before and after the intermediate. The deeper the well the more stable the intermediate.
Inversion	Used in respect to the configuration of a chiral centre, when in the course of a reaction the geometry of the chiral atom is reversed.
Iodoform	The reaction of a secondary alcohol which is adjacent to a terminal methyl group, which after oxidation to a ketone by a iodine/hydroxide mixture, then further reacts to from the carboxylic acid and CHI_3, the latter precipitates as a yellow solid. It is a test for the general structure: RCHOHMe or RC(=O)Me.
Iodonium ion	The iodine equivalent of a bromonium ion.
Ionic bond	A bond which is really a Coulombic attraction between two or more nuclei, which has resulted from the transfer of electrons from

one atom to another which gives rise to cations and anions and hence the electrostatic interaction.

Isochronous   Used for hydrogens which are indistinguishable in their NMR spectrum.

Isocyanate   The univalent $\leftarrow N=C=O$ group.

Isocyanide   *Vide* isonitrile.

Isoinversion   *Vide* isoracemization.

Isomer   A configuration of the molecule which corresponds to the molecular formula.

Isonitrile   The univalent $\leftarrow N=C$ group. Note that carbon is only divalent in this moiety.

Isoracemization   Used to describe the situation in which racemization occurs faster than isotope exchange on the chiral centre. The first step must be isoinversion.

Isothiocyanate   The univalent $\leftarrow N=C=S$ group.

Isotope   A variant of an element which has the same atomic number, but a different atomic mass, *i.e.* a different number of neutrons within the nucleus.

Isotope effect   Primary isotope effect is where the substitution of a deuterium atom for a hydrogen atom has an effect on the rate because that bond is broken in the rate determining step. Secondary isotope effect is where there is an effect on the rate but the C-H/D bond does not break in the reaction. This is subdivided into α and β effects. In the latter, substitution of a deuterium for a hydrogen β to the position of bond breaking slows the reaction due to the effect on hyperconjugation. In the former a replacement of a hydrogen by a deuterium at the carbon containing the leaving group effects the rate. The greater the carbonium ion character the greater the effect.

Jones reagent   A solution of chromic acid and sulfuric acid in water, used to readily oxidize secondary alcohols to ketones.

Ketal   Compound of the general type $R_2C(OR)_2$. *Vide* also acetal.

Ketene   *Vide* ketoketen.

Ketene acetal	The functional grouping $R_2C = C(OR)_2$.
Ketene aminal	The functional grouping $R_2C = C(NR_2)_2$.
Ketenimine	The functional grouping $R_2C = C = N-R$.
Ketenimmonium salt	The functional grouping $R_2C = C = NR_2{}^+$.
Ketimine	*Vide* imine.
Keto	The prefix to indicate the presence of a carbonyl group within the molecule which is joined to two carbon atom, Thus, for example, a keto acid is $RC(=O)CO_2H$.
Keto/enol tautomerism	The tautomerism between the keto and enol forms: $R_2CH-C(=O)R$ and $R_2C = C(-OH)R$. This type of isomerism requires at least one α hydrogen atom.
Ketone	A molecule with a non-terminal carbonyl group, *e.g.* $RC(=O)R$.
Ketoxime	The functional grouping $R_2C(=N-OH)$, where R is not hydrogen.
Ketyl radical	The $R_2C \equiv (-O^-)$ species.
Kiliani method	Used in carbohydrate chemistry to extend the length of the carbon chain by one. Following the formation of a cyanohydrin of the corresponding ketone, the nitrile grouping is then hydrolysed to give the carboxylic acid.
Kizhner	The Russian name for the Wolff-Kishner reaction.
Knoevenagel	The condensation reaction in which a methylene unit which has two adjacent -M groups such as carbonyl, ester or nitro groups, then reacts in the presence of base with an aldehyde or ketone, which usually do not have an α-hydrogen. Dehydration usually follows to yield the α-unsaturated carbonyl compound.
Koble	The electrolyic decarboxylation dimerization of the potassium salts of two carboxylate anions, used to prepare symmetrical R-R alicyclic compounds from $RCO_2^-K^+$.
Koble-Schmitt	The carboxylation of sodium phenoxides by carbon dioxide in the *ortho* position. The potassium salt forms mainly the *para* substituted compound.
Koch	The hydrocarboxylation of a double bond using CO and H_2O, in the presence of H^+ ions at elevated temperatures and pressures.

Kornblum's rule	As the nature of a reaction changes from SD1 to SN2, an ambident nucleophile becomes more likely to attack with its less electronegative atom.
Lactam	A cyclic amide.
Lactide	A cyclic dimer formed from two molecules of α-hydroxy acids.

Lactone	A cyclic ester.
Lanthanide shift	An organic complex of a lanthanide element which has the property of shifting the NMR signals of a compound with which it can form a coordination complex. Chiral lanthanide shifts reagents shift the peaks of the two enantiomers to different extents.
Lasso mechanism	The groups that are going to react together are placed next to each other and then encircled, *i.e.* "lassoed", to indicate that they have bonded, while the remaining truncated molecules fuse together.
Least motion	A theory proposed by Hine that those elementary reactions will be favored that involve the least change in atomic position and electronic configuration.
Levelling effect	If in any given solvent, two bases appear to be equally protonated within experimental error, then even though they may actually have very different pK_bs, they appear to be approximately the same in the given solvent, *i.e.* the solvent is said to level the effect of the bases. Equally applicable to acids.
Lewis acid/base	A species which is capable of accepting/donating a pair of electrons, *e.g.* H^+ or $AlCl_3$ versus NH_3.
Lewis structure	The representation of a molecule, ion or radical which shows the position of the

localized electrons in the bonding orbitals. If there are two or more such structures for any given compound, then they may also be called canonical forms, which when combined give the resonance structure.

Lindlar catalyst

Pd on $CaCO_3$, partly poisoned with $Pb(OAc)_2$. Used to partially, not partly, hydrogenate alkynes to alkenes.

Lone pair

A pair of electrons in the valence orbital, *e.g.* the non-bonding pair of electrons in NH_3.

Lossen

The O-acyl derivative of hydroxamic acid give an isocyanate when treated with a base or, sometimes, just heat.

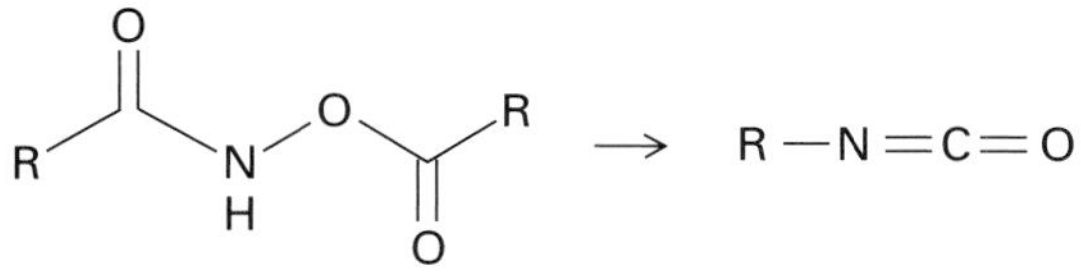

Lucas test

When an alcohol is mixed with concentrated HCl and $ZnCl_2$ at room temperature, tertiary alcohols form the chloride immediately, while secondary ones take about five minutes, and primary ones react far more slowly.

Maleic acid

Cis-butenedioic acid, the *trans* isomer is fumaric acid.

Malic acid

2-Hydroxypentandioic acid.

Malonic acid

A compound of the formula $RCH(CO_2H)_2$, *i.e.* α *geminal* carboxylic acid.

Mannich

The reaction of a ketone with formaldehyde in the presence of a secondary amine under acidic conditions to give the 1,3 amino carbonyl adduct, which is refered to as a Mannich base. After permethylation and heating with silver oxide the α-unsaturated carbonly compound may be formed.

Markovnikov

The addition of the positive end of the addition molecule to the carbon which had the most hydrogens on it. This may be remembered by the saying that "the rich become richer, while the poor become poorer."

Meervein-Ponndorf-Verley　The reduction of a ketone to a secondary alcohol using isopropanol and aluminium isopropoxide. The opposite of the Oppenauer oxidation.

Meisenheimer　The rearrangement of a tertiary amine oxide to give a substituted hydroxylamine. The migrating group is usually allylic or benzilic.

Mercaptal　The sulfur equivalent of an acetal, $CRH(SR)_2$ sometimes called thioacetal. The $CRH(OR)SR$ type compound tends to be called a hemimercaptol instead of the more correct hemimercaptal.

Mercaptan　A molecule contianing the ←SH functional group, *i.e.* the sulfur equivalent of an alcohol. Also called a thiol.

Mercaptide ion　The RS^- anion.

Mercaptol　Compound of the general type $CR_2(SR)_2$, also called mercaptole, dithioacetal or thioketal. The $CR_2(OR)SR$ type compound is called a hemimercaptol.

meso　A stereochemical term meaning that there is a plane of symmetry passing through the molecule and hence it is not optically active.

Mesoionic　A molecule which cannot be satifactorily represented by Lewis forms not involving charge separation. Nearly all known examples contain five membered rings of which the most common are the sydnones.

Mesomeric effect　The movement of π electrons in the π orbital system.

Mesylate　A compound of the type $p\text{-MeC}_6\text{H}_4\text{SO}_2\text{-O-R}$.

Metaformaldehyde　*Vide* paraformaldehyde.

Metallocene　Also called a sandwich compound, it usually contains two cyclopentadienyl anions which form a sandwich around a metallic ion, but may contain a tropylium ion instead of one of the cyclopentadienyl anions.

Metathesis

The reaction when two alkenes react in the presence of catalysts to interchange their alkylidiene groups ($R_2C=$). Also called dismutation and disproportionation of olefins. The resulting mixture is a usually the statistical spread of all possible products.

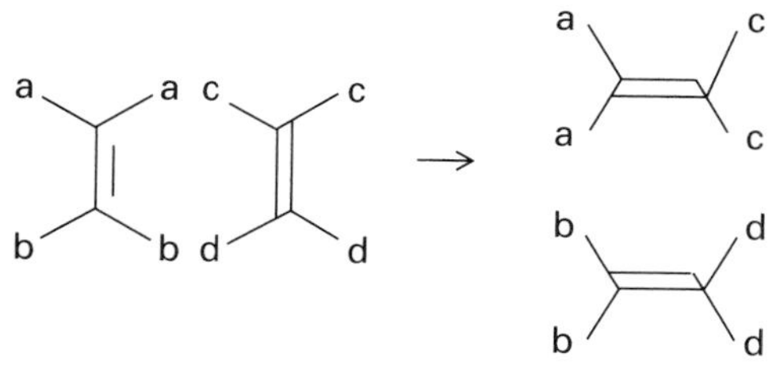

Methanonium ion

The CH_5^+ cation, very rare except in students' examination scripts.

Methoxide ion

The CH_3O^- anion.

Methylene unit

The $\leftarrow CH_2 \rightarrow$ unit.

Michael-type

The addition to an α,β-unsaturated carbonyl compound by a carbanion, in particular if the anion is stabilized by a carbonyl group, or other -M group such as nitrile or nitro.

Molecular formula

The formula which gives the actual number of each element in one molecule of the compound.

Molecular orbital theory

A theory used to describe the electronic structure of a molecule by the combination of atomic orbitals which give rise to molecular orbitals. The latter being orbitals which are centred on more than one nucleus.

Molozonide

The result of the first step of the 1,3 dipolar addition of ozone to an alkene, also called the initial or primary ozonide.

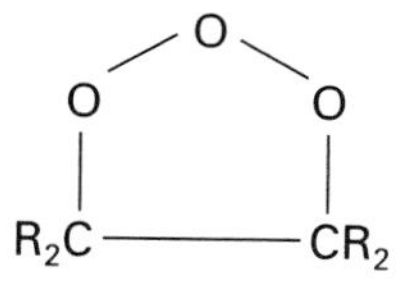

N-alkylation

Vide alkylation.

Naphthalene

The compound with the formula $C_{10}H_8$, a condensed aromatic.

Nef

Primary or secondary aliphatic nitro compounds may be hydrolysed respectively to aldehydes or ketones by treatment with firstly a base and then mild aqueous acid which yields the aci form of the nitro compound. This can then be hydrolysed by concentrated sulfuric acid to give the carbonyl compound.

Neopentane

The molecule $C(CH_3)_4$, whose radical neopentyl radical has the structure $Me_3CCH_2\equiv$. More correctly written as *neo*-pentane.

Newman projection

A notation used to indicate the geometry around an sp^3/sp^3 bond.

Nitration

The addition of a $\leftarrow NO_2$ group, *i.e.* nitro compounds.

Nitrene

The univalent $\leftarrow N$ species, *i.e.* the nitrogen equivalent of a carbene. The simplest nitrene NH, is also called nitrene, as well as imidogen, azene or imine.

Nitrenium ion

Species of the form RHN^+ or $R_2C=N^+$.

Nitrile

Compound of the general type $R\text{-}C\equiv N$, also called cyanide.

Nitrile oxide

The grouping $R\text{-}C\equiv N^+\text{-}O^-$.

Nitrilium ion

The $R\text{-}C\equiv N^+\text{-}R$ cation.

Nitrite ester

A molecule with the general grouping $R\text{-}O\text{-}N=O$.

Nitrite ion

The NO_2^- anion.

Nitro

The $\leftarrow NO_2$ group.

Nitrone

A molecule of the general form $R_2C=N^+\text{-}(\text{-}O^-)\text{-}R$.

Nitronic ester

The result of O-alkylation of the carbanion of the aci tautomer of a nitro compound and having the general formula $R_2C=N^+\text{-}(\text{-}O^-)\text{-}O\text{-}R$.

Nitronium ion

The NO_2^+ cation.

Nitrosation

The addition of the $\leftarrow NO$ group.

Nitroso

The $\leftarrow NO$ group.

Nitrosonium ion

The NO^+ cation.

Nitrous acid

The molecule HONO.

No-bond resonance

Vide hyperconjugation.

Non-alternant

Vide alternant.

Norrish I	The photolytic cleavage of an aldehyde or ketone to give $RCO^{\bullet}$ and $R^{\bullet}$.
Norrish II	The photolytic cleavage of $R_2CH\text{-}CR_2\text{-}CR_2COR$ to give $R_2C=CR_2$ and CHR_2COR.
Nosylate	A compound of the type $p\text{-}O_2NC_6H_4\text{-}SO_2\text{-}O\text{-}R$.
Nucleofuge	A leaving group which departs with an electron pair.
Nucleophile	A species which attacks areas of positive charge or electron deficiency. Usually negatively charged or possessing a lone pair.
O-alkylation	*Vide* alkylation.
Octant rule	A rule which attempts to predict the conformation of a carbonyl carbon by cutting it into octants using three perpendicular planes that bisect the carbonyl carbon, and then looking at the orientation of the substituents within each octant. Used mainly on cyclohexanone derivatives.
Olefin	A compound with a carbon/carbon double bond, more properly referred to as an alkene.
Oligomer	A polymer which contains only a few tens of the monomer, as opposed to several tens which would be a telomer, or hundreds of the monomer which would be a true polymer.
Onium salt	Formed when an Lewis base expands its valence, *e.g.* $Me_4N^+I^-$. Analogous to ate complexes.
Oppenauer	The oxidation of a secondary alcohol to a ketone using $(Me_2CHO)_3Al$, the reverse of the Meerwein-Ponndorf-Verley reduction.
Optically active	A compound which rotates the plane of polarized light. It possess a chiral centre.
Orbital	The solution of a Schrödinger wave equation which is used to describe the behaviour of an electron is called an eigenfunction. When a graphical representation of such an eigenfunction is made it is called an orbital.
Orbital symmetry	A property of orbitals, which results from the wave nature of the equations used to define the properties of the electron.

Order of a reaction	The summation of the indexes which govern the concentrations of the reagents in the rate equation.
Organometallic	A compound which contains an organic component which is covalently bonded to a metal atom.
Ortho carbonate	A molecule of the type $C(OR)_4$.
Ortho effect	In electrophilic substitution reactions of aromatic rings which already have two substitutent, one of which is a *meta* directing group, which is positioned *meta* to an *ortho/para* directing group, then the new substitutent tends to go to the *ortho* position.
Ortho ester	A molecule of the general formula $RC(OR)_3$.
Ortho/para ratio	The ratio of electrophilic substitution at the 2, as opposed to the 4, position of a mono-substituted aromatic ring.
Osazone	A molecule of the general formula $RC(=N\text{-}NHPh)\text{-}C(=N\text{-}NHPh)R$, particularly important in carbohydrate chemistry.
Out-in isomerism	Found in the salts of tricyclic diamines with the nitrogens at the bridgeheads. If the bridges are longer than six atoms then the N-H bond may be formed either inside the cage or outside.
Oxa-di-π-methane rea	The light induced rearrangement of β,γ-unsaturated ketones to give the acyl derivative of cycloproprane.

Oxalic acid	The molecule with the formula $(CO_2H)_2$.
Oxaphosphetane	The four membered phosphorous/oxygen ring formed as an intermediate in the Wittig reaction.

$$R_3P \longrightarrow O$$

Oxetane	A four membered ring comprising of three atoms of carbon, one of oxygen.

Oxidation	The gain of an oxygen atom or the loss of a hydrogen atom or an electron.
Oxidation number	The nominal loss or gain of electrons which has occurred on any given atom. *Vide* the appendix on oxidation numbers for more detail.
Oxime	The $R_2C = NOH$ grouping. Oxime ether A compound of the general formula $R_2C = N\text{-}O\text{-}R$.
Oxirane	Another name for an epoxide.
Oxo process	The commercial hydroformylation of an alkene by treatment with CO and H_2 over usually a cobalt carbonyl catalysis.
Oxonium ion	The $OR_3{}^+$ or $R_2C = O^+\text{-}R$ cation.
Oxy-Cope	The rearrangement of a 3-hydroxy-1,5-diene to the 1-hydroxy-1,5-diene which tautomerizes to the ketone or aldehyde.

Oxyamination	The formation of a carbon/oxygen and carbon/nitrogen bond from a carbon/carbon double bond.
Ozonide	The five membered ring which results from the ozonolysis of carbon/carbon double bonds.

Ozonolysis	The cleavage of a carbon/carbon double bond with ozone to give the aldehyde or ketone fragmentation derivatives.
π donor	A compound that donates a pair of electrons which were originally in a π-orbital, and so forms a π complex. An example of

which is when benzene complexes with an electrophile, initially forming a π complex which then rearranges to form a σ complex or Wheland intermediate.

π-orbital

A molecular orbital which has one nodal place.

p-orbital

An atomic orbital which has one nodal plane.

Paenecarbanion

An E2 elimination reaction whose transition state resembles the E1cb mechanism.

Paenecarbonium

An E2 elimination reaction whose transition state resembles the E1 mechanism.

Para-Claisen

Vide Claisen.

Paraformaldehyde

The linear polymer form of formaldehyde, $(CH_2O)_n.H_2O$. The cyclic trimer is called trioxane, trioxymethylene or metaformaldehyde. The trimer of acetaldehyde is called paraldehyde, while its tetramer is called metaldehyde.

Paramagnetic ring currents

Vide ring currents.

Paratropic

Used to describe a compound which substains paramagnetic ring currents.

Parrafin

The trival name for an alkanes.

Passerini

The formation of an α-acyloxyamide when an isontrile is treated with a carboxylic acid and an aldehyde or ketone.

Paterno-Büchi

The photochemical addition of a ketone to an alkene to give an oxetane.

Pechmann

Vide von Pechmann

Peptide

A polymer of amino acids.

Per

A prefix used to indicate that there is a maximum amount of the element under consideration, *e.g.* perchloroethane for C_2Cl_6.

Peracid

A molecule of the general formula RC(=O)-O-O-H.

Perester

A molecule of the general formula RC(=O)-O-O-C(=O)R.

Pericyclic

A reaction type in which the bond breakage and formation occur simultaneously without any charged or radical intermediates. There are three main sub-divisions, namely electrocyclic, sigmatropic and cycloadditions.

Perkin

The reaction of an acid anhydride with an aromatic aldehyde in the presence of base, which after dehydration results in the β-aryl-α,b-unsaturated carboxylate anion.

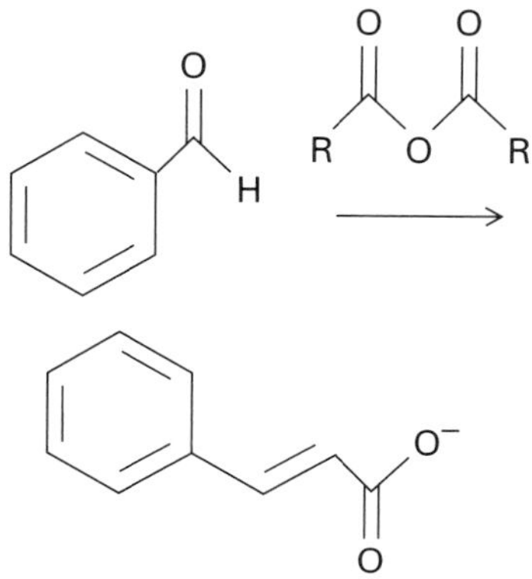

Peroxide

A molecule with the general formula R-O-O-R.

Peroxide effect

The use of peroxides to change the regio-stereochemistry from Markovnikov to anti-Markovnikov.

Peroxirane

An epoxide which has an oxygen atom attached to the oxygen in the ring, and so results in an ylid.

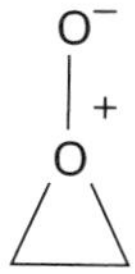

Peroxy acid

Vide peracid.

pH

Defined as $-\log_{10}[H^+]$. Strictly, may only be defined with water as the solvent, but is often used in a less rigid sense in different solvents.

Phantom atom

Used in the CIP notation to terminate the valency of all atoms except hydrogen, when the atom in question was involved in a multiple bond.

Phenol

The molecule C_6H_5OH.

Phenol-dienone

The reverse of the dienone-phenol rearrangement.

Phenonium ion

The bridged carbonium cation which results

from the departure of a leaving group which is β to the aromatic ring.

Phenoxide ion   The $C_6H_5O^-$ anion.
Phenoxy   The univalent group $C_6H_5O\rightarrow$.
Phenyl   The univalent group $C_6H_5\rightarrow$.
Phenyl cation   The $C_6H_5^+$ cation.
Phenylcarbene   The C_6H_5CH neutral molecule.
Phosgene   The molecule $COCl_2$.
Phosphine   A compound of the type R_3P.
Phosphonate   A compound of the general formula $(RO)_2P(=O)\text{-}R$.
Phosphonium ion   A compound of the type R_4P^+.
Phosphorane   An ylid of the type $Ar_3P^+\text{-}CH^-\text{-}R$.
Photochemistry   Reactions which proceed when exposed to light. Usually radical reactions or pericyclic reactions, as the reactions often occur in either the gas phase or in inert solvents which do not favor charged intermediates.
Photolysis   The cleavage of molecule induced by light.
Pinacol/Pinacolone The rearrangment of a 1,2 diol when treated with concentrated sulfuric acid. Named after the rearrangement of pinacol to pinacolone.

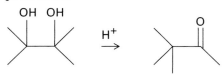

Pinner   The synthesis of the hydrochloride salt of an imidic ester when a mixture of a nitrile and an alcohol is treated with dry HCl. If hydrochloric acid is used the intermediate product is subsequently hydrolysed to give the ester.
Piperidine   A saturated five membered cyclic compound which contains one nitrogen and four carbon atoms.
Pitzer strain   The strain which results from eclipsed conformations, as opposed to transannular strain.

pK$_a$ value

A measure of the strength of an acid. The lower the value the stronger the acid. Negative values are possible for the strongest of acids. Defined as -log$_{10}$([A$^-$][H$^+$]/[HA]).

pK$_b$ value

A measure of the strength of a base. Defined as -log$_{10}$([BH$^+$][OH$^-$]/[B]).

Poisoning

A catalyst is said to be poisoned if it is working at less that maximum efficiency because of some chemical contamination, which is commonly sulfur.

Polarizability

The ease with which the electron cloud surrounding a group may be distorted by the approach of an electrical charge. When the electron cloud is easily distorted it is said to be very polarizable and so is also soft in nature.

Poly

A prefix meaning many.

Polymerization

The addition of many moners to each other to form a very long chain, which is called a polymer. Short polymers are called telomers, and even shorter ones are called oligomers.

Prevost method

The overall *anti*-hydroxylation of an alkene achieved by treating the alkene with a 1:2 molar ratio of iodine and silver benzoate, followed by hydrolysis. Overall *syn* hyroxylation may be achieved by the method of Woodward.Prilezhaev

The formation of an epoxide from the reaction of perbenzoic acid on an alkene.

Prochiral

A molecule of the general formula CABX$_2$, if one of the X groups was substituted for a group Y, then the molecule would become chiral.

Product spread

If a reaction gives rise to two isomeric products, then the greater the difference in yields of the two products the wider the product spread.

Propagation

A reaction in which the number of radicals formed equals the number of radicals comsumed.

Propargylic

The group HC$\equiv$C-CH$_2$-O$\rightarrow$.

Protic solvent

A solvent which is capable of forming hydrogen bonds with anionic solutes.

Prototropic

The isomerisation of a carbon/carbon bond,

usually to give the thermodynamic equilibrium mixture of products. Often achieved by the addition of a strong base and proceeds *via* a proton removal and an allylic rearrangement.

Puckered ring

When a ring system adopts a conformation which is not planar, it is said to be puckered, *e.g.* boat or chair conformations of cyclohexane.

Pyramidal inversion

The inversion of the configuration of a molecule which has a trigonal pyramidal geometry, such as simple nitrogen compounds. Also called the umbrella effect.

Pyran

A doubly unsaturated six membered ring compound which contains one oxygen and five carbons.

Pyrazole

A doubly unsaturated five membered cyclic compound which contains two nitrogen atoms.

3H–pyrazole

Pyrazoline

An unsaturated five membered cyclic compound which contains two nitrogen atoms.

1–pyrazoline   2–pyrazoline

Pyrazolone

An unsaturated five membered cyclic compound which contains two nitrogen atomst and a carbonyl group.

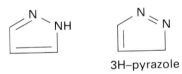

Pyridine

An aromatic six membered cyclic compound, which contains only one nitrogen. Often used as a solvent, or as a non-aqueous base.

Pyridinium ion

The $C_5NH_6^+$ cation.

Pyridone

A doubly unsaturated six membered ring compound with only one nitrogen and five carbons and one carbonyl group. Also known as pyrone.

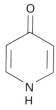

4–pyridone   2–pyridone

Pyrolytic

A elimination reaction which occurs on heating, requiring no other reagent.

Pyrrole

An aromatic five membered cyclic ring containing a nitrogen atom and four carbons.

Pyrrolidine

A saturated five membered cyclic ring four carbon atoms and a nitrogen.

Pyrroline

A five membered cyclic ring containing four carbon atoms and a nitrogen atom with one carbon/carbon double bond.

3–pyrroline

Pyruvic acid

A compound of the general formula, R-CH_2-CO-CO_2H.

Pyrylium ion

The six membered aromatic cation which contains one oxygen and five carbons.

Quasi racemate

A racemic compound formed between the *R* or *S* configuration of one compound and the *S* or *R* configuration of another, but closely related, compound.

Quinoline

Two condensed aromatic rings in which one carbon is replaced by a nitrogen atom.

Quinone

The *ortho* or *para* dicarbonyl derivative of cyclohexadiene.

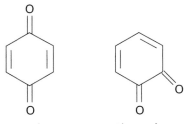

para–quinone   *ortho*–quinone

R configuration

The arrangement of the substituents around a sp³ hybridized centre, in which the group with the lowest priority according to the CIP rules, D, is furthest away from the viewer, and the remaining groups in order of decreasing priority, A > B > C, are arranged clockwise. If the arrangement is anticlockwise then it is the *S* configuration.

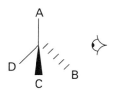

Racemic compound

A compound formed by the one to one mixture of the *R* and *S* enantiomers of compound, in which there is an unusually strong van der Waals interaction between the species. They are often characterized by the melting-point against molar ration curve.

Racemic mixture

A mixture which contains equal concentrations of both enantiomers and hence as a whole does not rotate the plane of polarized light.

Racemization

The process whereby a pure enantiomer is converted to a mixture which contains an

	increasing proportion of the opposite configuration, until the point is reached where the concentration of the enantiomers is equal, then the racemization process is complete.
Raney Nickel	An activated nickel catalyst.
Rate determining step	The slowest step in a multi step reaction, and hence the one which determines the overall rate of the reaction.
re face	Used to describe the configuration around a trigonal atom where the groups are arranged clockwise in the order $A > B > C$ according to the CIP priority rules. If the groups are arranged anticlockwise, then the configuration is said to be *si*.
Rearrangement	A reaction which involves a change in the configuration of the starting material, but with no overall change in the molecular formula.
Reduction	A reaction in which the oxidation level of the starting material is reduced, usually by the addition of a hydrogen or an electron, or the removal of an oxygen.
Reed	The chlorosulphonation using chlorine and SO_2 to give RSO_2Cl from RH.
Reformatsky	The reaction of an a-brominated ester in the presence of zinc with a ketone compound to give the a-unsaturated ester adduct.
Regiospecificity	When a reaction can potentially give rise to two or more structural isomers, but actually only produces one, then the reaction is said to be regiospecific.
Reimer-Tiemann	The formylation of aromatic rings using $CHCl_3$ and OH^-. Useful only for phenol and a few other heterocyclic compounds.
Relative atomic mass	The sum of the number of protons and neutrons in the nucleus of an element.
Relative configuration	Defines the molecular co-ordinates relative to an arbitrary standard.
Resonance	The representation of a real structure as the time-weighed average of two or more canonical forms.
Retention	Used to describe the situation when the relative geometry of the chiral centre under

investigation is the same at the end of the sequence of reaction steps as it was at the beginning. The opposite is inversion. If the geometry is scrambled then there is racemization. Where racemization occurs by proton exchange and the racemization took place faster than the related isotopic exchange process, then the process is called isoracemization.

Ring current
: The electric current which may be induced to flow around a system and thus characterizes it as being aromatic. The ability to substain such a ring current is called diatropicism.

Ring/chain tautomerism
: The tautomeric interconversion between the closed ring form and the open chain form of a molecule, *e.g.* in sugars, the interconversion between the open chain aldehyde and the closed ring hemiacetal.

Ritter
: The addition of an alcohol to a nitrile to give an amide, in the presence of a strong acid.

Robinson annulation
: The reaction of a cyclic ketone with an α,β-unsaturated methyl ketone to form another cyclic ketone. It is the cyclic version of the Michael reaction.

Rosenmund reduction
: The reduction of an acyl chloride to an aldehyde by hydrogenation using a palladium on barium sulphate catalyst.

Rotaxane
: A molecule in which a linear portion is threaded through a ring and cannot become free because of bulky end groups.

Ruzicha cyclization
: The reaction of a long chained dicarboxylic acid with ThO_2 to form a cyclic ketone. The reaction requires high temperatures and results in the overall pyrolytic elimination of CO_2 and water.

σ bond
: A molecular orbital which has no nodal plane.

σ complex
: *Vide* Wheland intermediate.

S configuration
: *Vide R* configuration.

S number
: The sum of the number of atoms in the bridges of a bicyclic system.

s orbital
: An atomic orbital which has no nodal planes.

Sandmeyer	The conversion of diazonium salts to aryl chlorides or bromides using cuprous chloride or bromide.
Sandwich compound	*Vide* metallocenes.
Saponification	The hydrolysis of an ester with an alkali to give the salt of the acid.
Saytzeff's rule	The German spelling of Zaitsev's rule. Also sometimes spelt Saytzev.
Schiff base	*Vide* aldimine.
Schlenk	*Vide* Grignard reagent.
Schmidt	The Beckmann reaction performed using hydrogen azide, HN_3, instead of using hyrdoxylamine *via* the oxime, which results in the overall conversion of a carboxylic acid to an amine with one carbon less.
Scholl	The coupling of two aromatic molecules by the treatment with a Lewis acid and a proton acid.
Schotten-Baumann	The formation of an ester from a carboxylic acid and an alcohol using an aqueous alkali.
Secoalkyation	The net addition of a $\leftarrow CR_2\text{-}CR_2\text{-}CO_2R$ group to a carbonyl group.
Secondary isotope	*Vide* isotope effect.
Semicarbazide	The molecule $NH_2NHCONH_2$.
Semicarbazone	Compound of the general type $R_2C= NNHCONH_2$.
Semidine	A molecule of the general formula *o-* or *p-*arylaminoanilines.
si face	*Vide re* face.
Side product	A product that is produced in minor amounts compared to the main product. However, this sometimes may be the desired product, or it may give very valuable information about the possible mechanistic pathways which have been followed.
Sigmatropic	*Vide* pericyclic.
Silyl enol ether	Enol ether in which the hydrogen of the hydroxyl group has been replaced by an alkylated silicon, *e.g.* the trimethylsilyl group (TMS).
Simmons-Smith	The reaction between an alkene and CH_2I_2 with a Zn-Cu couple leads to the cyclopropane adduct without the problem of having a free carbene in the reaction mixture.

Simonini | A 2:1 mixture of the silver salt of a carboxylic acid, RCOOAg, reacting with iodine gives the ester RCOOR. A 1:1 mixture gives the iodo product, RI, which is similar to the Hunsdiecker reaction.

Singlet | *Vide* triplet.

Skew boat | *Vide* twist conformer.

Skraup | The reaction between glycerol, sulfuric acid, a primary aromatic amine and an aromatic nitro compound to give a quinoline.

Small angle strain | The strain that results from the angles being much less than those resulting from normal orbital overlap. In three and four membered rings the bonds which form the ring are called bent or banana bonds.

Sodium amalgam | A solution of sodium in mercury, often used as a reducing agent.

Sodium borohydride | The compound with the formula $NaBH_4$.

Sodium copper cyanide | A mixed cyanide prepared from NaCN with CuCN, used to convert vinyl bromides to vinyl cyanides.

Sodium cyanoborohydride | The compound with the formula $NaBH_3CN$.

Soft acid/base | An acid or a base which is easy to polarize and hence tends to act as a nucleophile or electrophile respectively as oppose to a Brönsted-Lowry acid or base.

Solvated electron | In the reaction between sodium and liquid ammonia, free electrons are formed with are solvated by the ammonia molecules. Used as a reducing agent.

Solventolysis | The splitting of a large molecule usually into two fragments by the overall addition of a molecule of the solvent, which is often water and then the reaction is called hydrolysis.

Solvolysis | The substitution of a functional group with a molecule of the solvent.

Sommelet-Hauser | The rearrangement of a benzyl quaternary ammonium salt, $PhCH_2N^+R_3X^-$, when treated with an alkali metal amide to give the *ortho*-methylbenzyl tertiary amine, o-$MePHCH_2NR_2$. The tertiary amine may subsequently be alkylated and the reaction repeated until the *ortho* position is blocked.

Sonn-Müller method

The reduction of an aromatic amide to an aldehyde, in which the intermediate HCl salt of the imine is formed from the amide and PCl_5. This salt is then reduced to the aldehyde in a similar manner to the Stephen reduction.

Specific acid catalysis

In the two step reaction where the first step is the protonation of A by the conjugate acid of the solvent so as to form AH^+, and the second step is the reaction of AH^+ to form the products; if the first step is fast and the second step is rate determining, the the reaction is described as a specific acid catalysed.

Specific base catalysis

The same as specific acid catalysis, *mutatis mutandis*, but with the first step being the deprotonation of AH^+.

Specific rotation

The amount by which a chiral molecule rotates the plane of plane polarized light, $[\alpha]$, under standard conditions, and is defined for solutions as $[\alpha] = a/lc$ and for pure compounds as being equal to α/ld, where α is the observed rotation, l is the cell length in decimeters, c is the concentration in grams per millilitre and d is the density in the same units.

Spirane

A compound which has two number of rings joined in such a manner that there is only one carbon at the junction between the rings and so it has two bonds in one ring and two in the other, *i.e.* is a quaternary carbon.

Spiro dioxide

A compound which contains two epoxide groups which share a carbon atom bonded in a spiro manner.

Spiroannulation

The overall conversion of a ketone, RCOR, to a di-α-substituted cyclobutanone. The reaction may be performed by the addition of diphenylsulfoniumcyclopropylide with the ketone RCOR to give the oxaspiropentane which after treatment with a proton or a Lewis acid rearranges to give the cyclobutanone.

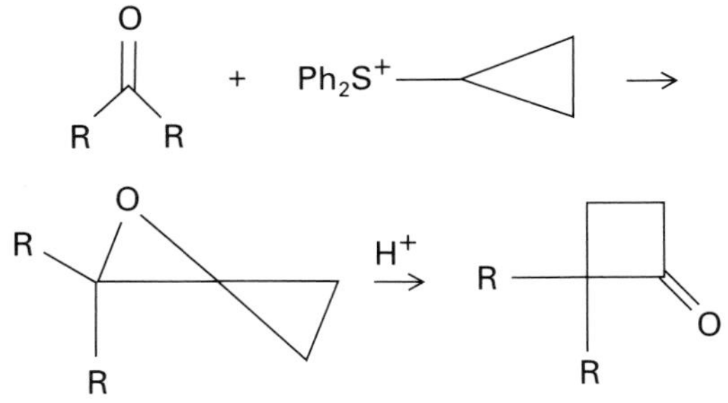

sp^n orbital

An atomic orbital which results from the hybridization of one s atomic orbital and n p atomic orbitals to give the hybrid orbital. sp is called diagonal hyrbidization, while sp^2 and sp^3 are called trigonal and tetrahedral respectively.

Square planar

The name given to describe the geometry in which a general molecule AB_4 exists with the A atom centrally placed between four B groups which are directed towards the corners of a square and all five atoms are in a single plane. This type of geometry may exhibit geometric, but not optical, isomerism.

Squaric acid

The four membered cyclobutane compound of the formula, $C_4O_4H_2$, which has two adjacent keto groups and two hydroxyl groups with a carbon/carbon double bond between them. The dianion is aromatic.

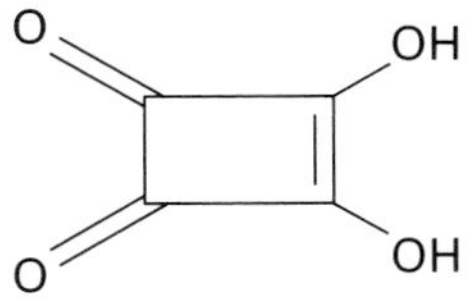

Staggered

Vide eclipsed.

Starting material

The principle carbon containing compound at the start of a synthetic route. If two or more carbon containing compounds are of approximately the same size then the designation is usually based upon which molecule is the nucleophile.

Stephen reduction

The reduction of a nitrile to an aldehyde using HCl, followed by anhydrous $SnCl_2$, which produces the intermediate imine, which is then hydrolysed.

Stereoheterotopic

A term which includes both enantitopic and diasteritopic atoms, groups and faces. Equivalent atoms, groups and faces would be homotopic.

Stereoisomerism

Those isomers that are differentiated by the relative or absolute position of each atom within a molecule to every other. Each isomer has exactly the same number of each type of bond. The opposite of structural isomerism.

Stereoselective

A reaction in which a preferred stereochemistry is selectively produced.

Stereospecific

A reaction in which the product depends on the stereochemistry of the starting material.

Steric crowding

The unfavorable steric interaction caused by the close proximity of large groups, or the increase of such interactions *e.g.* in going from sp^2 to sp^3 geometry.

Stobbe

The addition of a 1,4 ester, *i.e.* a succinate derivative, under basic conditions to a ketone, followed by an acidic work-up, resulting in a 1,3 double bond acid, and a 1,2 double bond ester functionalities.

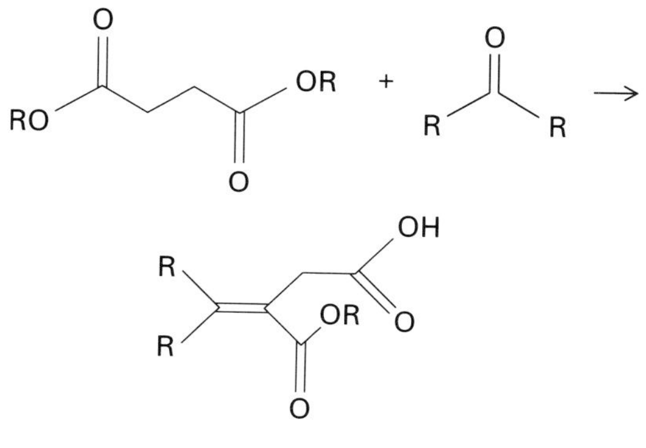

Stork emanime	The reaction of an emanime with an alkyl halide which results in the alkylation of the β-carbon. Usually followed by hydrolysis to give a ketone at the α-carbon.
Strecker	The treatment of a carbonyl compound with NaCN and NH_4Cl to give the α-aminonitrile derivative. It is a special case of the Mannich reaction. As the nitrile groups may be easily hydrolysed this is a convenient route to produce α-amino acids.
Structural isomerism	Isomers which are distinguished by having different bonds in the molecule.
Substitution	A reaction where a heteroatom or group, *i.e.* not carbon or hydrogen is replaced by another heteroatom or group.
Substrate	The molecule which supplies the carbon atom to the new bond. If a carbon/carbon bond is to be formed then the designation is arbitrary.
Succinic anhydride	The cyclic anhydride of butane-1,4-dicarboxylic acid.
Sulfate ester	A compound of the general type $(RO)_2SO_2$.
Sulfene	A molecule of the general formula $CR_2 = SO_2$.
Sulfenic acid	A molecule of the general formula RSOH, where R is an alkyl group.
Sulfenimide	A molecule of the general formula $(RS)_2NR$.
Sulfenyl chloride	A molecule of the general formula RSCl.
Sulfenylation	The addition of a RS→ group.
Sulfinic acid	A molecule of the general formula RSO_2H, where R is an alkyl group.
Sulfolane	A dipolar aprotic solvent, which comprises of a saturated five membered ring compound which contains one SO_2 group incorporated into a ring of four carbon atoms.

Sulfolene	A five membered ring compound which contains one SO_2 group incorporated into a ring of four carbon atoms, a single double

bond. In 3-sulfolene the double bond is furthest from the SO_2 group.

Sulfonamide	A molecule of the general formula $ArSO_2NHR$.
Sulfonation	The addition of a $\leftarrow SO_2OH$ group.
Sulfone	A compound of the general type R_2SO_2.
Sulfonic acid	A molecule of the general formula RSO_2OH, where R is an alkyl group.
Sulfonium ion	A cation of the general formula R_3S^+.
Sulfonyl azide	A compound of the general type RSO_2N_3.
Sulfonyl chloride	A molecule of the general formula RSO_2Cl.
Sulfonyl hydrazide	A compound of the general type RSO_2NHNH_2.
Sulfonylation	The addition of a $\leftarrow SO_2R$ group.
Sulfoxide	A compound of the general type $R_2S=O$.
Sulfoximine	A compound of the general type $R_2SO(=NR)$.
Sulfur dichloride	The molecule SCl_2.
Sulfurization	The addition of a $RS\rightarrow$ group.
Sulfuryl chloride	The molecule SO_2Cl_2.
Suprafacial	A cycloaddition reaction in which both of the new σ bonds are formed from the same face of the π system.
Sydnone	*Vide* mesoionic compound.
Syn	The addition or elimination of a molecule AB form the same side or face of another molecule.
Syn isomer	In the case of imines, oximes, and other $C=N$ compounds, if the substituent on the nitrogen is the same as one of the substituents on the carbon, then the *syn* isomer is the one which has the two identical substituents on the same side; if they are on opposite sides then it is the *anti* isomer. If all three substituents are different then the E/Z notation must be used.
Syn-anti dichotomy	An unusual result which occurs when alkenes are formed by an elimination reaction from large rings, in which the *cis* isomer is formed by *anti* elimination, but

the *trans* isomer is formed by *syn* elimination.

Synclinal	*Vide guache.*
Synfacial	The attaching species enters from the same side as the leaving groups departs, as opposed to apofacial.
Synperiplanar	*Vide anticlinal.*
Synthon	A structural unit within a molecule which can be formed or assembled by known or conceivable synthetic operations.
Taft equation	A structure/reactivity equation which correlates only field effects.
Target molecule	The desired compound at the end of a synthetic route.
Tautomer	The relationship which exists between, for example, the keto and enol forms of a carbonyl compound. These compounds are structurally distinct but are in rapid equilibrium. Usually the difference is due to the position of a proton, but occasionally may another element, such as carbon valence tautomerism.
Telomer	Short polymeric molecule, containing fewer monomeric units than a polymer, but more than an oligomer.
Termination	A reaction in which the product contains no radical species, but is formed from radical precursors.
Tetrahydrofuran	A saturated five membered ring compound in which there is one oxygen atom and four carbon atoms.
Tetrahydropyran	A saturated six membered ring compound in which there is one oxygen atom and five carbon atoms.
Tetrasulfide	A compound which contains four sulfur atoms in a chain, *e.g.* $R-S_4-R$.
Tetrazoline	A five membered cyclic compounds which contain four nitrogens and one carbon.

$$N=N$$
$$HN\diagdown\diagup NH$$

Thiirane	*Vide* episulfide.
Thiiranium ion	An episulfide which is substituted on the sulfur to give a bridged sulfur cation.

Thioacetal	*Vide* mercaptal.
Thio acid	*Vide* thionic acid.
Thioamide	A molecule containing the grouping RC(=S)NHR.
Thiocarbamate	A molecule containing the grouping RNHCOSR.
Thiocyanate	The univalent group ←S-C≡N.
Thiocyanogen	The molecule $(SCN)_2$.
Thiocyanogen chloride	The molecule ClSCN.
Thioether	The sulfur equivalent of an ether.
Thioisocyanate	The univalent group ←N=C=S.
Thiolic acid	Compound of the general type RCOSH. *Vide* also thioic acid.
Thiol	*Vide* mercaptans.
Thionic acid	Compound of the general type RCSOH. Thiolic and thionic acids are sometimes not distinguished and just called thio acids.
Thionyl chloride	The molecule $SOCl_2$.
Thiophene	The sulfur equivalent of a furan.

Thiosulfate ion	The $S_2O_3^{2-}$ anion.
Thiourea	Compound of the general type RNH-C(=S)-NHR.
Thorpe	The addition of the anion, which results from the deprotonation of a nitrile compound, to another nitrile group. The resulting imine may be hydrolysed to yield a β-keto nitrile. If the reaction is performed internally it is known as the Thorpe-Ziegler reaction.
Thorpe-Ziegler	*Vide* Thorpe.
Threo isomer	A stereochemical term used to describe the configuration of a $C_{abx}C_{aby}$ system. *Vide* also *erythro* configuration.

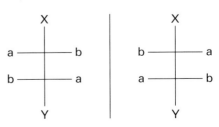

Tishchenko

When aldehydes are treated with aluminium ethoxide, one molecule is reduced while the other is oxidized and the product is the resulting ester. This reaction involves a hydride ion transfer. With more basic alkoxides, aldehydes with an a-hydrogen undergo the aldol reaction.

Tollens'

The reaction of an aldehyde or ketone which has an α-hydrogen, with formaldehyde, in the presence of $Ca(OH)_2$ to give initially a mixed aldol reaction, which then usually is reduced by a second molecule of formaldehyde to give a 1,3-diol, which is the result of a cross Cannizzaro reaction.

Topochemistry

The study of reactions which occur in the solid state.

Torsion effect

The interaction between non-bonded groups which are related by defining the dihedral angle between the groups.

Tosylate

A compound that contains the *p*-toluenesulfonate group, $\leftarrow OSO_2C_6H_4Me$.

Trans

A stereochemical term which indicates that the two similar groups which are attached at each end of a double bond are diametrically opposed, rather than on the same side as in the *cis* isomer. If there are more than two different groups then the CIP rules need to be invoked and the *E/Z* notation used instead.

trans   cis

Transacetalation

When acetals or ketals are treated with an alcohol of a higher molecular weight than the one already bonded, an exchange occurs and the heavier alcohol displaces the lighter one.

Transannular strain

The strain which occurs due to the interaction of the substitutents on C-1 and C-3 or C-1 and C-4, when a *gauche* conformation has been assumed by the molecule. It results from there being insufficient internal space to accommodate the groups in question.

Transesterification	The exchange of one alcohol group in an ester for another alcohol group, a reaction which may be catalysised by either acids or bases.
Transetherification	The exchange of one alkoxyl group for another.
Transition state	The moment of rearrangement of the reacting molecules at the peak of the reaction coordinate diagram which indicates the configuration that has the highest energy along the reaction pathway. The molecular species that corresponds to the transition state cannot be isolated.
Transoid	A stereochemical term used to describe the conformation of whereby four atoms form a plane with the first and fourth atoms being diametrically opposed, *i.e.* they adopt the same positions as would *trans* orientated groups around a double bond. The opposite of *cisoid*.

Transoximation	The conversion of a ketone to an oxime by treatment with another oxime.
Tresylate	A compound which contains the 2,2,2-tri-fluoroethanesulfon ate group, $\leftarrow OSO_2CH_2-CF_3$.
Triazene	The $R_2N-N=NR$ grouping.
Triazine	An aromatic six membered ring in which every other position is a nitrogen.

N
⫽ ⫽
N   N

Triazole	An aromatic five member ring which contains three nitrogen atoms and two carbon atoms. Isomeric with osotriazole.

triazole   osotriazole

Triflate

A compound which contains the trifluoromethanesulfonate group, $\leftarrow OSO_2CF_3$.

Trigonal hybrid

The combination of one s atomic orbital with two p atomic orbitals to give three degenerate, *i.e.* equal in energy, hybrid orbitals which are labelled sp^2. They project towards the corner of an equilateral triangle and so are all in the same plane.

Trioxane

A saturated six membered ring in which there are three oxygen atoms. *Vide* also paraformaldehyde.

1,3,5-trioxane

Triplet

When in a molecule, two unpaired electrons have the same spin. A molecule in which all the spins are paired is called the singlet state.

Tropone

Trival name for cycloheptatrienone.

Tropylium ion

The $C_7H_7{}^+$ aromatic cation. Also called the tropenium ion.

Twist confomer

A conformation of a six membered ring system which is half way between the chair and boat conformations. Also called the skew boat form.

Udenfriend

A reagent formed from a mixture of ferrous ions, oxygen, ascorbic acid, and ethylenetetraaminetetraacet ic acid (EDTA) which is used to oxidize aromatic rings to phenol.

Ugi

Also called the Ugi four-component condensation (4CC) in which an isonitrile is treated with a carboxylic acid and an aldehyde or a ketone in the presence of ammonia or an amine to give a bis amide. Related to the Passerini reaction.

Ullmann

The coupling of aryl halides with copper to give the di-aryl product.

Ullmann ether	The synthesis of diaryl ethers by the coupling of an aroxide group with an aryl halide in the presence of copper.
Umbrella effect	*Vide* pyramidal inversion.
Unsaturated	Indicates the presence of multiple bonds in a molecule, which can undergo addition reactions, in particular with hydrogen in order to saturate all the available valencies with single bonds.
Urea	The molecule of the formula $NH_2C=ONH_2$.
Ureide	The acyl derivatives of urea, $RCONH-CONH_2$.
Urethanes	*Vide* carbamates.
Valence isomer	A type of structural isomer which has the same basic skeleton, but differs in the exact bonding arrangement, *e.g.* the primanes, or the Ladenburg formula, and the bicyclo-[2.2.0]hexadienes, or the Dewar formula, are valence isomers of benzene.

Ladenburg   Dewar   Kekulé

Valence tautomerism	The rearrangement of the covalent bonds within a molecule, with the accompanying change in position of the nuclei which are involved in the bonds. Such molecules are said to be fluxional.

van der Waals forces	The electrostatic attraction between molecules which results from dipole/dipole interactions, whether permanent or induced.
Vilsmeier	The use of disubstituted formamides with $POCl_3$ in order to formylate an aromatic ring. Also know as the Vilsmeier-Haack reaction.
Vinologue	Related by the addition of a $\leftarrow CH=CH\rightarrow$ unit.

Vinyl ether

Vicinal

von Braun

von Pechmann

von Richter

W formation coupling

Wagner-Meervin

Walden inversion

Wheland intermediate

Wilkinson's catalyst

Williamson

Vide enol ether.

Used to describe the configuration when the two functional groups in question are on adjacent carbon atoms in the chain. Equivalent to *ortho* in aromatic systems.

There are two different reactions with this name. The reaction of N-alkyl substituted amides with PCl_5 to give the related nitrile and alkyl chloride; or, the cleavage of tertiary amines by BrCN to give an alkyl bromide and a disubstituted cyanamide.

The formation of a coumarin by the condensation of a phenol with a β-ketoester. The first step is a transesterification followed by a Friedel-Crafts ring closure. Sometimes simply called the Pechmann reaction.

The conversion of a *para* aromatic nitro compound with cyanide ions to give the *meta* carboxylic acid. This is a *cine* reaction.

A four bond coupling observed in the NMR spectrum, *i.e.* ^4J, *e.g.* between *meta* hydrogens in an aromatic ring.

The rearrangement of carbonium ions, usually from a secondary to a tertiary carbonium ion.

The inversion of the configuration at a chiral centre.

The addition of an electrophile to an aromatic ring gives rise to a cation of the general formula $C_6H_6E^+$, which is called a Wheland intermediate, or σ complex or an arenium ion.

A homogeneous catalyst with the formula chlorotris(triphenylphosphine)rhodium.

The formation of an ether by the treatment of an alkyl halide with an alkoxide anion. *gem*-Dihalides react with alkoxides to give acetals, while 1,1,1-trihalides give *ortho* esters.

Wittig	The reaction of a carbonyl with a phosphorous ylid (also called a phosphorane), $Ph_3P^+CR_2^-$ to give the unsaturated adduct.
Wohl-Ziegler bromination	The allylic bromination of an olefin using N-bromosuccinimde (NBS).
Wolff Kishner	The reduction of a carbonyl group using $NH_2NH_2H_2O$ and a base to give the methylene unit.
Wolff rearrangement	*Vide* Arndt-Eistert synthesis.
Woodward method	The *syn* hydroxylation of an alkene: the first step involves treating the alkene with iodine and silver acetate in a 1:1 molar ration in wet acetic acid. The hydrolysis of the *anti*-β-keto ester intermediate is *via* a normal SN2 reaction in which there is no neighbouring group effect because the ester function is solvated by the water. Overall *anti* addition may be achieved by the method of Prevost.
Woodward-Hoffmann rules	These rules are base upon the principle of the conversation of orbital symmetry, and are used to predict which concerted reactions are allowed and which are forbidden.
Wurtz	The coupling of two alkyl halides, RX, with sodium metal to give the R-R compound. Usually of almost no synthetic value as there are many side reaction. However the reaction of 1-bromo-3-chlorocyclobutane with sodium gives the bicyclobutane in about 95% yield.
Wurtz-Fittig	The coupling of an alkyl and an aryl halide with sodium to give the alkylated aromatic compound.
Xanthate	Compound of the general type $ROS(=S)$-SR.
Xylene	Compound of the general type $C_6H_4Me_2$, may exist in the *ortho*, *meta* or *para* isomeric forms.
Ylid	A molecule with a permanent positive and negative charge adjacent to each other in which the positive charge is on an atom from group V or VI of the periodic table, while the atom with the negative charge is

	carbon, *e.g.* a Wittig reagent. Also spelt ylide.
Ynamine	Compound of the general type $R-C\equiv C-NR_2$.
Ynediamine	Compound of the general type $R_2N-C\equiv C-NR_2$.
Z isomer	*Vide E* isomer.
Zaitsev's rule	The double bond being formed goes towards the most highly substituted carbon. Sometimes spelt in the German manner, Saytzeff, or even Saytzev.
Zig zag coupling	A five bond coupling observed in the NMR spectrum, *i.e.* ^5J, *e.g.* between the H1 and H5 of naphthalene.

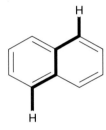

Zielger alkylation	The alkylation of a heterocyclic nitrogen compound using aklyllithium. It proceeds by the addition/elimination mechanism.
Zielger catalyst	A mixture of nickel complexes and alkylaluminum compounds which catalysis the addition of an alkene to another alkene.
Zinin reduction	The reduction of a nitro compound to an amine using a sulfide or polysulfide.
Zwitterion	A molecule which contains both positive and negative charges, but not on adjacent atoms as that is an ylid, *e.g.* $CHRNH_3^+CO_2^-$.

19

Abbreviations

19.1 Atomic, Group and Molecular Abbreviations

9-BBN	9-borobicyclo[3.3.1]nonane
Ac	acyl group, $CH_3CO\rightarrow$, group
acac	acetylacetone, $CH_3COCH_2COCH_3$.
AcOH	ethanoic acid (acetic acid), CH_3COOH
AO	atomic orbital
Ar	a general aryl group
B:	a general base, usually in the B-L sense
B⁻	a general base, usually in the B-L sense
B-L	Brönsted-Lowry acid/base
bis	two complex substituent groups joined to the main chain
Bu	butyl group, $C_4H_9\rightarrow$
sec	secondary carbon, *e.g.* *sec*-butyl cation, $CH_3CH_2CH^+CH_3$
Bs	brosylate, $p\text{-}BrC_4H_6SO_2O\rightarrow$, *para*-bromo-phenylsulfonyl
Bz	benzyl group, $C_6H_5CH_2\rightarrow$
CIP	Cahn-Ingold-Prelog rules
d	dextro rotatory
D	related to D glyceraldehyde
di	two substituent groups joined to the main chain
DCC	dicyclohexylcarbodiimide
DMSO	dimethylsuphoxide, $(CH_3)_2SO$
DMF	dimethylformamide, $HCON(CH_3)_2$
E	*cis* geometry at rigid double bond, using CIP system
E⁺	a general electrophile
Et	ethyl group, $C_2H_5\rightarrow$
FG	functional group
HA	general acid

353

hex	hexane, C_6H_{14}
HOMO	highest occupied molecular orbital
$\pm I$	inductive effect
IPA	*iso*-propyl alcohol, $CH_3CHOHCH_3$
i-Bu, ^iBu	*iso*-butyl group, $CH_3CH_2C(CH_3)H\rightarrow$
i-Pr, ^iPr	*iso*-propyl group, $(CH_3)_2CH\rightarrow$
ipso	1,1-disubstituted on a benzene ring
In•	a general initiator for radical reactions
l	levo rotatory
L	related to L glyceraldehyde
LG	a general leaving group
LUMO	lowest unoccupied molecular orbital
n-Bu, ^nBu	normal butyl group, $CH_3CH_2CH_2CH_2\rightarrow$
n-Pr, ^nPr	normal propyl group, $CH_3CH_2CH_2\rightarrow$
m	*meta* disubstituted, *i.e.* 1.3-
$\pm M$	mesomeric effect
$M\bullet^+$	radical cation in mass spectroscopy
Me	methyl group, $CH_3\rightarrow$
*m*CPBA	*meta*-chloroperbenzoic acid, $m\text{-}ClC_6H_4CO_2\text{-}OH$
MO	molecular orbital
Ms	mesyl group, $CH_3SO_2O\text{-}$
NBS	N-bromosucciamide
neo	primary carbonium ion that is adjacent to a quaternary carbon, *e.g.* $(CH_3)_3CCH_2^+$
Ns	nosylates, *para*-nitrobenzenesulfonat es, $p\text{-}O_2NC_4H_6SO_2O\rightarrow$
Nuc	a general nucleophile
Nu:	a general nucleophile
Nu⁻	a general nucleophile
o	*ortho* disubstituted, *i.e.* 1,2-
p	*para* disubstituted, *i.e.* 1,4-
Ph	phenyl group, C_6H_5, (also $\emptyset$ symbol)
PhH	benzene
Pr	propyl group, $C_3H_7\rightarrow$
R	a general alkyl group, distinguished by raised prefix, *e.g.* ^1R or ^2R
R	*recto* in the CIP rules
S	*sinister* in the CIP rules
SM	starting material
t-Bu, ^tBu	tertiary butyl group, $(CH_3)_3C\rightarrow$
THF	tetrahydrofuran
tri	three substituent groups joined to the main chain
tris	three complex substituent groups joined to the main chain

TM	target molecule
Ts	tosyl group, *para*-methylphenylsulphoni c acid, p-MeC$_6$H$_4$SO$_2$O$\rightarrow$
TS	transition state
X	any halogen, or any -I group
Y	any electrophilic group
Z	any -M group
Z	*trans* geometry at a rigid double bond, using the CIP system

19.2 Mechanistic Abbreviations

‡	concerned with the transition state
β	electron
e⁻	electron
:	lone pair of electrons
⁻	lone pair of electrons
hv	photon, *i.e.* light required
Δ	heat required
⇅	reflux
⌒	movement of one electron from the base of the arrow to the head of the arrow, *e.g.* a radical reaction
⌒→	movement of two electrons from the base of the arrow to the head of the arrow, *e.g.* a polar reaction
↜⌐	shorthand for a pair of electrons moving onto and then off an atom. Commonly found in carbonyl reactions
C$_a$=C$_b$	two electrons reacting through C$_b$
C$_a$---C$_b$---C$_c$	a partly broken C$_a$---C$_b$ bond in which those electrons are in the process of forming a new C$_a$---C$_c$ bond.
↔	resonance arrow linking two canonical forms
=	chemical equation, reaction goes form left to right
→	reaction goes essentially to completion
⇌	equilibrium arrow, *i.e.* significant concentration of all compounds at equilibrium
⇀↽	equilibrium in noticeably to one side, in this case to the right hand side

Symbol	Description

—o→   reaction proceeds with inversion of configuration at the chiral centre

C_a ⟲ C_b rotation about the C_a/C_b bond

⇒   synthon arrow, *i.e.* working backwards from the target molecule towards the starting material

C_a ⌇ X   undetermined stereo position

I I I I I (dashes)	stereo chemistry determined in the manner shown, *i.e.* it reveals the optical chemistry chiral centre or isotopic label
(solid wedge)	
—	single covalent bond
=	double covalent bond
≡	triple covalent bond
.....	partial bond, *i.e.* bond being made or broken. Or else a bond which is partially delocalized
H →— Cl	bond polarized towards the arrowhead, *i.e.* the chlorine is the negative end
+ /-	full units of charge
(-)/(+)	partial charges, which will add up to one or zero
δ + /-	small charges, which sum to zero, but do not add up to one unit
D	dissociation energy
E	bond energy
FGI	functional group interchange
DG	change in Gibbs free energy
DH	change in enthalpy
lp	lone pair of electrons
π	*pi* molecular orbital
rds	rate determining step
σ	*sigma* molecular orbital
DS	change in entropy
upe	unpaired electron

19.3 Reaction Type Abbreviations

4CC	Ugi four component condensation.
A1	*vide* SN1cA.
A2	*vide* SN2cA.

AAC1	Acid catalysed ester hydrolysis, acyl cleavage, unimolecular.
AAC2	Acid catalysed ester hydrolysis, acyl cleavage, bimolecular.
AAL1	Acid catalysed ester hydrolysis, alkyl cleavage, unimolecular.
AAL2	Acid catalysed ester hydrolysis, alkyl cleavage, bimolecular.
AdN-E	Tetrahedral mechanism of nucleophilic addition followed by elimination.
Ad3	Termolecular addition.
A-SE2	General acid catalysis of acetal in which the protonation of the substrate is the rds.
BAE1	Base catalysed ester hydrolysis, acyl cleavage, unimolecular.
BAC2	Base catalysed ester hydrolysis, acyl cleavage, bimolecular.
BAL1	Base catalysed ester hydrolysis, alkyl cleavage, unimolecular.
BAL2	Base catalysed ester hydrolysis, alkyl cleavage, bimolecular.
DA	Diels Alder reaction.
E1	Unimolecular elimination.
E1cB	Unimolecular elimination conjugate base.
E2	Bimolecular elimination in general.
E2C	Bimolecular elimination with the base interacting primarily with the carbon.
E2cB	Bimolecular elimination conjugate base.
E2H	Bimolecular elimination with the base interacting primarily with the hydrogen.
ECO2	Bimolecular elimination forming a $C=O$ bond.
Ei	pyrolytic intramolecular elimination reaction.
FC	Friedel Crafts.
FGI	Function group interchange.
HVZ	Hell-Volhard-Zelinskii.
SE1	Electrophilic substitution unimolecular.
SE1(N)	a SE1 pathway in which a nucleophile (which may be the solvent) assists in the removal of the electrofuge.
SE2	Electrophilic substitution bimolecular, front or back.
SE2(Co-ord)	*vide* SEC.
SE2(cyclic)	*vide* SEi.

S$_E$C	Electrophilic substitution in which there is an initial bond formation between the leaving group (before it becomes detached) and the electrophile and then subsequent bond-breaking step to yield the product. Also called S$_E$2(cyclic).
S$_E$i	Electrophilic substitution in which a portion of the electrophile assists in the removal of the leaving group, forming a bond with it at the same time that the new C-Y bond is formed. Also called S$_F$2 or S$_E$2(cyclic).
S$_E$i'	Electrophilic substitution internal, but at the allylic position.
S$_F$2	*vide* S$_E$i.
S$_H$1	Homolytic cleavage resulting in substitution.
S$_H$2	Bimolecular radical abstraction.
S$_N$1	Nucleophilic substitution unimolecular.
S$_N$1cA	Nucleophilic substitution unimolecular, conjugate acid, when the leaving group only departs after protonation. Also called A1.
S$_N$1cB	Nucleophilic substitution unimolecular conjugate base, when there is an initial deprotonation which then forms a carbene.
S$_N$2	Nucleophilic substitution bimolecular.
S$_N$2'	Nucleophilic substitution bimolecular at the allylic position.
S$_N$2Ar	*vide* S$_N$Ar.
S$_N$Ar	Nucleophilic aromatic substitution, also called S$_E$2Ar and intermediate complex mechanism.
S$_N$i	Substitution nucleophilic internal.
S$_N$i'	Substitution nucleophilic internal at the allylic position.
S$_N$2cA	Nucleophilic substitution bimolecular, conjugate acid, when the leaving group only departs after protonation. Also called A2.

20

Molecular Notations

20.1 Non-Structural Notations

20.1.1 Empirical Formula

This indicates the ratio of the constituent elements using the smallest possible integers, *i.e.* CH_2O and not $C_6H_{12}O_6$ for the glucose molecule. It conveys the least amount of information of any of the possible notations and hence it is of limited use. Normally only found as a step in the calculation of the molecular formula from the elemental percentage composition, in particular in examination questions.

20.1.2 Molecular Formula

This gives the actual numbers of each constituent atom in a molecule. It is derived from a knowledge of the empirical formula and the relative molecular mass (RMM), so *e.g.* C_2H_2 for ethyne and C_6H_6 for benzene. In this case each has the same empirical formulae of CH, but they different RMMs of 26 and 78 respectively. Of little use as it contains so little information, the most important of which is the RMM. But this ambiguity is used to advantage in two cases. Firstly, in examination questions when setting problems, *i.e.* what are the possible isomers which may correspond to a given molecular formula. Secondly, in indexes of organic compounds where it is useful to ignore the isomeric possibilities when trying to find a compound.

In indexes the compounds are listed firstly in order of increasing carbon number. The remaining elements are listed in alphabetical order. The symbols D and T are used for the isotopes of hydrogen, *i.e.* deuterium and tritium respectively. Other non-standard isotopes are indicated by a preceding superscript, and listed by increasing atomic mass, *e.g.* ^{17}O and then ^{18}O.

$CBrH_3$	bromomethane
CCl_2H_2	dichloromethane
CCl_3D	deuterochloroform
CCl_3H	trichloromethane

C_2BrH_5 bromoethane
C_2H_4 ethene
C_2H_6 ethane
C_2H_6O ethanol or dimethylether

20.2 Two Dimensional Structural Notations

All structural notations attempt to give some indication of the arrangement of the atoms within the molecule. The notations fall in two categories, the two dimensional and the three dimensional. The former project the molecule as though it were flat and leaves unresolved any explicit information about three dimensional relationships, except that which may be implied. The latter sets out to give as full a picture of the molecule as is possible on a flat sheet of paper.

20.2.1 Dot and Cross

This is a cumbersome form of structural notation and is very cumbersome. It is only used when it is necessary to illustrate the origin of each electron in a covalent bond. A dot is used to indicate an electron originating from one type atom, while a cross is used for an electron from another type. Only used for very simple molecules.

hydrogen, H_2 H ⦁ H

methane, CH_4

ethane, C_2H_6

× originally a H electron

• originally a C electron

ethene, C_2H_4 or

It can be used to represent the electronic configuration of a molecule which contains more than two different elements, but this can easily become confusing. Often the electrons which originate from the third element would be represented by a small triangle, or such like, or if it occurred only in an isolated part of the molecule then it may be represented by either a dot or a cross, whichever being the more convenient.

The use of a third symbol may also be employed in order to distinguish between the origin of electrons form different atoms of the same type, *e.g.* the two carbons in ethane.

ethane, H_3C_a —— C_bH_3

× originally a H electron
• originally a C_a electron
▪ originally a C_b electron

cyanoethane, CH_3CN

× originally a H electron
• originally a C electron
▪ originally a N electron

or

× originally a H or a N electron depending on context
• originally a C electron

Note that group abbreviations tend not to be used here as the object is to clearly delineate the bonding which is present, electron by electron. Useful in ensuring that no electrons have been omitted by mistake.

20.2.2 Line Notation

The chemical symbols are written on one line using the normal atomic symbols and relevant subscripts or superscripts. Each element is bonded directly to the one to the left, unless all the valences of that atom are satisfied in which case it is bonded to the next one to the left, and so on as necessary. Parentheses may be used to avoid ambiguities. Abbreviation may be used for common groups when certain points wish to be highlighted.

CH_4	methane
R^1R^2CO	generic ketone
$CH_3CH_2CH_3$	propane
CH_3COCH_3	propanone
$CH_3CHOHCH_3$	propan-2-ol
$CH_3CH_2CH_2OH$	propan-1-ol
$CH_2CHCHCH_2$	butadiene
CH_3COOH or CH_3CO_2H	ethanoic acid
$MeCOOH$ or $MeCO_2H$	ethanoic acid
$AcOH$	ethanoic acid
$(4\text{-}Me)C_6H_4CO_2H$	4-methylbenzoic acid
$X\emptyset Y$	disubstituted benzene ring

Useful for typescripts and manuscripts, as a lot of information is conveyed in a condensed manner. This notation suffers from the problem that it does not give a pictorial representation of the stereochemistry of the molecule, but optical detail may be included by using the Cahn-Ingold-Prelog notation, and geometrical detail may be included by using prefixes.

$(2R, 3S)$-$CH_3CHOHCHBrCH_3$ $(2R, 3S)$-3-bromobutan-2-ol
p-BrC_4H_4Me 4-*para*-bromotoluene
cis-$CH_3CHCHCH_3$ *cis*-but-2-ene

20.2.3 Expanded Line Notation

The line notation is sometimes expanded by using two or three lines to indicate double or triple bonds as appropriate, *e.g.* $R^1R^2C=O$ for a generic ketone. This notation highlights the position of the multiple bonds and so may be of use in distinguishing closely related compounds, *e.g.* $CH_3CH_2CH_2CH_3$ for butane, and $CH_3CH=CHCH_3$ - rather than $CH_3CHCHCH_3$ - for but-2-ene. The geometry of the double bond may be indicated by the appropriate prefix, namely *cis* or *trans*.

Sometimes a single bond is indicated to clarify the structure. Parentheses may also be used to ensure that there is no ambiguity:

$CH_3(=O)OH$ ethanoic acid
$CH_2=CH-CH=CH_2$ butadiene
$MeC\equiv N$ cyanomethane

This notation is useful when typing a compound, especially an aliphatic compound which has only a few substituents. However, multi-substituent alicyclic or aromatic compounds can rapidly become a little unclear because of the number of parentheses that must be used.

20.2.4 Stick Notation

Atomic symbols or group abbreviations joined by single, double or triple lines are used to represent single, double or triple covalent bonds. All atoms are included in a stylized manner, so right angles at a sp^3 centre, 120° at a sp^2 centre and 180° at a sp centre. This is the usual notation as it ensures that all the atoms are included and that the molecule as depicted on paper corresponds to the molecular formula of the desired compound. Its weakness are that it is time consuming and complicate structures may become extremely cluttered.

ethane, C_2H_6

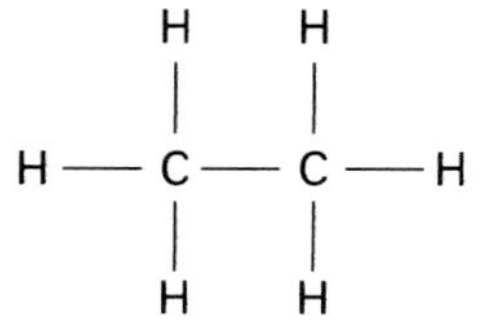

ethanonoic acid, CH_3CO_2H

3-bromobutan-2-ol, $CH_3CHOHCHBrCH_3$

20.2.5 2-D Skeletal Notation

This is the normal method used by organic chemists, because it is fast and accurate. The method uses the following conventions. Firstly, hydrogen atoms joined to carbons are not shown, they are assumed to be added as required to complete the normal valency of the carbon. Secondly, carbon atoms are not represented by the letter "C", but instead they are represented by either the end of a line, or the junction between two lines. Heteroatoms are represented with the normal symbol from the periodic table. Thirdly, covalent bonds are represented by lines which connect the atomic symbols of heteroatoms or carbons joined to heteroatoms, but with the above two provisos. Fourthly, the atomic symbol for hydrogen is included when it is bonded to a heteroatom or when it is bonded to a carbon atom and is involved in that mechanistic step, *e.g.* the enolization of a carbonyl group. This method is of little use to portray very simple molecules, for example, methane is just a dot "."; ethane is a line "-"; ethene is a double line "=", methanol, "-OH" and methanal, "=O".

The method is more valuable as the complexity of the molecule increases, *e.g.* hexane is represented by a zigzag of five parts: "∧∧∧". Aromatic rings are represented by a regular hexagon with a circle inside to indicate the delocalization. The method comes into its own for bigger molecules, *e.g.* 2-bromotoluene, hexatriene, cyclohexene, propylamine, ethylbutanionate, *cis* and *trans* butene. This notation is good for aliphatic and aromatic compounds.

2-bromotoluene, $2\text{-BrC}_6H_4CH_3$

hexatriene, C_6H_8

cyclohexene, C_6H_{10}

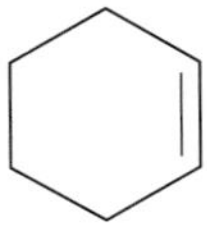

propylamine, $C_3H_7NH_2$

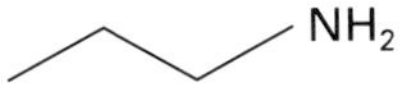

ethylbutanoate, $C_3H_7CO_2C_2H_5$

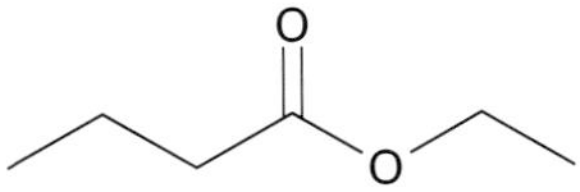

cis-butene, C_4H_8

trans-butene, C_4H_8

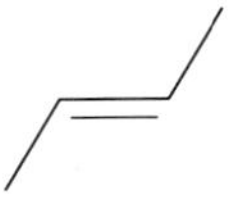

Many types of isomer may be easily illustrated, *e.g.* geometric and structural. No attempt, however, is made to indicated the optical conformation of the molecule, as the molecule is drawn as a flat geometric figure. Hence the stereochemistry of butan-2-ol is not shown. Similarly cyclohexane derivatives are represented by a flat, regular hexagon with side branches at appropriate junctions, with no attempt to indicate whether they are *cis* or *trans* to each other.

1,2-dibromocyclohexane, $C_6H_{10}Br_2$

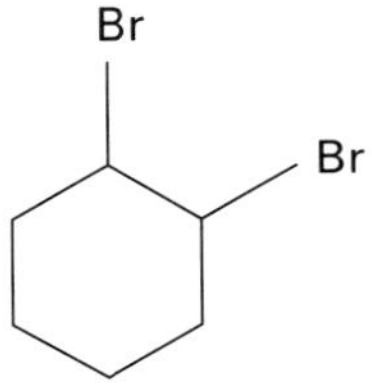

1-bromo-1-methylcyclohexane, $C_6H_{10}BrCH_3$

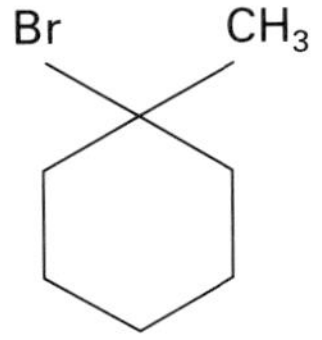

There is a variation that is used in some cases for fused ring systems, such as decalin derivatives and steroids. In this variation even though the rings are drawn as symmetrical polygons, a hydrogen which projects up from the page at the ring junction is represented by a small solid circle over the carbon at that junction, and similarly a hydrogen which projects downwards is represented by an open circle. Usually only one of the hydrogens is portrayed as the orientation of the other may be deduced.

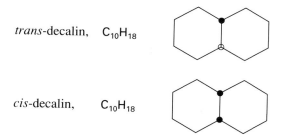

trans-decalin, $C_{10}H_{18}$

cis-decalin, $C_{10}H_{18}$

This variation is of very limited use, and in order to convey more fully the three dimensional structure of the molecule more sophisticated notations must be employed.

20.3 Three Dimensional Structural Notations

20.3.1 Haworth Notation

In this notation a six membered ring is represented by a regular hexagon, but as viewed from the side and slightly above. The bonds which represent the bonds which project towards the reader are thickened and the substituents are positioned above and below the flat ring joined with normal thickness lines. Alternatively, the bonds within the ring are represented by lines of equal thickness and the groups above and below the ring are represented by solid wedges or dashed lines respectively. There is no attempt to indicate whether a bond is in front or behind another bond which it crosses from the view point of the reader. Examples of this notation may be found in the literature, but there are clearer notations which may be used.

D-glucose, $C_6H_{12}O_6$

20.3.2 Three Dimensional Skeletal Notation

This is an extension of the skeletal notation. It is very good for ring compounds and the cyclic transition states of aliphatic compounds. For example, 1,4-dichlorocyclohexane is shown as a puckered ring and the positions of the chlorine atoms are indicated by placing them in conventional positions relative to each other. Vertical lines are axial, while lines at a slight angle to the horizontal are equatorial. This clearly shows whether the chlorines are *cis* or *trans* to each other. If a bond crosses another because of the view point of the reader, then it is convenient to introduce a break in the hindermost bond to as to assist with the visual interpretation of the diagram. Both optical and geometrical isomers may be easily distinguished using this notation. For most purposes this is the best notation to use as it is quick and easy to draw by hand, it is easily understood, and finally it clearly indicates the relative positions of the different parts of the molecule and so helps in writing mechanisms.

cis-1,4-dichlorocyclohexane, $C_6H_{10}Cl_2$ Cl_2

one chlorine atom is axial,
while the other is equatorial

trans-1,4-dichlorocyclohexane,

equivalent to

both chlorine
atoms are axial

both chlorine
atoms are equatorial

20.3.3 Stereo Projection

This is another extension of the two dimensional skeletal notation, but with the addition of dashed and solid wedge lines to indicate bonds that are either going away, or coming towards, the reader. Bonds shown by the usual thickness of line are in the plane of the paper. Hence the stereochemistry at a chiral centre may be shown clearly and as such it is particularly good for displaying the change in geometry of a chiral centre in substitution reactions. For example, in a simple S_N2 reaction, the inversion of the carbon which is substituted may be clearly seen.

There are two common variations on this notation. The first is that sometimes at a chiral centre, only three bonds may be shown, two which are of

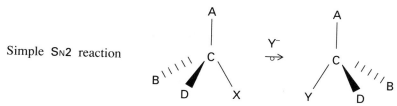

Simple S$_N$2 reaction

normal thickness and third which is either dashed or a solid wedge; the fourth bond is to an unshown hydrogen and is assumed to be orientated as necessary to complete the usual geometry of the chiral centre. The other common variation is that three bonds to the chiral centre are represented with lines of normal thickness, and only the fourth uses a solid wedge or dashed line to indicate the stereochemistry.

20.3.4 Sawhorse Projection

This method is used to give a clear picture of the geometry along a carbon/carbon single or double bond, in particular one around which an addition or elimination reaction is proceeding.

E2 mechanism

It is valuable when trying to illustrate the geometry of the formation of a three membered ring, as in a Darzen's condensation.

epoxide formation in the
Darzen's condensation

Lastly it may be used when describing the geometry of conformers in aliphatic compounds.

conformers of butane

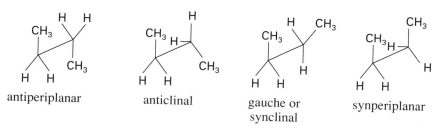

antiperiplanar   anticlinal   gauche or   synperiplanar
                                             synclinal

20.3.5 Newman Projection

In this projection the reader looks down the single bond which is adjacent to the doubly bonded carbon that is of interest. The α-substituents are placed on lines emanating from the rim of the circle which represents the α-carbon, which is away from the reader. The carbon of interest is represented by a dot in the centre of the circle and is towards the reader. Issuing from this central point are the double lines representing the double bond to the neighbouring atom, which is often a carbonyl oxygen, and a single line which is the bond to the last remaining group.

It is used in particular to determine which side an incoming nucleophile will attack a carbonyl group, based upon Cram's rule that the side of attack is the one which offers less steric hindrance. Hence the sterochemistry of the product may be deduced.

attack of a nucleophile on a carbonyl group

S, M, L : Small, medium and large groups attached to α-carbon

20.3.6 Fisher Projection

This method was developed to illustrate the optical chemistry of small biological molecules, in particular monosaccharides and amino acids. It rapidly becomes cumbersome for larger molecules. The vertical lines represent bonds that are going into the page, away from the reader; while, the horizontal bonds represent bonds that are coming out of the page, towards the reader. For multichiral centred molecules the arrangement must be vertical on the page, which means that it takes up a lot of space.

a general chiral module C_{abcd}

21

Stereochemical Terminology

21.1 Introduction

The elemental composition of a molecule may be unambiguously described using the molecular formula. In inorganic chemistry this is often completely sufficient to identify the species in question, because there is usually only one possible arrangement of the constituent elements that are combined in that particular ratio. One obvious exception to this general rule would be the elements that may exist in various allotropic forms. For example, to fully identify the type of phosphorus that is being used it is necessary to indicate the color, *i.e.* white, yellow, red or black. These terms refer to various forms of phosphorus which differ in their relative spatial orientation of the component atoms within the lattice. Another major exception in inorganic chemistry would be the need to characterize the possible isomers of the coordination compounds of the transition metals, for these may exhibit geometric or optical isomerism.

In organic chemistry the situation is very different, because, for all but the very simplest of molecules, for any given ratio of the constituent elements there are many possible spatial arrangements that those atoms may assume relative to one another. So even though the molecular formula conveys some useful information it is not enough to completely characterize any particular compound in question.

The problem may be easily illustrated by considering the molecules which have the molecular formulae $C_4H_{10}O$. There are five isomers which have this molecular formulae. Four of these isomers may be distinguished by writing out the compounds using the line notation: $CH_3CH_2CH_2CH_2OH$, $CH_3CH_2CH(OH)CH_3$, $CH_3CH_2OCH_2CH_3$, $CH_3CH_2CH_2OCH_3$. The first two compounds are alcohols, butan-1-ol and butan-2-ol; the last two compounds are ethers, propylmethyl ether and diethyl ether. However, there is a further ambiguity that has not yet been revealed by the line notation. For butan-2-ol there are actually two possible structures, which differ only in the way the groups are arranged in space relative to one another, *i.e.* there are two stereoisomers.

369

the enantiomers of butan -2 - ol, $CH_3CHOHCH_2CH_3$

mirror
plane

In order to distinguish between these two possibilities further information must be included in the name of the compound, *i.e.* the introduction of stereochemical terminology is needed.

In the example given above there were five possible arrangements of the constituent atoms that corresponded to the same molecular formula. Each one of these possible arrangements is called an isomer. Isomers may be divided into two fundamental classes. The first type consists of those isomers that differ in the exact number of each type of bond that occurs in the various molecules. These are called structural isomer, because the structure is different in each one. They are also said to have different configurations. In the second type, the isomers all have exactly the same number of the each type of bond, but differ only in the precise position that each group occupies relative to all the others in the molecule. These are called stereoisomers.

21.2 Structural Isomerism

Structural isomers may be divided further into five sub-types: skeletal, positional, functional, tautomeric and meta.

There are also generic terms that are used to describe, and distinguish between, certain classes of alcohols, amines and amides, and we will consider these first.

Alcohols are classified according to how many carbon chains are attached to the carbon to which the hydroxyl group is bonded. A primary alcohol has one carbon chain attached to the carbon of the COH group; while a secondary alcohol has two; and a tertiary has three. Notice that methanol has no carbon chains attached to the carbon of the COH group and so is not even a primary alcohol, although it is often erroneously included in the study of primary alcohols.

primary   secondary   tertiary

For the amines the classification depends on the number of carbon chains that are directly attached to the nitrogen, so the general formula is R_nNH_{3-n}, where n is the level of the amine. R_4N^+ is a quaternary ammonium ion. Notice that ammonia is not even a primary amine, *cf.* methanol.

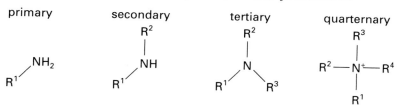

For amides again the classification refers to the number of carbon chains that are attached to the nitrogen, but now the general formula is $RCONH_{2-n}R_n$, so if there are no carbon chains joined to the nitrogen, the molecule it is a primary amide and so on.

21.2.1 Skeletal Isomerism

This type of isomerism is concerned with differences in the carbon backbone of the molecule, whether it is straight or branched. It is also sometimes called chain isomerism. There must be at least four carbon atoms before this type of isomerism can occur.

skeletal isomers
of butane   *normal*   *iso-*

skeletal isomers
of pentane   *normal*   *iso-*   *neo*-pentane

The best way to distinguish between these isomers is by naming them according to the IUPAC system, which based upon the longest continuous carbon chain.

butane   2-methylpropane

pentane   2-methylbutane   2,2-dimethylpropane

However, it is still common to find references to an older system of nomenclature, where each type of configuration of a carbon chain was given a particular prefix. A straight chained molecule is given the prefix "normal", which is abbreviated to "*n*-", *e.g.* *n*-butane for $CH_3CH_2CH_2CH_3$. A compound which contains the $(CH_3)_2CH$ group is called "*iso*", which is abbreviated to "*i*-", *e.g.* *i*-pentane for $CH_3CH_2CH(CH_3)CH_3$. If the compound contains a quaternary carbon then the prefix is "*neo*", which is not abbreviated, *e.g.* *neo*-pentane for $(CH_3)_4C$.

This old system is also used to describe carbonium ions and radicals. Thus primary, secondary, and tertiary carbonium ions are given the respective label, "*normal*", "*secondary*" (abbreviated to "*sec*-" or just "*s*-"), and "*tertiary*" (abbreviated to "*t*-"). The term "*neo*" may be used to describe a primary carbonium ion that is adjacent to a quaternary carbon, *e.g.* $(CH_3)_3CCH_2{}^+$.

n-pentyl   *sec*-pentyl   *t*-pentyl   *neo*-pentyl

21.2.2 Positional Isomerism

On a long straight chained molecule that has only one functional group, then this group may be placed at a number of different positions on the carbon chain. This is referred to a positional isomerism. At least three carbons are needed in the chain before this type of isomerism may occur.

Again, as with all structural isomers, the best way to distinguish between closely related compounds is to name them according to the IUPAC system.

However, it is still common to find references to an older system of nomenclature where the point of attachment of the functional group to the alkyl chain is indicated by a prefix, such as iso or *tertiary*. This bears a relationship to the prefix system that is used for distinguishing between the different structural isomers of alkanes, and have been included in parenthesis under the examples given.

possible alcohol isomers of C_3H_7OH

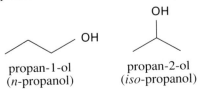

propan-1-ol   propan-2-ol
(*n*-propanol)   (*iso*-propanol)

possible alcohol isomers of C$_4$H$_9$OH

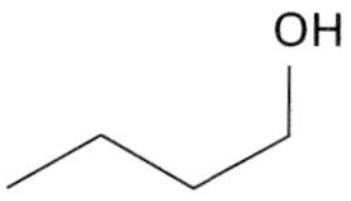

butan-1-ol
(*n*-butanol)

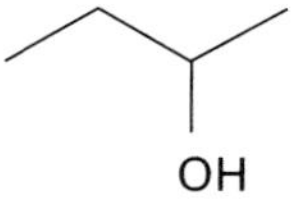

butan-2-ol
(*iso*-butanol)

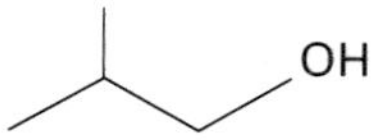

2-methylpropan-1-ol

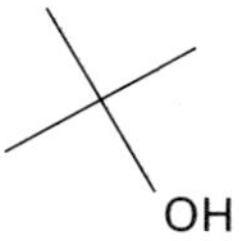

methylpropan-2-ol
(*t*-butanol)

possible alcohol isomers of C$_5$H$_{11}$OH

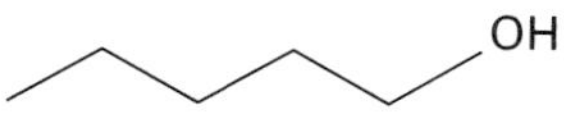

pentan-1-ol
(*n*-pentanol)

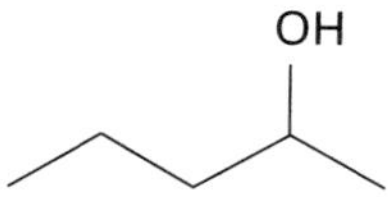

pentan-2-ol
(*iso*-pentanol)

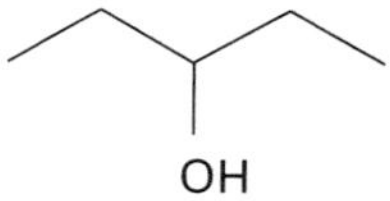

pentan-3-ol

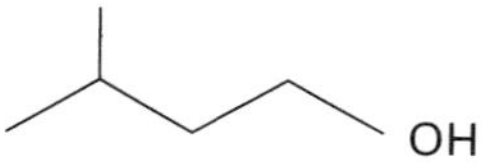

3-methylpentan-1-ol

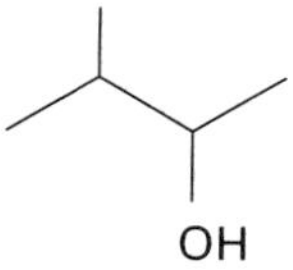

3-methylpentan-2-ol

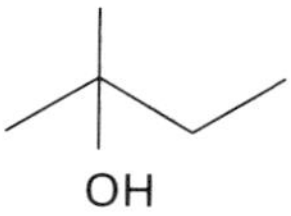

2-methylpentan-2-ol
(*t*-pentanol)

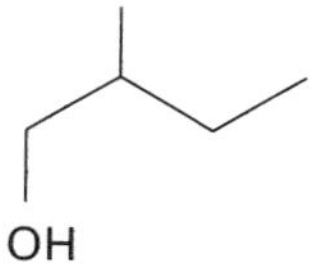

2-methylpentan-1-ol

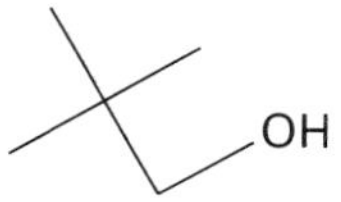

2,2-dimethylpentan-1-ol
(*neo*-pentanol)

It is also important to be able to refer to a particular carbon which is adjacent to the functional group in question. This is often done using small Greek letters, which may be illustrated by using 1-bromooctane as an example.

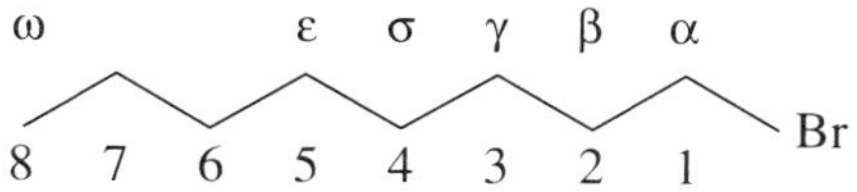

The w position always refers to the terminal carbon no matter how long the chain actually is.

There there are two identical function groups that are bonded to the same carbon atom, then they are said to be *geminal* to each other, or abbreviated to "*gem*", *e.g.* 1,1-dibromoethane, H$_3$C-CHBr$_2$, contians *geminal* bromine atoms. If, however, the two functional groups in question are bonded to adjacent carbons, then they are said to be *vicinal* to each other, abbreviated to "*vic*", *e.g.* 1,2-dibromoethane, H$_2$BrC-CH$_2$Br.

In elimination reactions if both groups are lost from the same atom, then it is called a 1,1- or α-elimination; if the groups are lost from adjacent atoms, then it is called a 1,2- or β-elimination; and if the groups are lost from atoms which are separated by another atom, then it is called a 1,3- or γ-elimination. The chain in which the elimination does not occur is indicated by a prime symbol, " ' ". For example, 4-bromopent-2-one gives pent-3-en-2-one on the elimination of HBr, and so the C5 methyl carbon has the β'-hydrogens, while the C3 methylene carbon has the β-hydrogens.

When the functional group is one that includes a carbon which is fully bonded to heteroatoms then the lettering starts from the closest carbon.

In disubstituted aromatic systems, the position of the second substituent with respect to the primary functional group is indicated either by a number, with the primary group being numbered 1, or by the prefix *ortho*, *meta* or *para* indicating the 2, 3 or 4 positions respectively. These prefixes are often abbreviated to *o*, *m* or *p*. If a mono-substituted aromatic compound is attacked at the ring carbon which already bears the substituent, then this is called *ipso* attack.

If the nitrogen of an amine group is substituted, then the position of this substituent is indicated by the prefix N. If there are two relevant nitrogens which are substituted, then the second is indicated by the prefix N'. Similarly oxygen substitution may be indicated by an O.

21.2.3 Functional Isomerism

It is very common for a given molecular formula to be capable of forming molecules which have different functional groups. A simple example is C_2H_6O, which may exist either as an alcohol or an ether.

ethanol, C_2H_5OH dimethylether, CH_3OCH_3 or MeOMe
or EtOH

21.2.4 Tautomeric Isomerism

This is a special type of functional isomerism, in that even though different functionalities are present in the tautomers, they are often in equilibrium with each other. There are several types, but by far the commonest type involve a proton shift between an heteroatom and a carbon. The two commonest examples of this type are the enol/keto and the aci/nitro tautomeric systems.

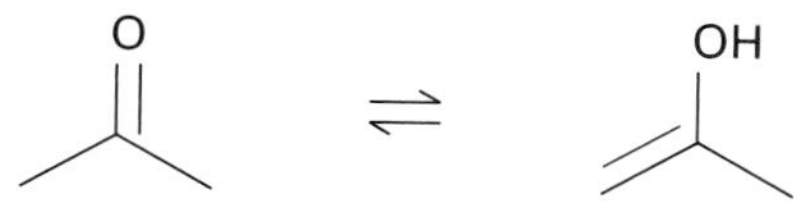

keto   enol

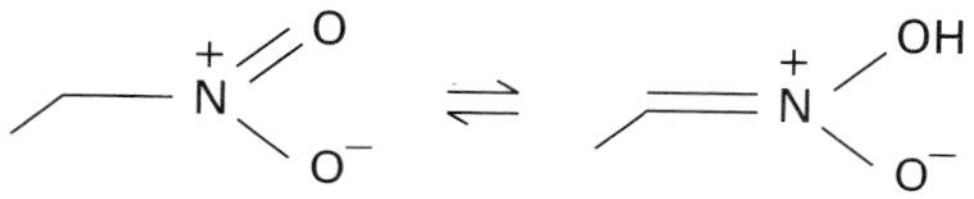

nitro   aci

These are usually named according to the structure of the more stable tautomer, that is the keto and nitro versions respectively. If reference is made to the other tautomer it is just called the enol or aci tautomer of the corresponding keto or nitro compound.

The enol tautomers may be isolated and trapped but this is quite rare, because they are very easy to make in situ.

There are other types of proton shift tautomers, such as phenol/keto, nitroso/oxime and imine/enamine, but these are less often encountered. There are also valence tautomers which exhibit fluxional structures, which undergo rapid sigmatropic rearrangements. Again molecules which exhibit this type of isomerism are very interesting but are not often encountered. An example of a valence tautomer is the Cope system.

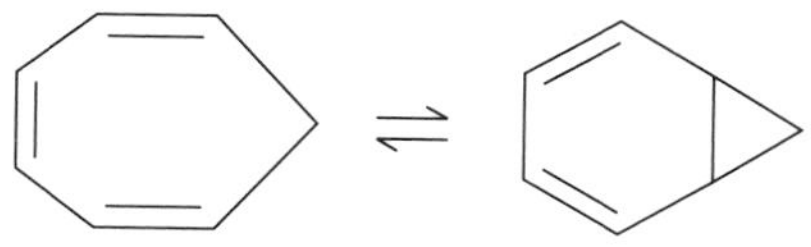

Sugars exhibit a different type of type of tautomerism called ring/chain, so named because it reflects the change from the cyclic hemialdol derivative to the open chain aldehyde derivative.

21.2.5 Meta Isomerism

An asymmetrical functional group like an ester functionality may have a different group on each end. If these groups are then exchanged, the two isomers are said to be meta-isomers of each other. This is a classification that is not used very much.

methylpropanoate　　　　　　　　　ethylmethanoate

21.3 Stereoisomerism

While structural isomers differed in the number and type of bonds to be found in each isomer, in stereoisomers the only difference is the relative position of the groups with regard to each other. There are three main sub-types of stereoisomer: geometric, optical and conformational.

When discussing this type of isomerism the Cahn-Ingold-Prelog system is used to rank the various groups which are attached to a particular centre.

1. Each group is ranked in order of decreasing atomic number of the atom directly joined to the carbon or group in question.

2. Where two or more of the atoms connected to the system in question are the same, then the atomic number of the second atom determines the order. The first difference is taken so (O.H.H.) ranks higher than (C.C.C.), even though it has an overall lower summation of atomic numbers. If this still does not resolve the issue then continue until there is a difference.

3. All atoms except hydrogen are formally given a valency of four. Where the actual valency is less than four, then phantom atoms are added which are assigned an atomic number of 0, which is represented by a subscripted zero.

4. Higher isotopes take precedence over lower ones.

5. All multiple bonds are split into single bonds, and phantom atoms are used to terminate the bonds.

Group	CIP representation	
— CH_3		(H,H,H)
— CH_2 — CH_3		(C,H,H)

$— CH(CH_3)_2$

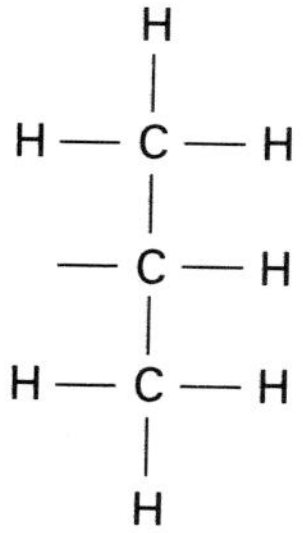

$$H—\underset{\underset{\displaystyle H}{|}}{\overset{\overset{\displaystyle H}{|}}{C}}—H$$
$$—\underset{|}{C}—H$$
$$H—\underset{\underset{\displaystyle H}{|}}{C}—H$$

(C,C,H)

$— CH_2OH$

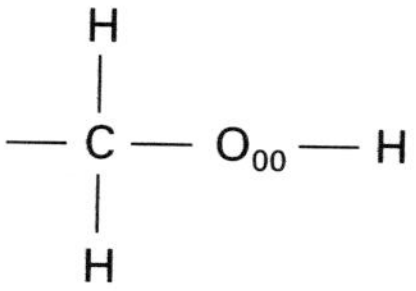

$$—\underset{\underset{\displaystyle H}{|}}{\overset{\overset{\displaystyle H}{|}}{C}}—O_{00}—H$$

(O,H,H)

$— CH = O$

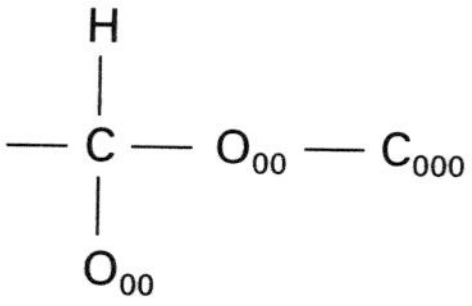

$$—\underset{\underset{\displaystyle O_{00}}{|}}{\overset{\overset{\displaystyle H}{|}}{C}}—O_{00}—C_{000}$$

(O,O,H)

$— CH = CH_2$

$$—\underset{\underset{\displaystyle C_{000}}{|}}{\overset{\overset{\displaystyle H}{|}}{C}}—\underset{\underset{\displaystyle H}{|}}{\overset{\overset{\displaystyle H}{|}}{C}}—C_{000}$$

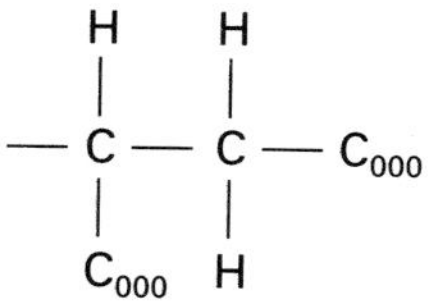

(C,C,H)
or more fully
(C$_{CHH}$, C$_{000}$, H)

$— CH \equiv CH$

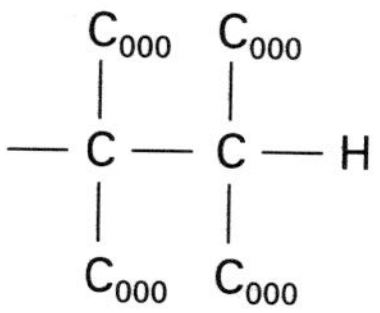

$$—\underset{\underset{\displaystyle C_{000}}{|}}{\overset{\overset{\displaystyle C_{000}}{|}}{C}}—\underset{\underset{\displaystyle C_{000}}{|}}{\overset{\overset{\displaystyle C_{000}}{|}}{C}}—H$$

(C,C,C)
or more fully
(C$_{CCH}$, C$_{000}$, C$_{000}$)

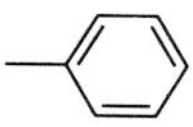

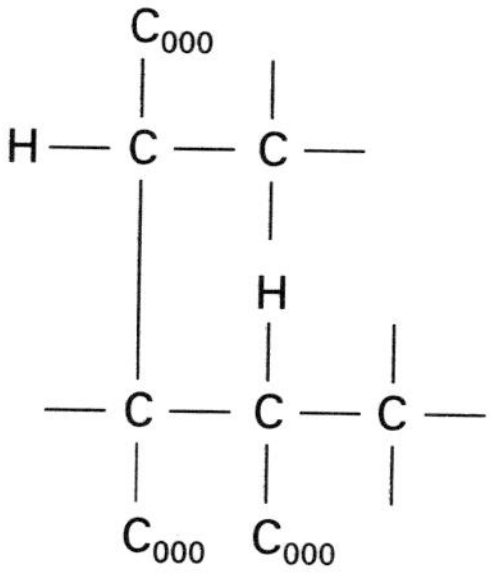

$$H—\underset{|}{\overset{\overset{\displaystyle C_{000}}{|}}{C}}—\underset{\underset{\displaystyle H}{|}}{C}—$$
$$—\underset{\underset{\displaystyle C_{000}}{|}}{C}—\underset{\underset{\displaystyle C_{000}}{|}}{C}—\underset{|}{C}—$$

(C,C,C)
or more fully
(C$_{CCH}$, C$_{CCH}$, C$_{000}$)

21.3.1 Optical Isomerism

If a compound rotates the plane of polarized light it is said to be optically active. It has been found by experiment that all molecules which are optically active are not superimposable on their mirror image. This property is called chirality. There are three requirements for a molecule to be chiral. It must not have a centre of symmetry; it must not have a plane of symmetry and it must not have an alternating axis of symmetry (i.e. rotation followed by reflection in the plane perpendicular to this axis of rotation). In all examples that will be encountered in common practice, only the first two types of symmetrical transformation need to be considered.

For a compound which is chiral, there are two molecules that are related by the fact that one is superimposable on the mirror image of the other. These two molecules are enantiomers of each other, sometimes called enantiomorphs. Enantiomers have exactly the same physical and chemical properties except in two important respects. Firstly, they rotate plane polarized light by opposite, but equal, amounts. The compound that rotates the plane to the left is called the *levo*, *l*, or (-) isomer, while the other is the *dextro*, *d*, or (+) isomer. The second difference is that they may react with other optically active compounds at different rates.

For a normal tetrahedral carbon it will be chiral if there are four different groups around it, *i.e.* C_{abcd}.

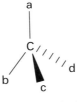

general chiral
molecule

The stereochemistry at every chiral centre may be identified by using the Cahn-Ingold-Prelog rules to rank the groups. Position the lowest ranking group furthest away from the observer, then if the remaining groups are arranged in a clockwise manner when going from highest to lowest priority group, that isomer is labelled the *R* isomer, while the opposite arrangement in which the groups are arranged in an anti-clockwise manner is labelled the *S* isomer.

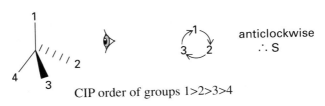

CIP order of groups 1>2>3>4

If instead of the general chiral molecule, C_{abcd}, two groups are the same, *i.e.* C_{abc2}, then the carbon under consideration will no longer be chiral, because it has a plane of symmetry which passes through the central carbon atom and between the "a" and "b" groups.

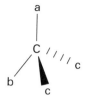

prochiral
molecule

However, if one of the "c" groups is substituted for a fourth different group the molecule will become chiral again. The original molecule, C_{abc2}, is said to be prochiral, and the two "c" groups are said to be enantiotopic, because the replacement of either of them would result in an enantiomer being formed. The two "c" groups will behave differently in an asymmetrical environment, and may be labelled *pro-R* and *pro-S* to distinguish them.

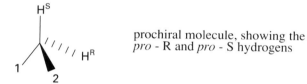

prochiral molecule, showing the
pro - R and *pro* - S hydrogens

The idea of enantiometric molecules may be extended to a faces in trigonal molecules, by an extension of the Cahn-Ingold-Prelog system. Thus when looking at one face of a trigonal system, if the groups are arranged in a clockwise manner when moving from the highest to the lowest, then that face is called the *re* face. If the groups are arranged in an anti-clockwise manner, then that face is the *si* face.

If there is more than one chiral centre in a molecule then the number of possible *d,l* pairs of stereoisomers increases, and is given by the formulae 2^{n-1}, where n is the number of chiral centres. So if there are two chiral centres there are four possible stereoisomers, comprising of two *d,l* pairs. Stereoisomers are called diastereomers if their relationship to one another is not enantiomeric. Where two diastereomers differ by the stereochemistry at only one chiral centre then they are called epimers. If the two "c" groups in the prochiral given above, where in such positions that if they were replaced then diastereomers were formed, then the "c" groups would be called diatereotopic.

If there are only two chiral centres, which instead of having the totally general formula of $C_{abc}C_{xyz}$, have the symmetrical formula of $C_{abc}C_{abc}$, then instead of the existence of two *d,l* enantiomeric pairs, there is only one *d,l* enantiomeric pair, as well as a molecule which has an internal plane of symmetry and is thus not optically active. This optically inactive compound is called the *meso* isomer.

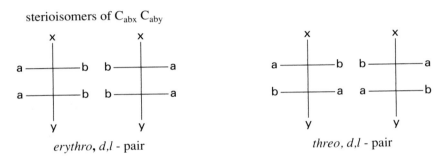

sterioisomers of $C_{abc} C_{abc}$

mirror plane   *meso* - isomer

d, l - pair

If there are two groups which are the same on each carbon, while the third is different, *i.e.* $C_{abx}C_{aby}$, then these compounds form the *threo* and *erythro* *d,l* pairs.

sterioisomers of $C_{abx} C_{aby}$

erythro, *d,l* - pair   *threo*, *d,l* - pair

This type of geometry is common in biological molecules and that is why it is often illustrated by using the Fischer projection.

If there is an equal mix of the enantiomers of a compound then the resultant mixture will not rotate the plane of polarized light because one isomer will cancel out the effect of the other. This is called a racemic mixture and sometimes has different properties from either of the enantiomers, *e.g.* different melting point and solubility.

The tetrahedral geometry is found most commonly at a normal sp^3 carbon centre. However, it may be found in other situations, and there is no need for the chiral centre to be a carbon atom, *i.e.* other tetravalent asymmetrical centres such as quaternary nitrogen salts, or sulfones with two different isotopes of oxygen can be chiral centres. The latter element provides an example which illustrates that only a very small difference in the four groups is necessary, *i.e.* $R^1R^2S^{18}O^{16}O$ is optically active.

Any trisubstituted nitrogen is chiral, but because at room temperature there is a rapid inversion of the molecule, the time averaged view means that such a molecule does not exhibit optical activity. This inversion is called umbrella effect.

The inversion process is slowed down sufficiently in R^1R^2SO compounds for the activity to be measured, *e.g.* $[\alpha]_{280} = +0.71°$ for (+) $Ph^{12}CH_2$-$SO^{13}CH_2Ph$.

It is not necessary to have a simple tetrahedral shape. For example, adamantane is a caged structure which projects four bonds to the corners of an expanded tetrahedron, and so suitably substituted adamantanes may be optically active.

Other structures are possible as well, *e.g.* those that possess perpendicular disymmetric planes. Optical activity will occur if there are two even numbered rings joined in a *spiro* manner, with each ring substituted at the carbon furthest from the *spiro* junction. In this case an extended tetrahedron is formed.

In order to be chiral the groups at each end must be different from one another, but they do not need to be different from those at the other end. This extended tetrahedral geometry may be found where two double bonds are joined directly together as in an allene. Thus $_{ab}C=C=C_{ab}$ will be chiral, but $_{ab}C=C=C_{cc}$ will not. This is because the latter has a plane of symmetry.

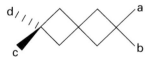

chiral   achiral

In general an even number of double bonds joined directly together gives rise to a chiral molecule, while an odd number gives rise to a molecule which displays geometric isomerism, which will be discussed in the next section.

In addition to the two ways described above of creating an extended tetra-hedral structure, it is possible to mix then, *i.e.* to have a double bond pro-jecting from an even numbered ring. This type of double bond is called an *exo* cyclic double bond, in contrast to an *endo* cyclic double bond.

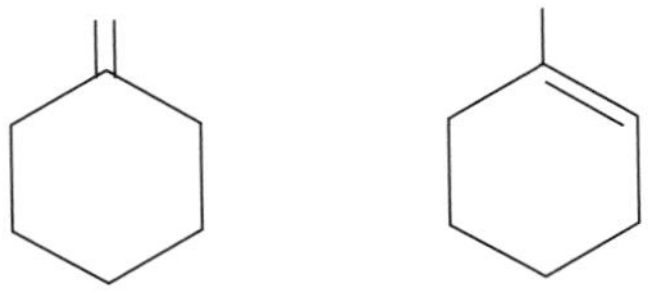

exo   *endo*

It is possible to have a chiral centre about a single bond so long as there is restricted rotation. An example of this hindered rotation occurs in biphenyl compounds, which have the same extended tetrahedral geometry as the *spiro* and allene compounds above. Isomers which may only be resolved because of hindered rotation about a single bond are called atropisomers.

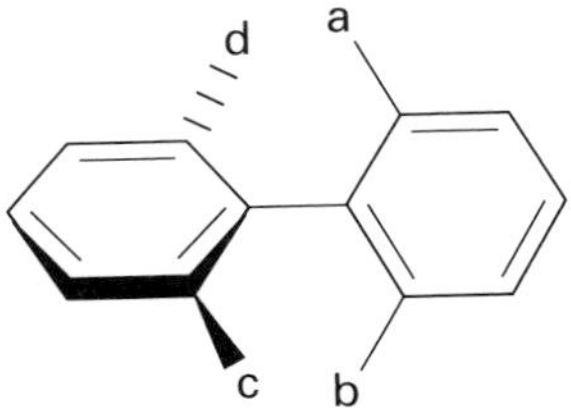

By using phase lagged x-crystallography is it possible to discover the absolute position of one group with respect to each of the others around a chiral centre. This was first done in 1951 by Bijvoet on a crystal of sodium rubidium tartrate. Up until this time it had been impossible to assign the absolute configuration to any molecule and so an arbitrary relative system had been suggested by Rosanoff. He proposed that the isomer of glycer-aldehyde which was *dextro* rotatory be assigned the configuration shown below, and called the D isomer Glyceraldehyde was chosen because it is related to many sugars. It turned out that his suggested configuration was correct. This configuration would now be called the R isomer.

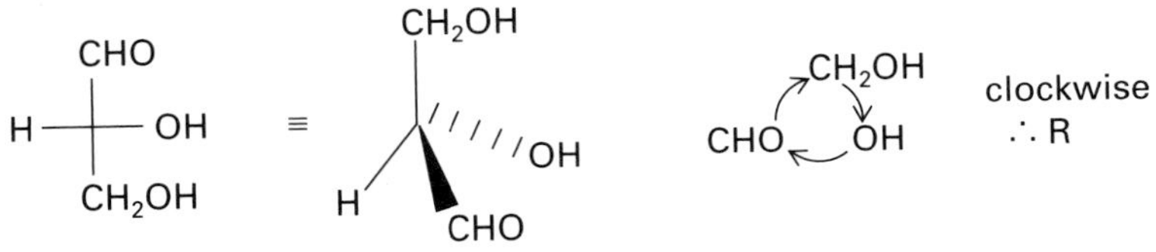

By relating one compound to another by synthetic routes, which had a known stereochemical consequence, all optically active molecules could be designated either D or L. This system did have it problems in that a few compounds could be related to either D or L glyceraldehyde by using different synthetic routes! So now all compounds are named according to the Cahn-Ingold-Prelog rules for each chiral centre.

A reaction in which only one set of stereoisomers is formed exclusively or predominantly is termed a stereoselective synthesis. In a stereospecific reaction, however, the particular isomer of the target molecule that is formed is determined by which isomer of the starting material that was used initially. All stereospecific reactions must be stereoselective, but the converse is not true.

It is possible for other geometries to give rise to optical activity within a molecule, *e.g.* helical structures and knotted compounds. These geometries are not often encountered.

21.3.2 Geometrical Isomerism

Compounds in which rotation is restricted may exhibit geometric isomerism. These compounds do not rotate the plane of polarized light (unless they also happen to be chiral) and the properties of the isomers are different.

Taking the commonest case of a carbon/carbon double bond which has two identical groups, one attached at each end of the double bond, there are two possible isomers:

cis *trans*

The isomer in which the two groups are on the same side is called the *cis* isomer, while the other one where they are on opposite sides is called the *trans* isomer. This type of isomerism used to be called *cis/trans* isomerism. However, this nomenclature system fails if there are four different groups on the double bond. Using the Cahn-Ingold-Prelog rules the groups may be ranked, and if the groups with the highest priority at each end of the double bond are diametrically opposed then that is the *E* isomer, otherwise it is the *Z* isomer.

Z *Z* *E*

The idea may be extended to other rigid double bond systems such as the $C=N$ unit or the $N=N$ system, in which the terms *syn* and *anti* are used instead of *cis* or *trans*.

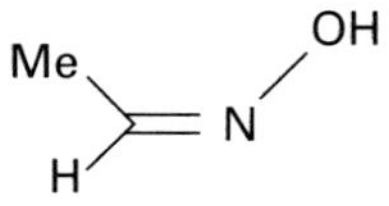

syn 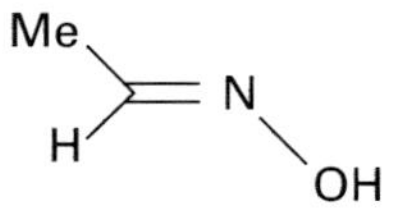 · anti 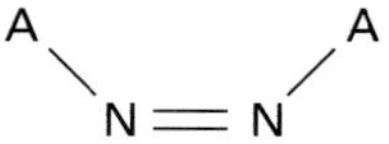 · syn 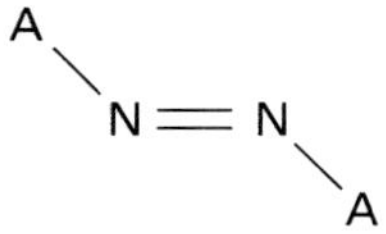 · anti

Hindered rotation may occur where formally there is no double bond. In such case sometimes it is possible to draw at least one canonical structure which has a double bond along the axis about which there is restricted rotation.

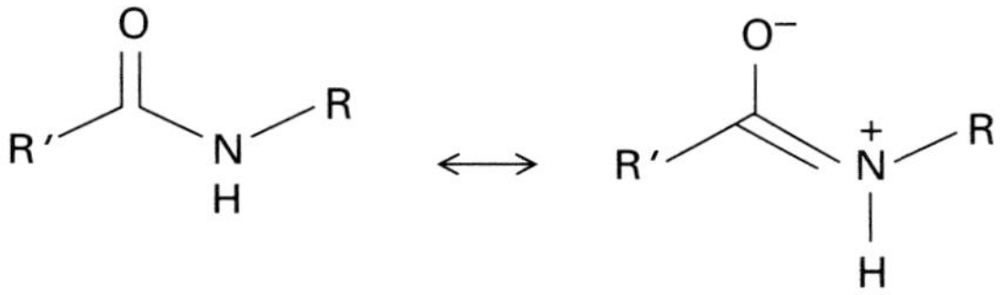

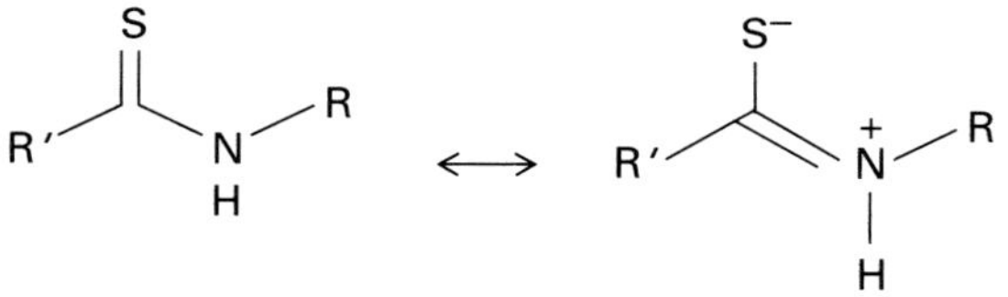

A further example of hindered rotation about a single bond occurs in conjugated dienes. The simple case of buta-1,3-diene exists in two forms, which are described as *transoid* and *cisoid*. The *transoid* form is the thermodynamically more stable form as there is less steric interaction between the terminal methyl groups.

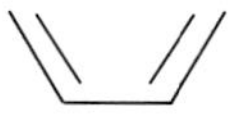

cisoid 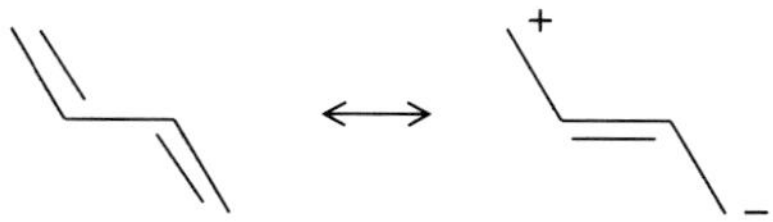   transoid

In other cases the restriction on free rotation is imposed by the formation of an ring, for example, geometric isomerism may be shown by the three membered ring of cyclopropane.

Note that in this example the *trans* isomer also exhibits optical isomerism, while the *cis* one does not because it has a plane of symmetry, *i.e.* is *meso*.

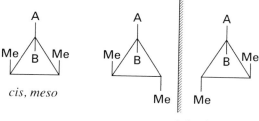

cis, meso

trans, d, l-pair

In general where there are two substituted carbons in the ring, *e.g.* C_{ab} and C_{cd}, then there will be four possible isomers if a $\neq$ b $\neq$ c $\neq$ d, since neither the *cis* nor the *trans* isomers is superimposable on its mirror image. This is true regardless of the ring size or which two carbons in that ring are substituted, except that in even numbered rings, where the two substituted carbons are opposite each other, then both the *cis* and the *trans* forms are *meso*.

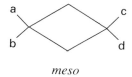

meso

When a = c and b = d, then the *cis* isomer is always meso, while the *trans* isomer will have a *d,l* pair, except as above.

cis, meso

trans, d, l-pair

Lastly, there are two type of multiple ring systems that must be considered. The first is the fused ring system, *e.g.* of decalin, which has two isomers, *cis* and *trans*.

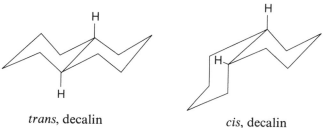

trans, decalin   *cis*, decalin

These isomers are rigid and cannot interchange.

The second type of multiple ring system, is the bridged ring system. If there is a substituent on one of the bridges, and it is closer to the longer of the unsubstituted bridge then that is the *endo* isomer; while if it is closer to the shorter bridge then that is the *exo* isomer. If the two unsubstituted bridges are of the same length, then this definition can not be strictly interpreted, but it still may be used if there is another functional group on only one of the other bridges.

| *exo*-2-norborneol | *endo*-2-norborneol | *endo*-7-methyl-2-norcamphor | *exo*-7-methyl-2-norcamphor |

The terms *endo* and *exo* are also used when describing the different orientations adopted by the reacting species in the Diels-Alder reaction just prior to the concerted bond formation.

The terms *syn* and *anti* are used when referring to the geometry of the neighbouring groups that are about to undergo α,β-elimination. In *syn* elimination the two groups are on the same side, while in *anti* they are geometrically opposed.

21.3.3 Conformers

If two different three dimensional arrangements in space of the constituent atoms are interconvertible by free rotation about bonds they are called conformations: if not, they are called configurations, *i.e.* to interconvert between them bonds must be broken and reformed. Isomers which have different configurations may be separated, while isomers which are conformations represent molecules that are rapidly interconverting at room temperature and are thus cannot be separated. Such isomers are called conformers. We have already seen an example of conformers in the previous section, where the *cisoid* and *transoid* structures of dienes were mentioned.

Conformation analysis is particularly important in alicyclic systems. Taking ethane as an example, there are obviously an infinite number of possible dihedral angles which the two methyl groups may adopt. However, two of the conformers are of particular interest in that they represent the states of lowest and highest interaction between the hydrogens on each methyl group: these are the staggered and eclipsed conformers respectively.

staggered   eclipsed

For the 1,2-disubstituted ethane, $XCH_2\text{-}CH_2X$, there are now four extreme conformers: antiperplanar, anticlinal, synclinal and synperplanar.

antiperiplanar

About a sp^3-sp^2 bond there are also four extreme conformers, which are divided into eclipsing and bisecting.

eclipsing   eclipsing   bisecting

In cycloalkane systems, completely free rotation is obviously no longer possible, however, there is still sufficient movement for a range of conforms to exist.

The cyclobutane ring exists in a non-planar open book conformation, which may flex further open or closed. The cyclopentane ring exists in two puckered conformations, the half chair, and the envelope, both of which are non-planar. The puckering may rotate around the ring, and this is called pseudorotation.

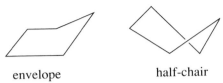

envelope   half-chair

The cyclohexane ring adopts three main conformers: the chair, the boat and the skew boat or the twist. Each of these conformers is puckered, *i.e.* not flat. This is important, because there are consequently implications on the possible angle of attack that an incoming reagent must follow.

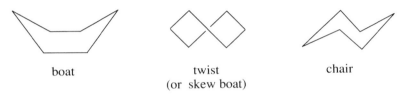

boat   twist   chair
                   (or skew boat)

The ring normally adopts a chair shape as this is the most stable state because the interactions are minimized. There are, however, two chair conformers which may interconvert *via* the boat or twist conformer. In each chair conformer there are six axial bonds and six equatorial bonds, but the orientation of each bond changes on the conversion from one chair conformer to the other.

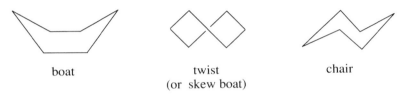

At room temperature this transformation between one chair conformer and the other is rapid, assuming that there is no bulky group that locks the molecule. This means that the cyclohexane system may be treated as if it were flat when trying to find how many stereoisomers exist for any given derivative.

For *cis*-1,2-disubstituted cyclohexanes one group must be axial while the other is equatorial. For *trans*-1,2-disubstituted cyclohexanes both groups may be axial or equatorial. The same analysis is true for 1,4-disubstituted cyclohexanes. However, for 1,3-disubstituted cyclohexanes the reverse is true:

in the *cis* isomer the groups may be both axial or equatorial, while in the *trans* isomer one group must be axial and the other group equatorial.

For alkyl substituents the most stable position is the equatorial one, which minimizes the axial/axial group interactions. Care needs to be exercised as the energy difference between the axial and equatorial states may be very small. *Trans*-1,2-dihalocyclohexanes exist predominantly in the axial/axial conformer. The ring may be locked into one conformer by using a very bulky group like *t*-butyl which must adopt the equatorial position and thus may be used to hold another group in the axial position. In such a derivative the two chair isomers do not rapidly interconvert.

When an oxygen is introduced into the ring, an alkyl substituent on the α-carbon is still more stable in the equatorial position. However, if the substituent is polar, such as an alkoxide group, the reverse is true. This accounts for the greater stability of α-glucosides over β-glucosides. This is known as the anomeric effect.

β-glucoside α-glucoside

22

Oxidation Numbers

22.1 Introduction

The oxidation number of an atom refers to the number of electrons which that atom has gained or lost. For any given atom this number must be an integer, because it represents the hypothetical number of whole electrons that have been transferred to or from that atom. Thus it corresponds to the whole number of electrical units of charge which that atom bears. The complete transfer of an electron only takes place in the most ionic of univalent compounds, *e.g.* KF, and as such is quite rare. As the degree of ionic nature of the bond decreases the artificial nature of the concept of oxidation numbers becomes more apparent. In covalent compounds, where the bonds between the nuclei are formed from electrons which are being shared more or less equally, the oxidation number bears no relationship to the charge that resides on any particular atom. However, so long as the limitations of the idea are remembered then the concept may still be used to advantage in covalent compounds, because the relative oxidation states of the same atom in different molecules may indicate a possible synthetic link.

22.2 Oxidation Numbers in Ionic Compounds

Inorganic chemistry may be divided into four types of reaction, acid/base; precipitation; complexation and redox. It is when considering redox reactions that oxidation numbers are used to great advantage. In particular when ensuring that a reaction equation has been properly balanced, *i.e.* that the same amount of reduction as oxidation has taken place.

In simple ions the oxidation number represents the notional whole charge that resides on that ion. So in NaCl the sodium has an oxidation number of $+1$, while the chlorine has an oxidation number of -1. In $MgCl_2$ the chlorine still has an oxidation number of -1, but the magnesium has an oxidation number of $+2$.

In Fe_2O_3, the oxygen has an oxidation number of -2, while the iron has an oxidation number of $+3$. Here, though, the artificial nature of the concept becomes apparent, because any ion with a $+3$ charge would have such a high charge density that it would polarize the electron rich cloud around the oxygen to some extent and hence reduce its own charge.

In complex ions such as nitrate, NO_3^-, and sulfate, SO_4^-, this charge polarization is acknowledged and there is no suggestion that the central atom would bear a $+5$ or $+6$ charge respectively. These numbers, however, are the values of the oxidation numbers assigned to nitrogen and sulfur.

In inorganic chemistry the oxidation number is calculated by applying a number of simple rules:

1. All atoms in their elemental state have an oxidation number of zero.

2. Fluorine always has an oxidation number of -1 when it is combined with other elements.

3. Hydrogen usually has an oxidation number of $+1$, except when found in metallic hydrides when it has an oxidation number of -1.

4. Oxygen usually has an oxidation number of -2, except when found as a peroxide when it has an oxidation number of -1, or when in combination with fluorine, when it has an oxidation number of $+2$.

5. Electropositive elements usually have positive oxidation numbers, *i.e.* lose electrons readily; while electronegative elements usually have negative oxidation numbers, *i.e.* gain electrons readily.

6. In complex ions the difference between the sum of the positive oxidation numbers and the sum of the negative oxidation numbers of all the constituent atoms gives the charge that resides on the complex ion.

7. It is usually assumed that within a complex ion all the atoms of the same element have the same oxidation number.

In a complex ion where all the atoms of the same element occupy the same environment then it is reasonable to assume that they have the same electron distribution and hence the same oxidation number. So in the peroxide anion, O_2^{2-}, it is reasonable to assume that each oxygen has an oxidation number of -1. Or in the sulfate anion, it reasonable to assume that each oxygen has the same oxidation number, namely -2.

However, in anions such as the superoxide, O_2^-, or the azide, N_3^-, one arrives at fractional oxidation numbers of -1/2 and -1/3 respectively for each of the constituent atoms.

The problem is further highlighted by such complex anions as the thiosulfate anion, $S_2O_3^{2-}$, where if it is assumed that each sulfur bears the same oxidation number one arrives at a value of $+2$ for each sulfur atom.

The problem becomes even more acute in the persulfate anion, $S_2O_8^{2-}$. In this example one could assume that each of the oxygens has an oxidation number of -2 and then calculate the oxidation number of the sulfur, which would then be $+7$. Alternatively one could fix the oxidation number of the sulfur first at say $+6$, and then calculate the value required for the oxygens, which would then be -1.75. Which answer is correct, or even useful?

In the four examples that have been given in the last three paragraphs all the problems may be sided-stepped by realizing that the simple rules that were employed initially, even though they worked well for simple ions, are now no longer sufficient to cope with more advanced examples. Hence a new working definition is needed for molecular species which are bound together by covalent bonds.

22.3 Oxidation Numbers in Covalent Species

The new working definition for calculating oxidation numbers in species which contain covalent bonds may be stated simply. Every heteroatomic bond which is polarized towards the atom in question contributes -1 unit to the redox number of that atom. Conversely, every heteroatomic bond which is polarized away from the atom in question contributes $+1$ unit to the redox number of that atom. For covalent species which a have an overall charge then if an extra electron resides on the atom then -1 unit is added to the oxidation number, and if an electron is missing then $+1$ is added to the oxidation number. The emphasis is now on finding the oxidation number for each atom individually and not on calculating an average for all the atoms of the same element within a molecule.

Only heteroatomic bonds or any gain or lose of electrons are of concern when calculating the oxidation number, because it is only the gain or loss of electrons when compared to the electronic situation which is to be found in the elemental state which is of interest. So diatomic oxygen, O_2, has an oxidation number of zero, as does elemental carbon, C_n.

In methane, CH_4, each hydrogen is involved in a single heteroatomic bond that is polarized away from it, because the carbon atom has the greater electronegativity, and hence each hydrogen has an oxidation number of $+1$. The carbon has an oxidation number of -4, because it is involved in four heteroatomic bonds each of which is polarized towards it.

In ethane, C_2H_6, each hydrogen has an oxidation number of $+1$, while each carbon has an oxidation number of -3, because each is involved in three heteroatomic bonds that are polarized towards itself, and also one homoatomic bond that does not contribute to the overall oxidation number, because there is no overall polarization of that bond.

In ethene, C_2H_4, and ethyne, C_2H_2, the oxidation number of the carbon atoms is respectively -2 and -1, because they are now involved in only two heteroatomic bonds, or only one. The other bonds are homoatomic and so do not contribute to the oxidation state.

In longer chain alkanes the oxidation number may be easily calculated for each atom, so for example in propane, $CH_3CH_2CH_3$, the terminal carbons have an oxidation number of -3, while the central one has an oxidation number of -2. Similarly for 2-methylpropane, $CH_3CH(CH_3)CH_3$, the methyl carbons have an oxidation number of -3, while the central carbon has an oxidation number of -1. In Me_4C, the central carbon has an oxidation number of zero, which is perfectly reasonable when it is remembered that it is in a similar environment to elemental carbon, C_n.

In organic compounds which contain oxygen the oxidation number for each atom may be calculated in a similar manner. In methanol, CH_3OH, each hydrogen has an oxidation number of $+1$, while, the oxygen has an oxidation number of -2, and the carbon has one of -2, giving an overall sum of zero. The oxidation number of the carbon is the result of having 3 bonds which are polarized towards it, *i.e.* the 3 C-H bonds, and 1 bond which is polarized away from it, *i.e.* the C-O bond, giving a total of -2 units.

The oxidation numbers of the carbon atoms in the following oxygen containing compounds is: methanal, CH_2O (zero); methanoic acid, HCO_2H ($+2$); carbon dioxide, CO_2 ($+4$); ethanol, CH_3CH_2OH (-3, -1); ethanal, CH_3CHO (-3, $+1$); ethanoic acid, CH_3CO_2H (-3, $+3$).

Hence, it may easily be seen that as ethanol is converted to ethanal and then ethanoic acid and finally to carbon dioxide, each step involves an oxidation of the carbon atom to which the oxygen is becoming more attached. Similarly the addition of hydrogen to ethene to give ethane, may readily be seen to be a reduction as the oxidation number is reduced from -2 to -3.

If we now return to the compounds which caused problems under the previous definition we see that by analysing the oxidation state of each atom individually, and only considering heteroatomic bonds or any gain or loss of electrons, the problems are resolved.

Starting with diatomic oxygen, as there are no heteroatomic bonds and no electrons have been gained or lost, the oxidation number is zero. If one electron is added to form the superoxide anion, O_2^-, then considering this addition in two steps: firstly, the double bond between the oxygen atoms breaks in a homolytic manner to give the diatomic diradical held together by a single bond; secondly, an extra electron is added to one of these oxygens. This results in one oxygen having two lone pairs, one homoatomic bond and one unpaired electron, which means that it has an share in exactly the six electrons in the valence shell that it would have had in the elemental state and so its oxidation number is zero. The other oxygen has three lone pairs and one homoatomic bond. Thus it has a share in seven electrons, which is one more than the elemental state, and so it has an oxidation state of -1.

$$O = O \quad \rightarrow \quad :\overset{\cdot\cdot}{O} - \overset{\cdot\cdot}{O}: \quad \overset{e^-}{\rightarrow} \quad \left[:\overset{\cdot\cdot}{\underset{\cdot\cdot}{O}} - \overset{\cdot\cdot}{O}: \right]^{-}$$

When a further electron is added to form the peroxide anion, O_2^{2-}, now both oxygen atoms have three lone pairs and one homoatomic bond and so each atom has an oxidation number of -1.

$$\left[:\overset{\cdot\cdot}{\underset{\cdot\cdot}{O}} - \overset{\cdot\cdot}{O}: \right]^{-} \quad \overset{e^-}{\rightarrow} \quad \left[:\overset{\cdot\cdot}{\underset{\cdot\cdot}{O}} - \overset{\cdot\cdot}{\underset{\cdot\cdot}{O}}: \right]^{2-}$$

The azide anion, N_3^-, is more complicated. This will be considered three steps. Firstly, a neutral nitrogen atom which has five electrons accommodated in its valence shell, and hence an oxidation number of zero, gains a single electron to give it six electrons accommodated in three lone pairs. Thus now it has an oxidation number of -1.

$$:\overset{\displaystyle ..}{\underset{\displaystyle .}{N}} \quad \overset{e^-}{\longrightarrow} \quad \left[:\overset{\displaystyle ..}{\underset{\displaystyle ..}{N}} \right]^-$$

Secondly, a triple bonded dinitrogen molecule, in which each nitrogen has an oxidation of zero, datively bonds to the nitrogen anion to give rise to one of the canonical structures of the azide anion, namely, one terminal nitrogen has one lone pair and three homoatomic bonds and so has an oxidation number of zero. The central atom has four homoatomic bonds, but one of which is polarized away from it, *i.e.* the dative bond, and so it has an oxidation number of +1. The other terminal nitrogen started with one extra electron and now also has a homoatomic bond which is polarized towards it and so now has an oxidation number of -2. This nitrogen atom has three lone pairs.

$$\left[:\overset{\displaystyle ..}{\underset{\displaystyle ..}{N}} \right]^- \quad :N \equiv N: \quad \longrightarrow \quad \left[:\overset{\displaystyle ..}{\underset{\displaystyle ..}{N}}{}^{2-} \overset{+}{N} \equiv N: \right]^-$$

Thirdly, there is a rearrangement of the electrons in the π molecular orbitals to give another canonical structure, where the central nitrogen still has an oxidation number of +1, but now both the terminal nitrogens have an oxidation number of -1.

$$\left[:\overset{\displaystyle ..}{\underset{\displaystyle ..}{N}}{}^{2-} \overset{+}{N} \equiv N: \right]^- \quad \longleftrightarrow \quad \left[:\overset{\displaystyle ..}{N}{}^- = \overset{+}{N} = \overset{\displaystyle ..}{\underset{\displaystyle ..}{N}}{}^- \right]^-$$

In the thiosulfate anion, $S_2O_3^{2-}$, it is important to know how each sulfur is bonded.

$$\left[\begin{array}{c} S \diagdown \quad \diagup O \\ S \\ O^- \quad O^- \end{array} \right]^{2-}$$

The sulfur which occupies a position similar to that of the oxygens has two covalent homoatomic bonds to the central sulfur and so its oxidation number is zero. The central sulfur has four heteroatomic bonds to oxygen atoms and in each case they are polarized away from the sulfur and hence its oxidation number is +4. So instead of each sulfur being given the same oxidation number of +2, as was the result above, a distinction is drawn between them which reflects the different chemical environments which they occupy.

The persulfate anion, $S_2O_8{}^{2-}$, has the structure:

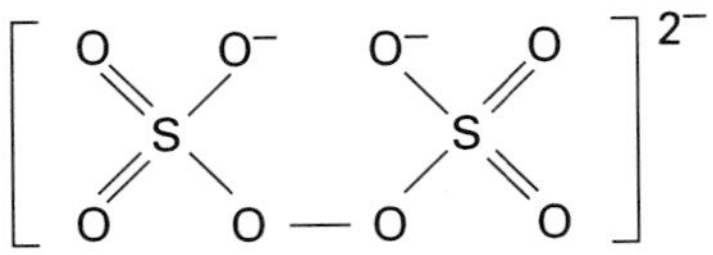

$$
\left[
\begin{array}{c}
\overset{\displaystyle O}{\underset{\displaystyle O}{\diagup}}S\overset{\displaystyle O^-}{\diagdown}O-O\overset{\displaystyle O^-}{\diagup}S\overset{\displaystyle O}{\underset{\displaystyle O}{\diagdown}}
\end{array}
\right]^{2-}
$$

Here each sulfur has a similar position and may be given an oxidation number of $+6$. The six terminal oxygens may be assigned an oxidation number of -2, while the two bridging oxygens, which form a peroxide type link between the two sulfurs, are only involved in one heteratomic bond each and so have an oxidation number of -1. It is this peroxide bond that is broken when the persulfate reacts as an oxidizing agent, and concomitantly these oxygens are reduced to -2 as the two sulfate anions are produced.

So, in summary, it may be clearly seen that by examining the electronic environment of each atom in turn within a covalently bonded molecule a sensible oxidation number may be reached which reflects the different chemistry of each atom.

23

Skeletal Index

This index is arranged to help you find the name which is used to describe the common functional groups and other common combinations of atoms in organic compounds. It is arranged in order of increasing complexity as perceived visually. Hydrocarbons are listed first, then in the subsequent sections, compounds which contain oxygen, nitrogen, sulfur and phosphorus are introduced. Complexity is partly subjective, but each section is arranged approximately along the following lines: compounds which contain a greater the degree of unsaturation, more heteroatoms, larger rings, or more rings systems appear latter within each section.

In this index, the old or trivial name for some compounds has been used, *e.g.* nitrous acid instead of nitric (III) acid for HNO_2. This is deliberate, because firstly that is the name used in practice, and secondly that is the name by which the compound is known in the older literature, which is were the most assistance is required.

23.1 Hydrocarbon Compounds

C_nH_{2n+2}	saturated hydrocarbon, alkane: methane (1), ethane (2), propane (3), butane (4), pentane (5), hexane (6), heptane (7), octane (8), nonane (9), decane (10).
$R_2C=CR_2$	alkene
	diene
	triene, and so on to a polyene
$R_2C=C=CR_2$	allene
$R_2C=C=C=CR_2$	cumulene
$R\text{-}C\equiv C\text{-}R$	alkyne
$R\text{-}C\equiv C^-$	acetylide
$R_2C:$	carbene
R_3C^+	carbonium ion
R_3C^-	carbanion

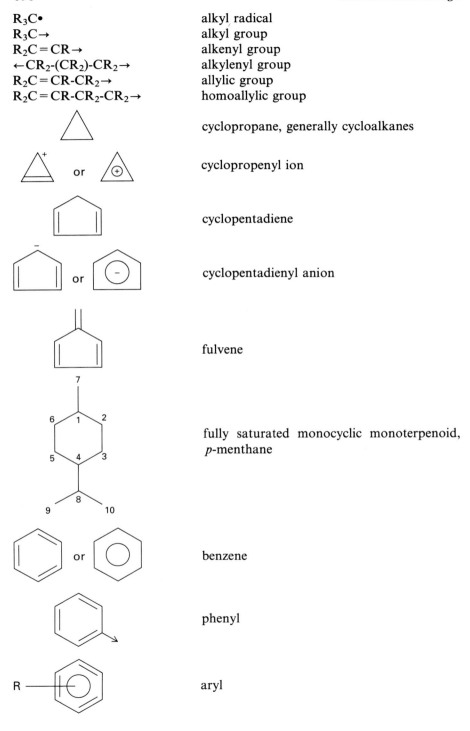

R₃C•   alkyl radical
R₃C→   alkyl group
R₂C=CR→   alkenyl group
←CR₂-(CR₂)-CR₂→   alkylenyl group
R₂C=CR-CR₂→   allylic group
R₂C=CR-CR₂-CR₂→   homoallylic group

cyclopropane, generally cycloalkanes

or   cyclopropenyl ion

cyclopentadiene

or   cyclopentadienyl anion

fulvene

fully saturated monocyclic monoterpenoid, *p*-menthane

or   benzene

phenyl

R—   aryl

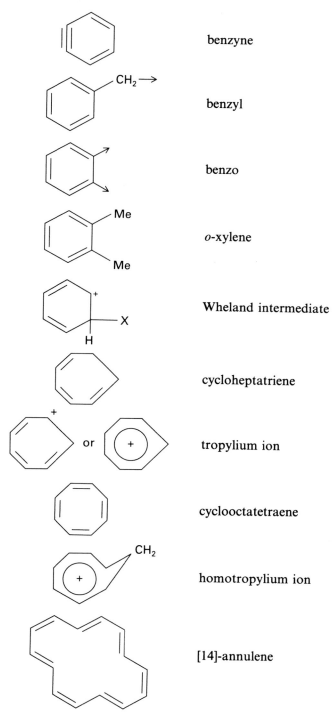

benzyne

benzyl

benzo

o-xylene

Wheland intermediate

cycloheptatriene

tropylium ion

cyclooctatetraene

homotropylium ion

[14]-annulene

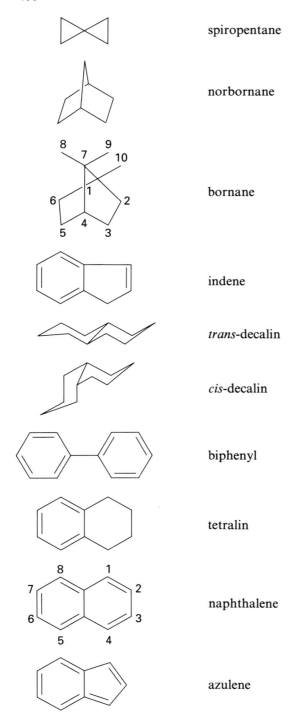

spiropentane

norbornane

bornane

indene

trans-decalin

cis-decalin

biphenyl

tetralin

naphthalene

azulene

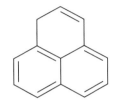

anthracene

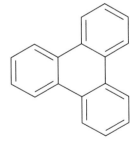

phenanthrene

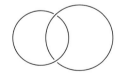

phenalene

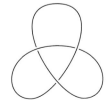

triphenylene

catenanes

trefoil

23.2 Oxygen Containing Compounds

RCH_2OH	primary alcohol
R_2CHOH	secondary alcohol
R_3COH	tertiary alcohol
$HO \rightarrow$	hydroxyl group
$RO \rightarrow$	alkoxyl group
HO^-	hydroxide anion
RO^-	alkoxide anion
ROR	ether
H_3O^+	hydroxonium ion
R_3O^+	oxonium ion
H_2O_2	hydrogen peroxide
$ROOH$	hydroperoxide
$ROOR$	peroxide
$RHC(OR)OH$	hemiacetal
$R_2C(OR)OH$	hemiketal
$RHC(OR)_2$	acetal
$R_2C(OR)_2$	ketal
$RC(OR)_3$	orthoester
$C(OH)_4$	orthocarbonic acid
$C(OR)_4$	orthocarbonates

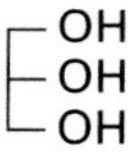

$$\begin{array}{l} -OH \\ -OH \\ -OH \end{array} \qquad \text{glycerol}$$

CO carbon monoxide

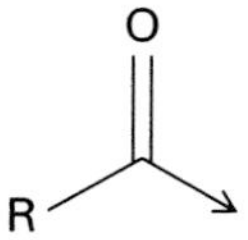

acyl group

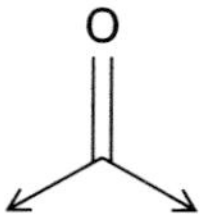

carbonyl group

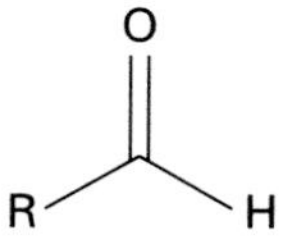

or RCHO   aldehyde

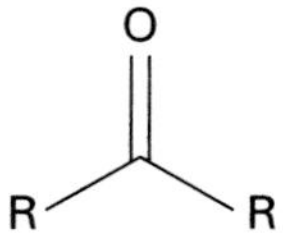

or RCOR   ketone

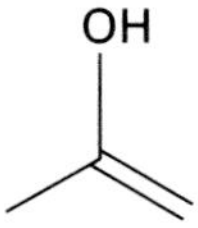

enol

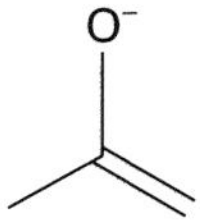

enolate

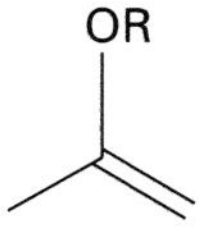

enol ether

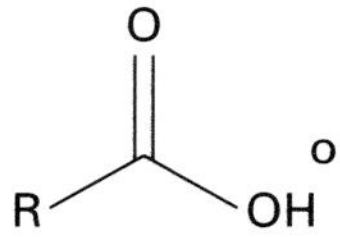

or RCO₂H carboxylic acid

RCO₂⁻ carboxylate anion

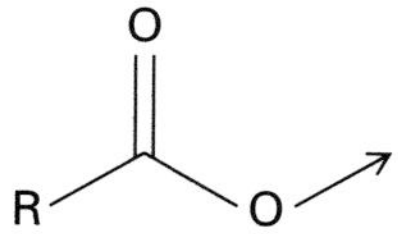

alkanoate group

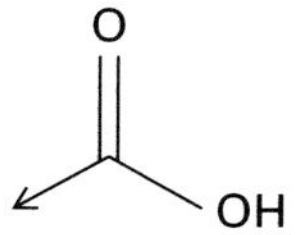

carboxyl group, or a carboxylic acid

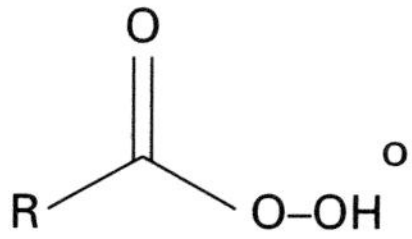

or RCO₃H percarboxylic acid

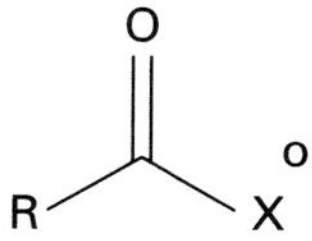

or RCOX where X is a halide, acid halide

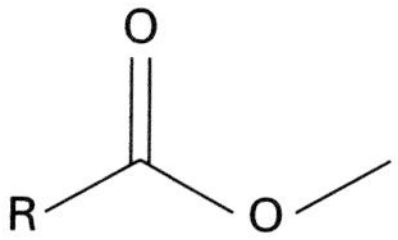

R or RCO₂R ester, if cyclic then a lactone

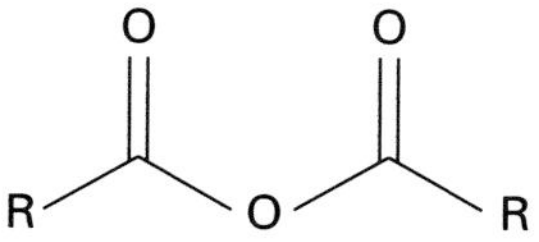

acid anhydride
or RCOOCOR

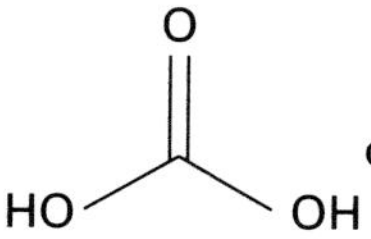

or H₂CO₃ carbonic acid

Structure	Name
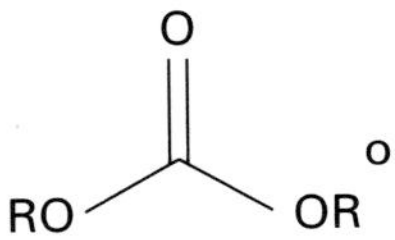 $\underset{\text{RO}}{}\overset{\displaystyle O}{\underset{}{\|}}\underset{\text{OR}}{}$ or $O = C(OR)_2$	carbonates
CO_2	carbon dioxide
$RHC = C = O$	aldoketene
$R_2C = C = O$	ketoketene
$O = C = C = C = O$ or C_3O_2	carbon suboxide

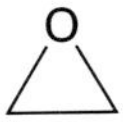

epoxide

oxirene

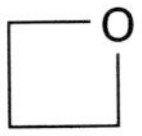

oxetane

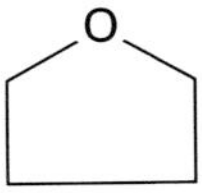

tetrahydrofuran

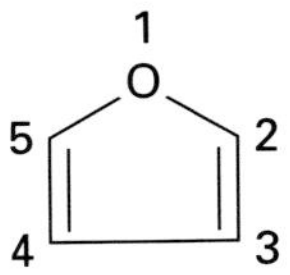

furan

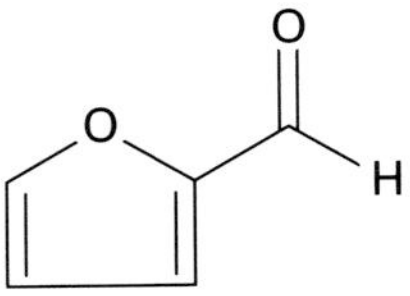

furfural

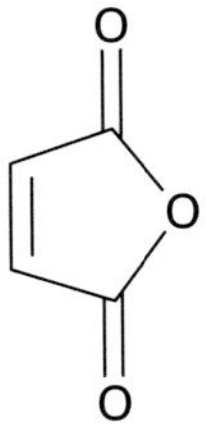

maleic anhydride

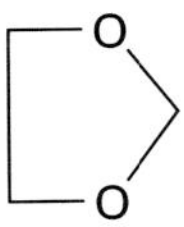

1,3-dioxolane

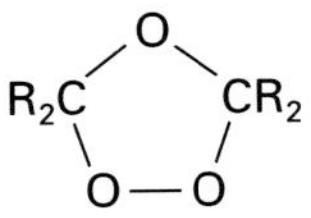

ozonide

molozonide

tetrahydropyran

pyran

pyrylium cation

1,4-dioxane

 2-pyrone

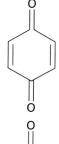 *p*-quinone

lactide

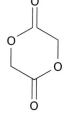 bisperoxide

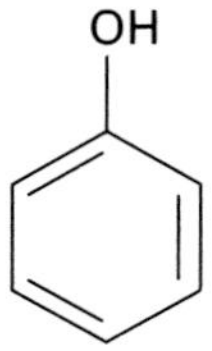

phenol

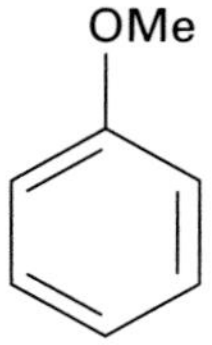

anisole

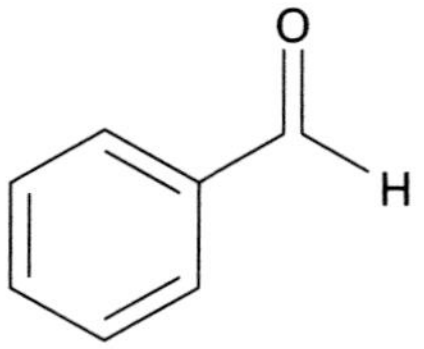

benzaldehyde

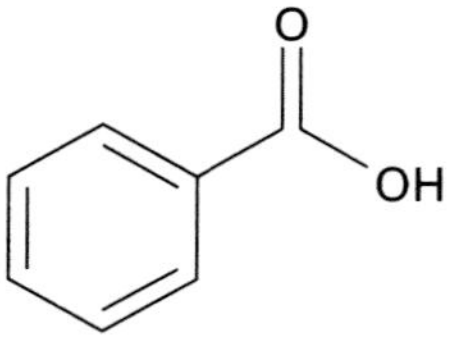

benzoic acid

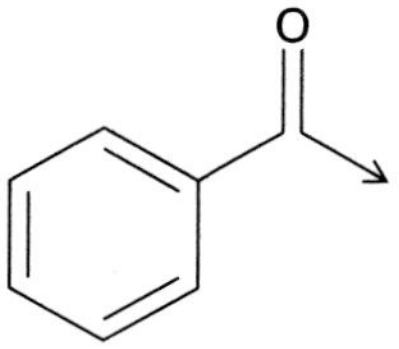

benzoyl

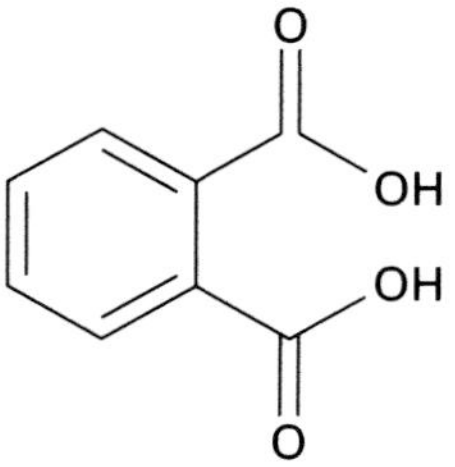

phthalic acid

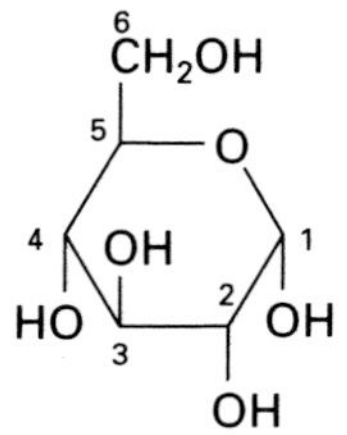

pyranose

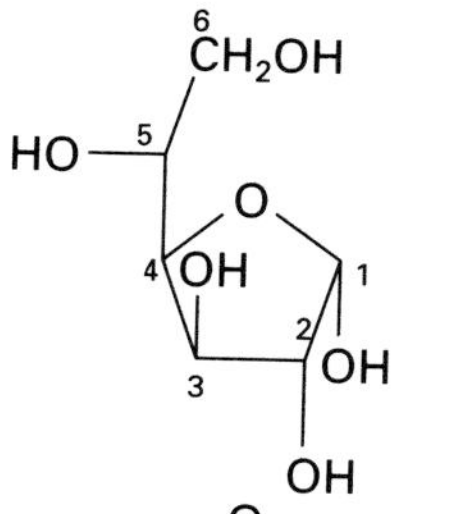

furanose

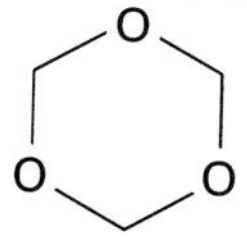

1,3,5-trioxane

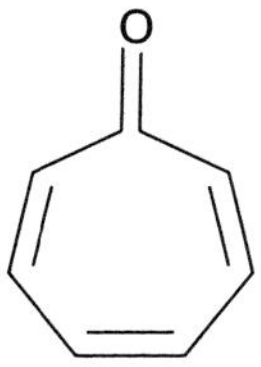

tropanone

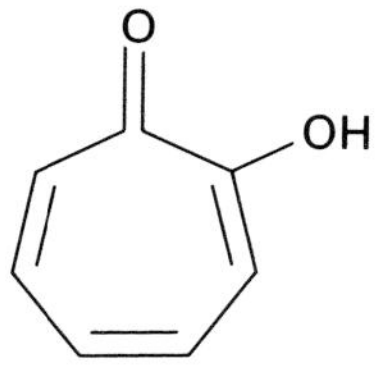

tropolone

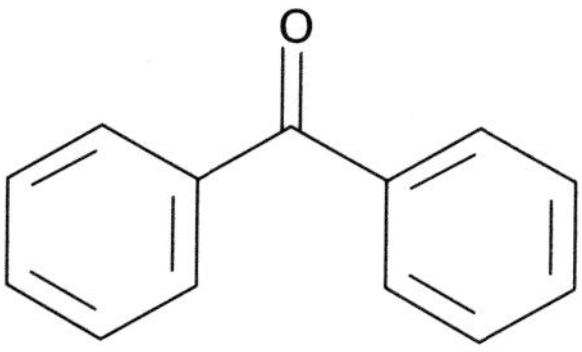

benzophenone

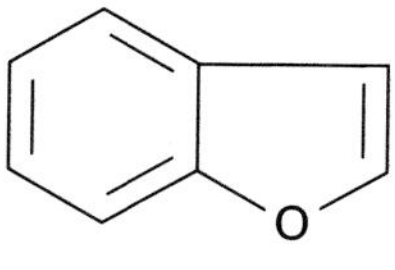

benzofuran

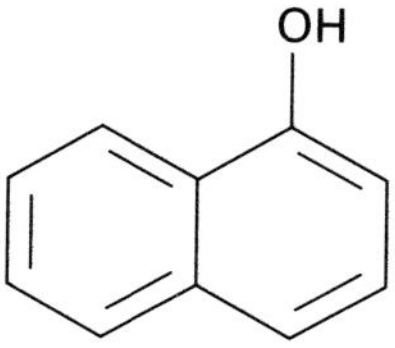

naphthol

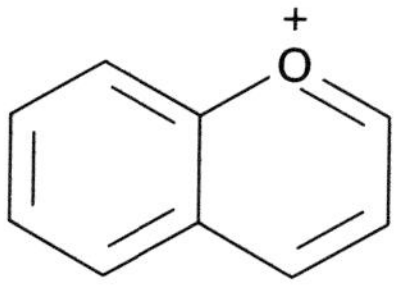

benzopyrylium cation

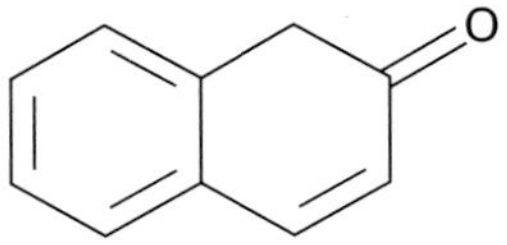

coumarin

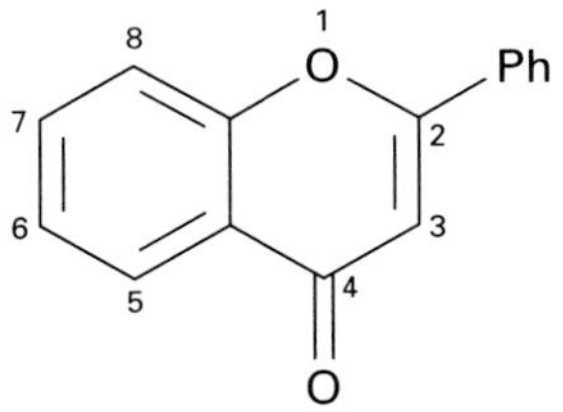

flavone

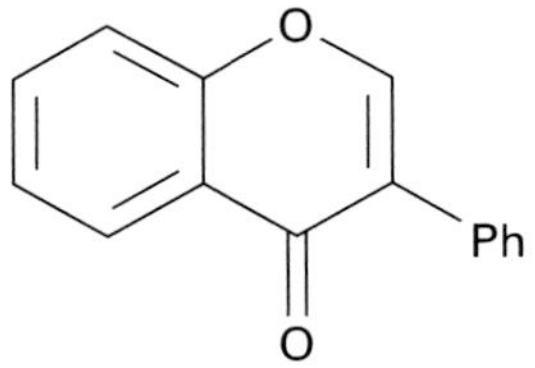

isoflavone

23.3 Nitrogen Containing Compounds

NH_3	ammonia
NH_4^+	ammonium salt
$R-NH_2$	primary amine
R_2NH	secondary amine
R_3N	tertiary amine
$R_4N^+X^-$	quaternary ammonium salt
RN:	nitrene

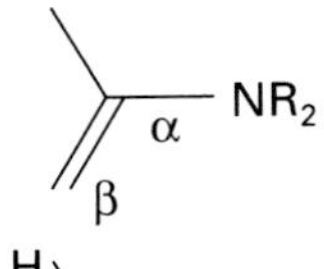

enamine

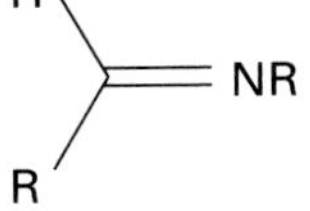

aldimine

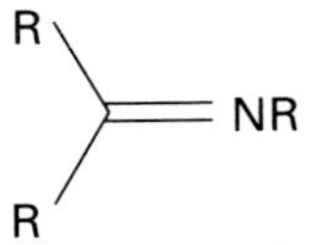

imine

$R_2C = NR_2^+$	iminium cation
$R-C \equiv C-NR_2$	ynamine
$R-N = C$	isonitrile
$R-N = CX_2$	where X is a halide, alkyliminocarbonyl halide

HCN	hydrocyanic acid, or hydrogen cyanide
R-CN	nitrile

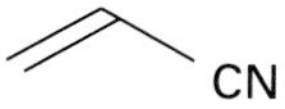

acrylonitrile

H_2N-NH_2	hydrazine
$HN=NH$	diimine
R-NH-NH-R	hydrazo compounds
R-$N=N$-X	one nitrogen bonded to a non-carbon atom, diazo
R-$N=N$-R	both nitrogens bonded to a carbon atoms, azo
R_2N-$C\equiv C$-NR_2	ynediamines
R_2N-$N:$	aminonitrene
R-$N^+\equiv N$ or R-N_2^+	diazonium cation
$R_2C=N$-NH_2	hydrazone
$R_2C=N$-NH^-	hydrazide anion
$R_2C=N^+=N^-$	diazoalkane
$R_2C=N$-$N=CR_2$	azine
R-$N=C=N$-R	carbodiimide
H_2N-CN	cyanamide
NC-CN	cyanogen

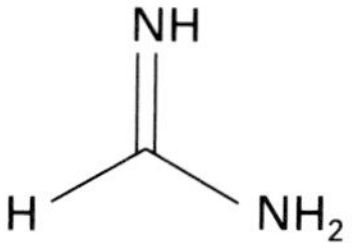

formamidine

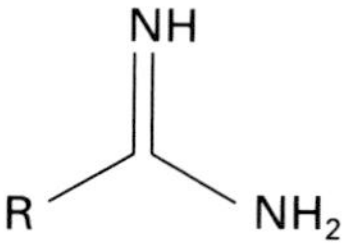

amidine

HN_3	hydazoic acid
N_3^-	azide anion
R-N_3	azide

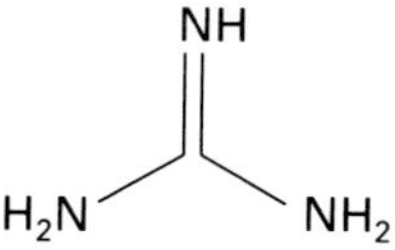

guanidine

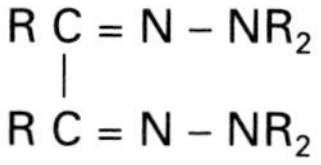

osazone

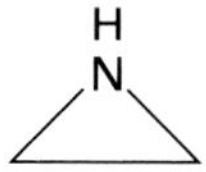

aziridine

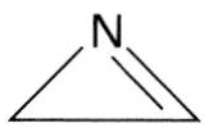

1-azirine

(structure)	diaziridine
(structure)	pyrrolidine
(structure)	3-pyrroline
(structure)	pyrrole
(structure)	imidazole
(structure)	pyrazolidine
(structure)	1-pyrazoline
(structure)	2-pyrazoline
(structure)	pyrazole
(structure)	3H-pyrazole
(structure)	osotriazolidine
(structure)	osotriazoline

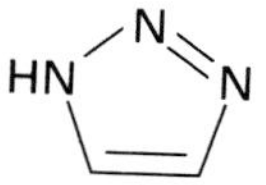

osotriazole

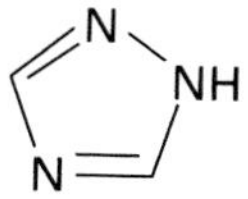

triazole

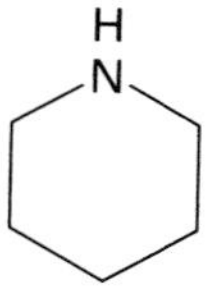

piperidine

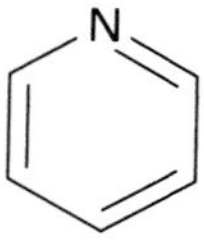

pyradine

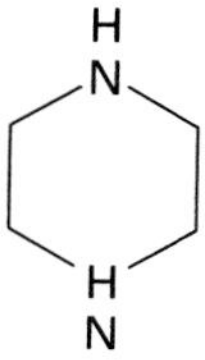

piperazine

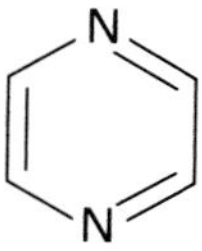

pyrazine

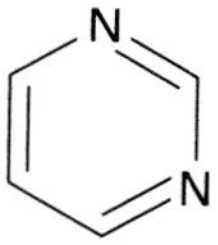

pyrimidine

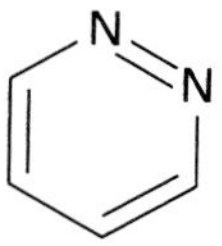

pyridazine

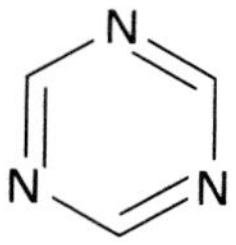

triazine

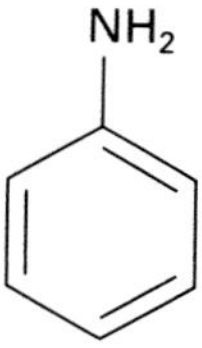

aniline

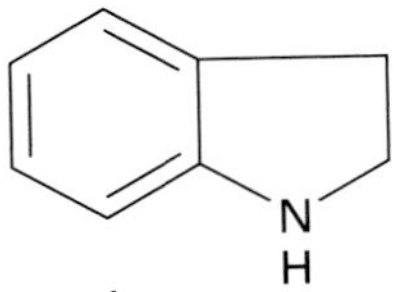

indoline

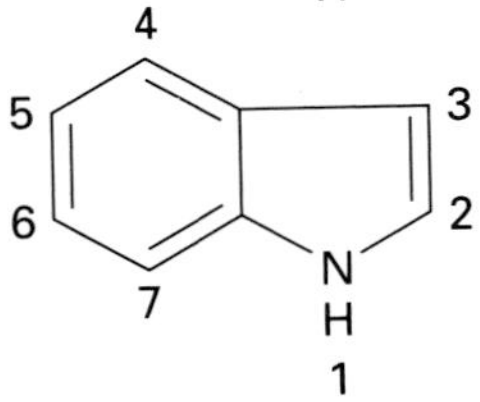

indole

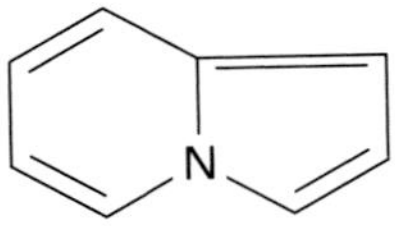

indolizine

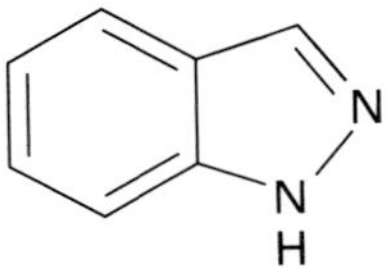

indazole

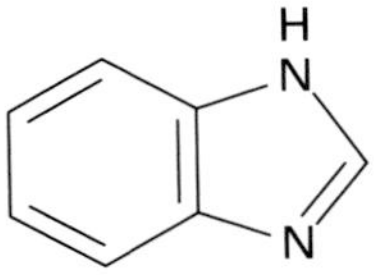

benzimidazole

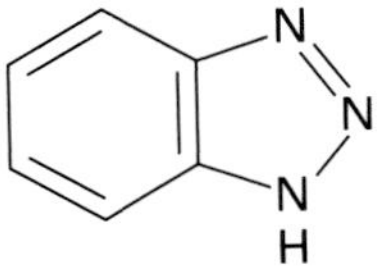

benztriazole

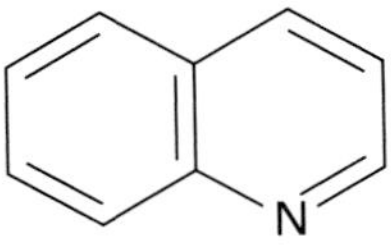

quinoline

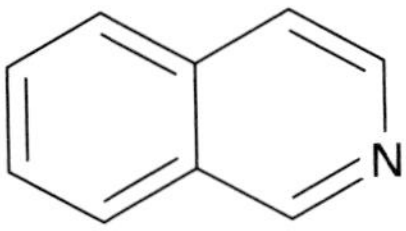

isoquinoline

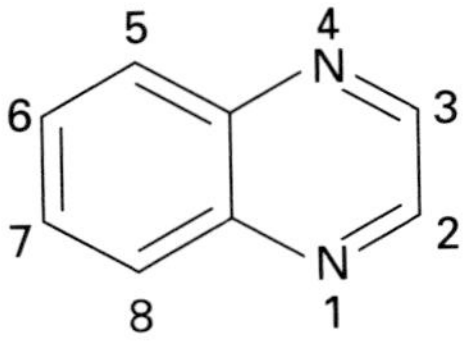

quinoxaline

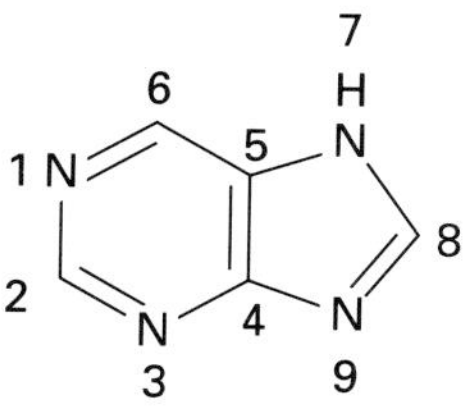

purine

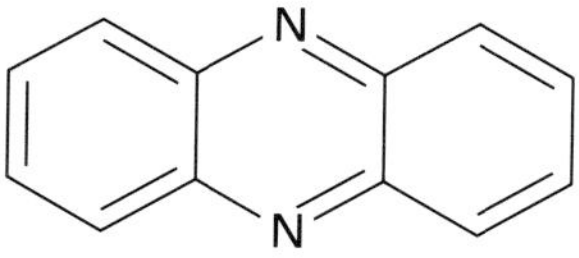

phenazine

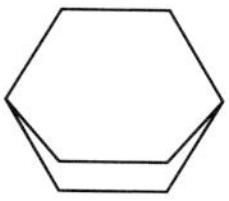

quinuclidine

23.4 Nitrogen and Oxygen Containing Compounds

N_2O	nitrous oxide
NO	nitrogen oxide
N_2O_3	dinitrogen trioxide
NO_2	nitrogen dioxide
N_2O_4	dinitrogen tetraoxide
N_2O_5	dinitrogen pentaoxide
$H\text{-}O\text{-}N=O$ or HNO_2	nitrous acid
$H\text{-}P\text{-}NO_2$ or HNO_3	nitric acid
NH_2OH	hydroxylamine
$H\text{-}N=C\text{-}O$	cyanic acid
$R\text{-}N=C\text{-}O$	isocyanate
$R\text{-}O\text{-}C\equiv N$	cyanate
$R\text{-}N=O$	nitroso

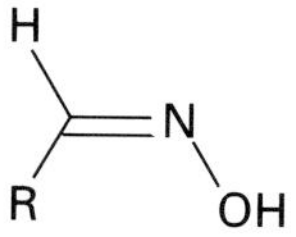

aldoxime

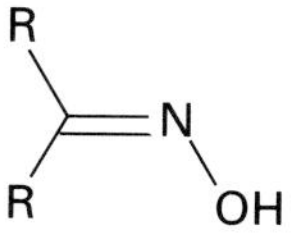

oxime

$R_3N^+\text{-}O^-$ amine oxide

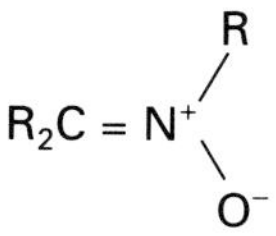

nitrone

$R-O-N=O$ — nitrite ester

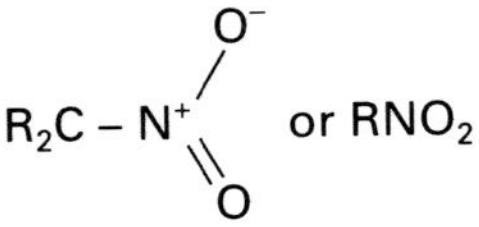

$R_2C-N^+\!\!\begin{array}{c}O^-\\ \\O\end{array}$ or RNO_2 — nitrate

$RCH=N^+\!\!\begin{array}{c}OH\\ \\O^-\end{array}$ — nitronic acid

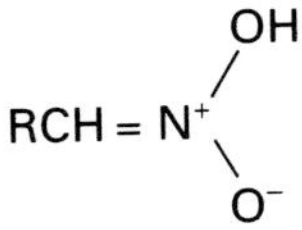

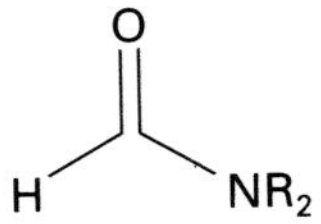

formamide

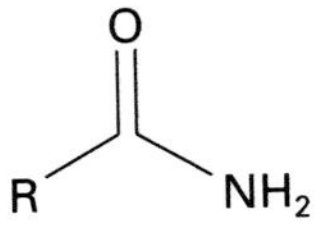

$$\underset{H}{\overset{O}{\|}}\underset{}{C}-NR_2$$

primary amide

$$R-\underset{}{\overset{O}{\|}}C-NH_2$$

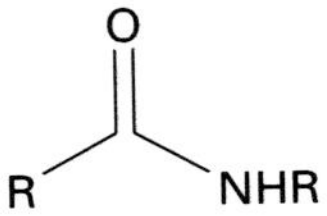

secondary amide, if cyclic then a lactam

$$R-\underset{}{\overset{O}{\|}}C-NHR$$

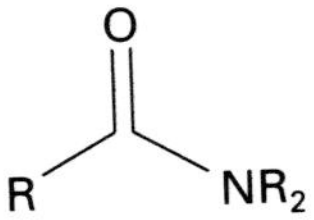

tertiary amide, if cyclic then a lactam

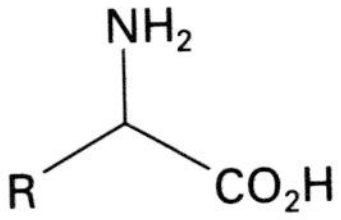

$$R-\underset{}{\overset{O}{\|}}C-NR_2$$

α-amino acid

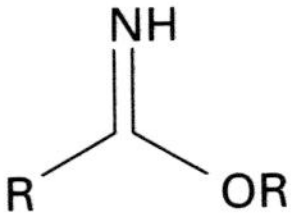

$$R-\underset{NH_2}{\overset{}{C}}H-CO_2H$$

imidic ester

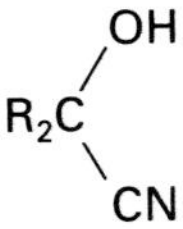

$$R-\underset{}{\overset{NH}{\|}}C-OR$$

cyanohydrin

$$R_2C\begin{array}{c}OH\\CN\end{array}$$

R_2N-NO — nitrosamine

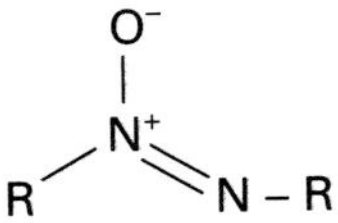

$$R-\underset{}{\overset{O^-}{\underset{}{N^+}}}=N-R$$

azoxy compound

$$R-CH\begin{array}{c}NO\\NO_2\end{array}$$

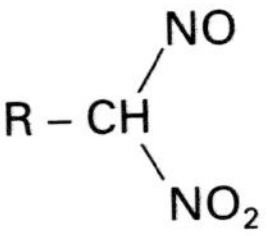

pseudonitrole

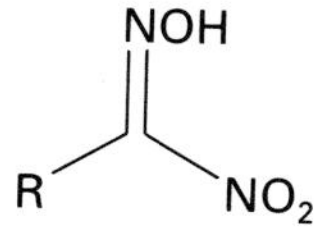

nitrolic acid

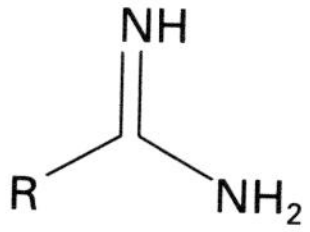

amidoximes

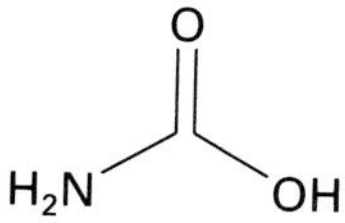

carbamic acid

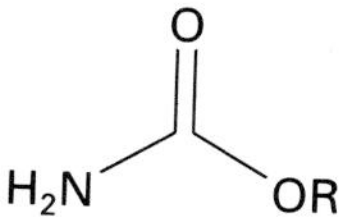

carbamate

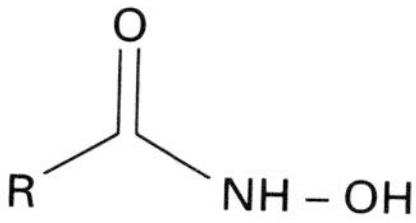

hydroxamic acid

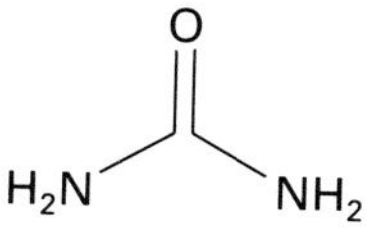

urea

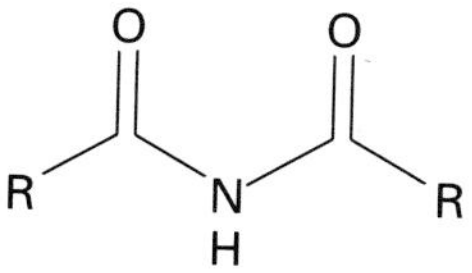

imide

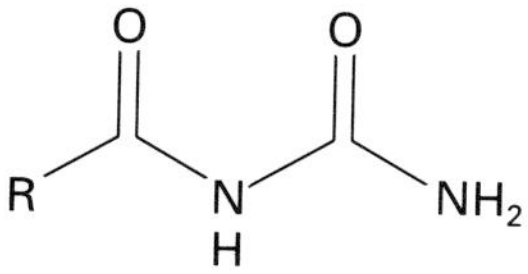

ureide

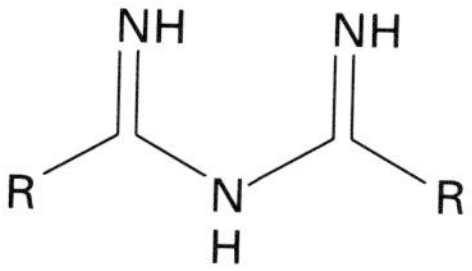

imidine

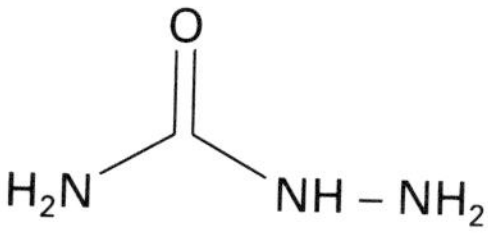

semicarbazide

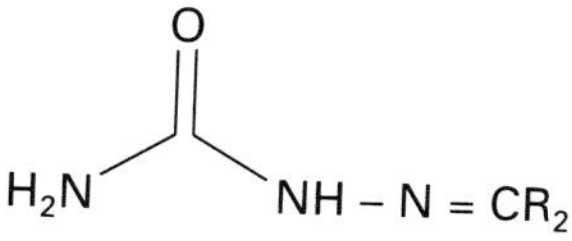

semicarbazone

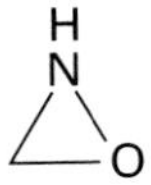

oxazirane

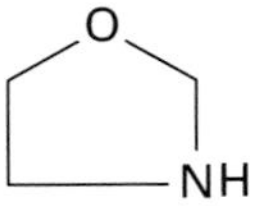

oxazolidine

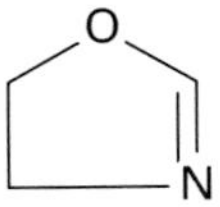

oxazoline

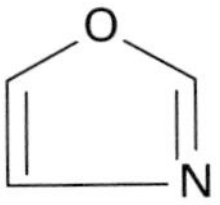

oxazole

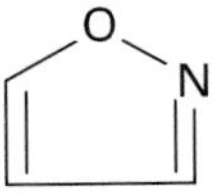

isoxazole

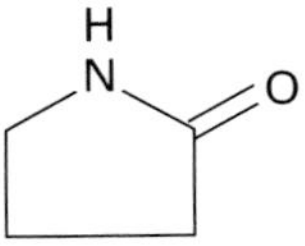

2-pyrrolidone

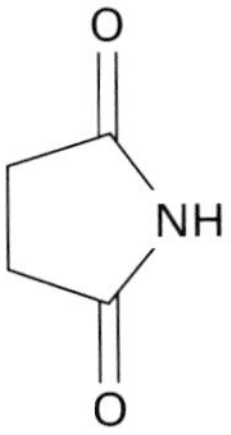

succinimide

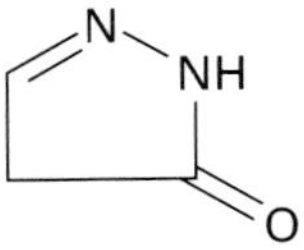

pyrazolone

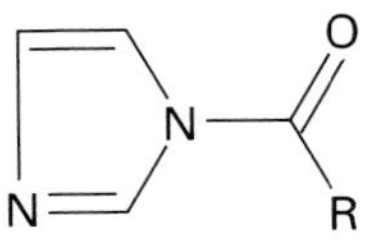

imidazolide

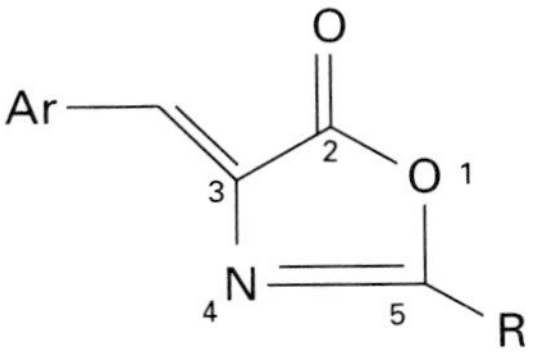

azlactone

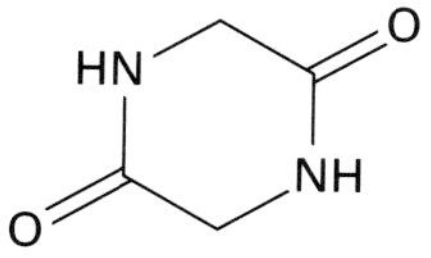

diketopiperazine

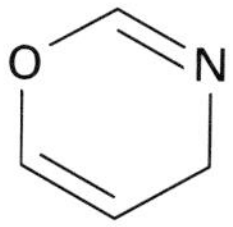

1,3-oxazine

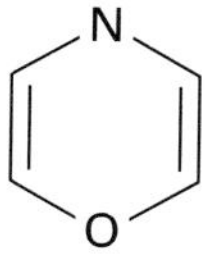

1,4-oxazine

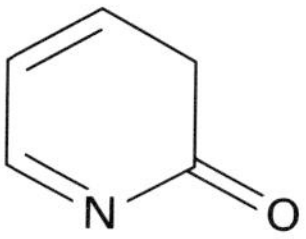

2-pyridone

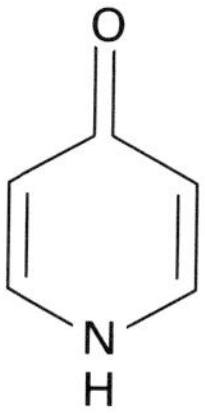

4-pyridone

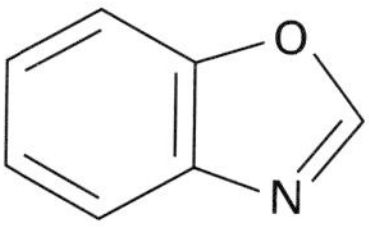

benzoxazole

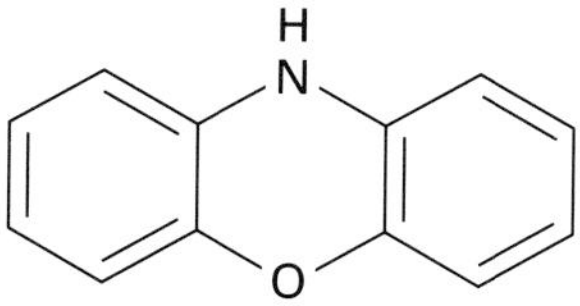

phenoxazine

23.5 Sulfur Containing Compounds

RSH	mercaptan
RS⁻	sulfide
RSR	thioether
R_3S^+	sulfonium cation
RS-OH	sulfenic acid
R-S-S-R	disulfide
R-(S)$_n$-R	polysulfide

$RHC(SR)_2$	mercaptal
$R_2C(SR)_2$	mercaptol
$R_2C(OR)SR$	hemithioacetal

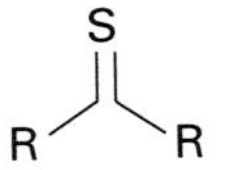

or RCSR — thiocarbonyl

$R_2S = CR_2$ or $R_2\overset{-}{S} - \overset{+}{C}R_2$ — sulfonium ylid

$R_2S = O$ — sulfoxide

$RSOCH_2^-$ — methylsulfinyl carbanion

$R_2S = O_2$ — sulfone

$R\text{-}S\text{-}CN$ — thiocyanate

$R\text{-}N = C = S$ — isothiocyanate

$R\text{-}N = N\text{-}S\text{-}R$ — diazosulfide

$R_2C = S = O$ — sulfine

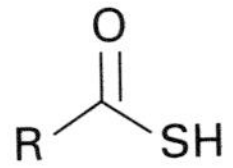

thiolic acid

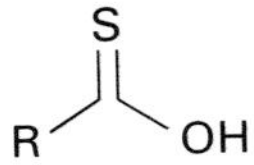

thionic acid

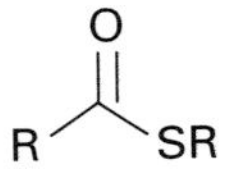

thioester

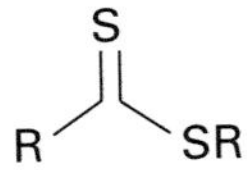

dithioester

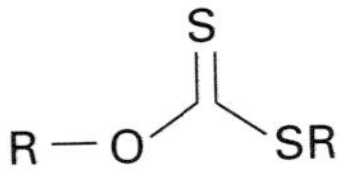

xanthate

$R\text{-}SO_2H$ — sulfinic acid

$R\text{-}OSO_2H$ — alkyl hydrogen sulfite

$R\text{-}SO_2OH$ — sulfonic acid

$R\text{-}OSO_2OH$ — alkyl hydrogen sulfate

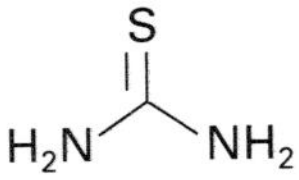

thiourea

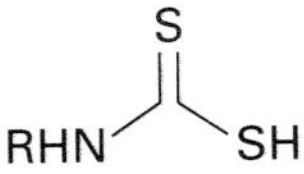

dithiocarbamic acid

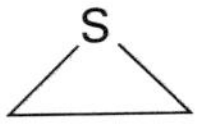

episulfide

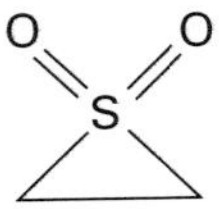

episulfone

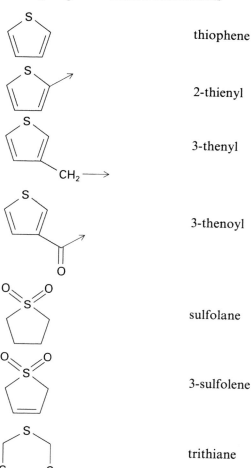

thiophene

2-thienyl

3-thenyl

3-thenoyl

sulfolane

3-sulfolene

trithiane

23.6 Phosphorus Containing Compounds

R_3P	phosphine
R_4P^+	phosphonium cation
H_3PO_3 or $P(OH)_3$	phosphorous acid
$P(OR)_3$	phosphite
$HP(OH)_2$	phosphonous acid
$HP(OR)_2$	phosphonite
$H_2P(OH)$	phosphinous acid
$H_2P(OR)$	phosphinite
H_3PO_4 or $O=P(OH)_3$	phosphoric acid
$O=P(OR)_3$	phosphate

$HP=O(OH)_2$ phosphonic acid
$H_2P=O(OH)$ phosphinic acid
$R_3P=CR_2$ phosphorane

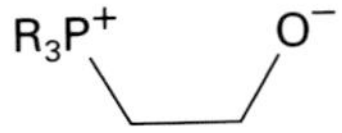

R_3P^+ O^-

betaine

R_3P-O

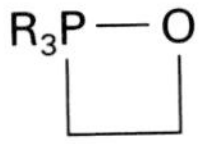

oxaphosphetane